METHODS IN MOLECULAR BIOLOGY™

Series Editor
John M. Walker
School of Life Sciences
University of Hertfordshire
Hatfield, Hertfordshire, AL10 9AB, UK

For other titles published in this series, go to
www.springer.com/series/7651

METHODS IN MOLECULAR BIOLOGY™

Bacteriophages

Methods and Protocols, Volume 1: Isolation, Characterization, and Interactions

Edited by

Martha R. J. Clokie

University of Leicester, Leicester, UK

Andrew M. Kropinski

Public Health Agency of Canada, Guelph, Ontario, Canada

Humana Press

Editors
Martha R. J. Clokie
University of Leicester
Leicester, UK
mrjc1@le.ac.uk

Andrew M. Kropinski
Public Health Agency of Canada
Guelph, Ontario, Canada
Andrew_Kropinski@phac-aspc.gc.ca

Series Editor
John M. Walker
University of Hertfordshire
Hatfield, Herts
UK

ISSN 1064-3745
e-ISSN 1940-6029
ISBN 978-1-58829-682-5
e-ISBN 978-1-60327-164-6
DOI 10.1007/978-1-60327-164-6

Library of Congress Control Number: 2008939449

Printed on acid-free paper

springer.com

Preface

We are increasingly aware of the many and varied roles that bacteriophages play in microbial ecology and evolution. The implications of bacteriophage–bacteria interactions range from the evolution of pathogenicity to oceanic carbon cycling. However, working with bacteriophages can be difficult due to their small size and specific bacterial host requirements. Written by top international bacteriophage researchers, these volumes pull together a vast body of knowledge and expertise, including almost forgotten classical methods as well as state-of-the-art molecular techniques. It is designed to be a valuable reference for experienced bacteriophage researchers as well as an accessible introduction to the newcomer to the subject.

The books are designed to be modular and are organised in the order in which one would carry out the work. A wide range of projects can be built from these modules by selecting appropriate chapters from each section. Volume 1's Section 1 concerns the isolation of phages from a range of environments. Sections 2 and 3 describe their morphological and molecular characterisation, and present methods for the investigation of their interaction with bacteria. Volume 2's Sections 1 – 3 are concerned with bacteriophage genomics, metagenomics, transcriptomics and proteomics. It concludes with chapters on applied bacteriophage biology (Section 4).

Martha R. J. Clokie
Andrew M. Kropinski

Contents

Contributors

STEPHEN T. ABEDON • *Department of Microbiology, The Ohio State University, Mansfield, OH, USA*
HANS-W. ACKERMANN • *Department of Medical Microbiology, Faculty of Medicine, Laval University, Quebec, QC, Canada*
SHELLIE R. BENCH • *Ocean Sciences Department, University of California, Santa Cruz, CA, USA*
CORINA P. D. BRUSSAARD • *Department of Biological Oceanography, Royal Netherlands Institute for Sea Research, Texel, The Netherlands*
DONALD L. COURT† • *Gene Regulation and Chromosome Biology Laboratory, Center for Cancer Research, National Cancer Institute at Frederick, Frederick, MD, USA*
BORRIES DEMELER • *Department of Biochemistry, The University of Texas Health Science Center at San Antonio, San Antonio, TX, USA*
TREVOR DOUGLAS • *Department of Chemistry and Biochemistry, Center for BioInspired Nanomaterials, and Thermal Biology Institute, Montana State University, Bozeman, MT, USA*
LOUIS-CHARLES FORTIER • *Département de microbiologie et d'infectiologie Faculté de médecine et des sciences de la santé, Université de Sherbrooke, Sherbrooke, QC, Canada*
KRISTYN FRANKLIN • *Public Health Agency of Canada, Laboratory for Foodborne Zoonoses, Guelph, Ontario, Canada*
DAVID I. FRIEDMAN • *Department of Microbiology and Immunology, University of Michigan, Ann Arbor, MI, USA*
JENNIFER FULTON • *Department of Microbiology and Thermal Biology Institute, Montana State University, Bozeman, MT, USA*
STEPHEN C. HARDIES • *Department of Biochemistry, The University of Texas Health Science Center at San Antonio, San Antonio, TX, USA*
SHIRLEY J. HAYES • *Department of Biochemistry, The University of Texas Health Science Center at San Antonio, San Antonio, TX, USA*
ELVIRA M. HÉBERT • *Cerela-Conicet, S.M. de Tucumán, Tucumán, Argentina*
REBEKAH R. HELTON • *Delaware Biotechnology Institute, University of Delaware, Newark, DE, USA*
PAUL HYMAN • *MedCentral College of Nursing, Mansfield, OH, USA*
ROGER P. JOHNSON • *Public Health Agency of Canada, Laboratory for Foodborne Zoonoses, Guelph, Ontario, Canada*
ANDREW M. KROPINSKI • *Public Health Agency of Canada, Laboratory for Foodborne Diseases, Guelph, Ontario, Canada; Department of Molecular and Cellular Biology, University of Guelph, Guelph, Ontario, Canada; Department of Microbiology and Immunology Queen's University, Kingston, Ontario, Canada*
ELIZABETH KUTTER • *The Evergreen State College, Olympia, WA, USA*

CHRISTOPHER N. LAROCK • *Department of Microbiology and Immunology, University of Michigan, Ann Arbor, MI, USA*
ERIKA LINGOHR • *Public Health Agency of Canada, Laboratory for Foodborne Diseases, Guelph, Ontario, Canada*
JONATHAN LIVNY • *Department of Microbiology, Tufts University School of Medicine, Boston, MA, USA*
STANLEY MALOY • *Center for Microbial Sciences, San Diego State University, San Diego, CA, USA*
AMANDA MAZZOCCO • *Public Health Agency of Canada, Laboratory for Foodborne Diseases, Guelph, Ontario, Canada*
ANDREW D. MILLARD • *Department of Biological Sciences, University of Warwick, Coventry, UK*
SYLVAIN MOINEAU • *Département de Biochimie et de Microbiologie, Faculté des Sciences et de Génie; Groupe de Recherche en Écologie Buccale (GREB), Faculté de Médecine Dentaire; and Félix d'Hérelle Reference Center for Bacterial Viruses, Université Laval, Québec City, Québec, Canada*
AMOS B. OPPENHEIM • *Department of Molecular Genetics and Biotechnology, The Hebrew University-Hadassah Medical School, Jerusalem, Israel*
ALICE C. ORTMANN • *Department of Plant Sciences and Plant Pathology, Montana State University, Bozeman, MT, USA*
GERARDO PEREZ • *Center for Microbial Sciences, San Diego State University, San Diego, CA, USA*
RAÚL R. RAYA • *Cerela-Conicet, S.M. de Tucumán, Tucumán, Argentina*
PHILIP SERWER • *Department of Biochemistry, The University of Texas Health Science Center at San Antonio, San Antonio, TX, USA*
CURTIS A. SUTTLE • *Departments of Earth and Ocean Sciences, Botany, and Microbiology and Immunology, University of British Columbia, Vancouver, British Columbia, Canada*
ANNE THIERAUF • *Department of Microbiology, University of Illinois, Urbana, IL, USA*
JULIE THOMAS • *Department of Biochemistry, The University of Texas Health Science Center at San Antonio, San Antonio, TX, USA*
LYNN C. THOMASON • *Gene Regulation and Chromosome Biology Laboratory, Center for Cancer Research, National Cancer Institute at Frederick, Frederick, MD, USA; SAIC-Frederick, Inc., Gene Regulation and Chromosome Biology Laboratory, Basic Research Program, Frederick, MD, USA*
ROHAN VAN TWEST • *Department of Molecular and Cellular Biology, University of Guelph, Guelph, Ontario, Canada*
ROBERT VILLAFANE • *Department of Biological Sciences, Alabama State University, Montgomery, AL, USA*
THOMAS E. WADDELL • *Pro-Lab Diagnostics, Richmond Hill, Ontario, Canada*
KURT E. WILLIAMSON • *J. Craig Venter Institute, Rockville, MD, USA*
DANIELLE M. WINGET • *Delaware Biotechnology Institute, University of Delaware, Newark, DE, USA*
K. ERIC WOMMACK • *Delaware Biotechnology Institute, University of Delaware, Newark, DE, USA*

JOHN YIN • *Department of Chemical and Biological Engineering, University of Wisconsin-Madison, Madison, WI, USA*
MARK YOUNG • *Department of Plant Sciences and Plant Pathology, Center for BioInspired Nanomaterials, and Thermal Biology Institute, Montana State University, Bozeman, MT, USA*

Introduction

Andrew M. Kropinski and Martha R.J. Clokie

The discovery of viruses specific to bacteria (referred to variably as bacteriophages, phages, and bacterial viruses in this volume) is credited to an English bacteriologist, Frederick William Twort *(1)* in 1915 and to a French-Canadian scientist, Felix d'Herelle *(2)* in 1917. It is the latter scientist who probably more accurately recognized what he was dealing with and is responsible for naming these agents of bacterial death. He realized these organisms propagated at the expense of bacteria so named them bacteriophages, which translates as bacterial eaters, "phages" coming from the Greek "phagein" meaning "to eat." He is also responsible for recognizing their potential clinical significance *(3)*.

The first golden age of bacteriophage research ran from the 1930s through to the 1970s and resulted in major discoveries such as the identification of DNA as genetic material and the subsequent deciphering of the genetic code and the discovery of messenger RNA; these breakthroughs led to the birth of the new science of Molecular Biology. This work is described in the book by John Cairns et al. "*Phage and the Origins of Molecular Biology*" *(4)* and on the excellent American Society for Microbiology Division M (Bacteriophage) homepage (http://www.asm.org/division/M/M.html (thanks to Susan Godfrey, Roger Hendrix, Eric Miller). A summary of some of the major discoveries made during this period is detailed in **Table 1** (the authors apologize for the omission of the impact of many eminent phage biologists).

Following on from this golden age was the 1980s and 1990s, where phages and phage-derived products were essential to the major biotechnological revolution that occurred. Recombinant DNA techniques were developed in which phage played a significant part as primary vectors (filamentous phage *(5)*, λ insertional and replacement vectors *(6)*) or parts of vectors (promoters [expression vector *(7–9)*], packaging signals [cosmids *(10, 11)* and phagemids *(12)*], integrative signals [integrative vectors *(13–15)*], replicons [phagemids *(16)*], P1-derived vectors *(17)*]). In addition, they contributed a great variety of enzymes which are employed in today's molecular biology laboratory, including integrases, polynucleotide kinases, DNA ligases, DNA polymerases, RNA polymerases, recombinases, single-stranded DNA binding proteins (SSB), endo- and exonucleases, and even methylases and restriction endonucleases *(18)*.

A good indicator for the amount of interest in bacteriophage research is the number of papers published per year that contain the word 'bacteriophage' in their title. This rose steadily from 1950 to 1965 (**Fig. 1**). There was then a sharp burst of phage publications from 1970 to 1975 followed by a precipitous drop in number. This was due to the unfortunate lack of interest and funding for phage biology where many eminent scientists gave up working on phages for more lucrative eukaryotic projects.

Recently due to an increased awareness of their importance, an interest in bacteriophages has been re-kindled, and an insight into the scale of this renewed enthusiasm can be seen from the huge increase in the number of sequenced phage genomes (**Fig. 1**).

Table 1
Significant experimental observation using bacteriophages

Grouping	Discovery	Year & Reference
Plaque assays		Felix d'Herelle 1917 *(50–52)*
Structure and taxonomy	First EM pictures of phages	T.F. Anderson 1942 *(53)*
	CryoEM	1992 *(54–56)*
	Development of modern taxonomic schemes	1962 *(57–59)*
Composition – general	Phages are composed of protein and DNA	1948 *(60)*
	Isolation of: first lipid-containing phage: PM2	1968 *(61)*
Nucleic acids	Genes are made of DNA	1952 *(62)*
	Introns: type I – T4	*(63–65)*
	Inteins	1998 *(66–69)* [http://www.neb.com/inteins.html]
	Genetic code	1961 *(70)*
	Modified bases: T4 (5-hydroxymethylcytosine)	1953 *(71)*
	tRNA encoding genes: T4	1972 *(72–75)*
	Restriction and modification: λ	
	a) Phenomenon	1953 *(76)*
	b) Mechanism	1962 *(77)*
	Novel genomes:	
	a) Single-stranded (ss) DNA - øX174	1959 *(78)*
	b) Single-stranded (ss) RNA - f2	1961 *(79)*
	c) Segmented double-stranded RNA - ø6	1973 *(80, 81)*
	d) Phage with terminal proteins - ø29	1971 *(82)*
	Sequence of first:	
	a) ssRNA virus	1976 *(83)*
	b) ssDNA virus	1977 *(84)*
Mutation	rII experiments – T4	1955 *(85, 86)*
	T1 resistance in *E.coli*	1943 *(87)*
Lysogeny and integration	Discovery of lysogeny	1934 *(88–90)*
	Isolation of phage λ	E.M. Lederberg 1951 *(91)*
	Induction	1950 *(92)*

(continued)

Table 1 (continued)

Grouping	Discovery	Year & Reference
	Integration:	
	a) Model	1962 *(93)*
	b) Site-specific recombination	1968 *(94, 95)*
	Repression:	
	a) Model	1961 *(96, 97)*
	b) Experimental evidence	1967 *(98, 99)*
	Integration of phage Mu causes host mutations	1963 *(100)*
	Not all temperate phages integrate:	
	a) P1	1951 *(101, 102)*
	b) Linear prophages - N15	N.V. Ravin 1964 *(103)*
	Lysogenic conversion: a) Toxigenicity – *Corynebacterium diphtheriae* phage B	1951 *(104, 105)*
	b) Serotype: *Salmonella* Anatum phage ε15	1955 *(106, 107)*
Genetic exchange - transduction	P22 and *Salmonella*	1952 *(108)*
	P1 and *Escherichia coli*	1955 *(109)*
	Specialized transduction: λ	1957 *(110, 111)*
	Origin of host DNA in P22 transducing particles	1972 *(112)*
Adsorption and injection	a) penetration of capsule	1979 *(113)*
	b) λ & LamB liposomes	1983 *(114)*
	c) T4 and spheroplasts	1983 *(115)*
	d) T7 DNA uptake requires transcription	2001 *(116)*
Intracellular development	One-step growth curve:	
	a) latent period & burst size	1939 *(117)*
	b) burst size from single cells	1945 *(118)*
	c) eclipse phase	1948 *(119, 120)*
	DNA replication:	
	a) DNA ligase	1967 *(121)*
	b) øX174 – rolling circle	1968 *(122, 123)*
	c) T4 – Okazaki fragments	1969 *(124)*
	d) M13 – RNA primers	1972 *(125)*
	e) T7 – visualization & formation of concatemers	1972
	General recombination – λ	1961 *(126–129)*

Table 1 (continued)

Grouping	Discovery	Year & Reference
	Transcription:	
	a) mRNA	1956 *(130–133)*
	b) antitermination	1969 *(134)*
	Protein synthesis:	
	a) SDS gels	1969 *(135)*
	b) discontinuous buffer system	1970 *(136)*
	c) slab gel	1973 *(137)*
	d) ribosomal slippage	1993 *(138)*
	Morphogenesis:	
	a) role of chaperonins	1972 *(139–143)*
	b) cross-linked capsid proteins	1995 *(144, 145)*
	c) packaging of φ29 require a small RNA molecule	1987 *(146)*
Phage therapy		F. d'Herelle 1917 *(147)*

From the very steep slope of the graph of the number of phage genome sequences per year it is quite apparent that we are in a new exponential phase of phage research.

There are three main reasons for this renewed interest in bacteriophages. The first is a result of bacterial genome sequencing projects which have revealed that most bacteria are lysogenic for at least one bacteriophage and that phages have played a major role in host genome evolution *(19–24)*. The first project of this kind was the sequencing of the *Haemophilus influenzae* in 1995 which was shown to contain a Mu-like prophage *(25)*. Some bacteria contain many phages in their genomes and pathogenicity is often linked to phage carriage for example the *Streptococcus* group C contain up to six phage or prophage-like elements *(26)*. Phages have been shown to encode a range of toxins and gene products that influence their bacterial cells or even the host in which the bacterium lives (reviewed in *(27)*). An example of the complexity of these interactions can be seen from phages which infect aphid gut bacteria which encode toxins that help bacteria defend the aphid from other invading bacteria *(28)*.

The second reason for the renewed bacteriophage interest is that phage ecologists have shown that soil and water contain between 10 and 100 times more phage particles than bacterial cells, leading to the speculation that the global abundance of phages is probably in the order of 10^{31} *(29)*. Diversity even within phages which infect one bacterial host is also high, for example genomic studies on mycobacterial phages have shown that not only do mycobacterial phage encode genes which are unlike all other genes sequenced thus far, they are also not present in the different phages *(30–35)*. Metagenomic viral studies focusing on the phage in the oceans have also demonstrated the enormous scale of phage genetic diversity *(29, 36, 37)*. With a raised awareness of phage abundance and diversity has come an appreciation for the consequence of phage action in influencing bacterial population dynamics and evolution and in maintaining essential biogeochemical cycles

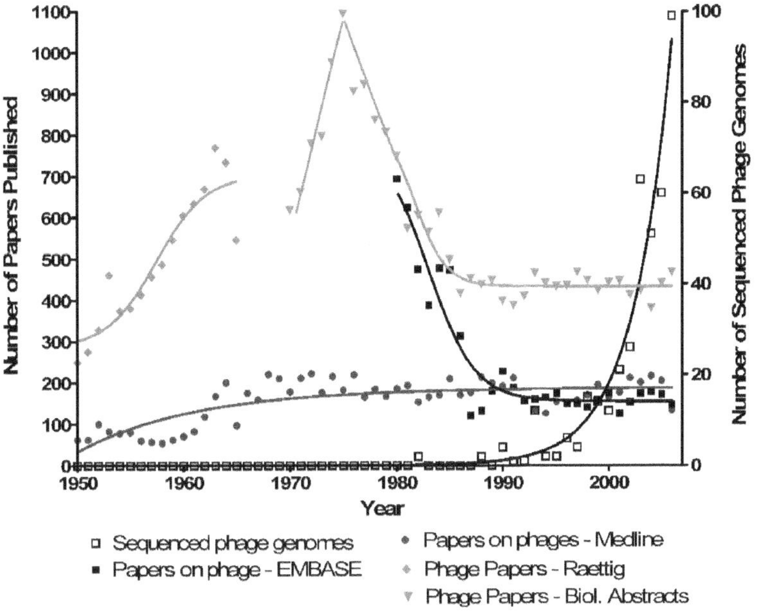

Fig. 1. Publications of bacteriophages and the appearance of phage sequences in GenBank as a function of year. The publication data was derived from four sources: Hansjürgen Raettig for data from 1950 to 1965 *(148, 149)* brought to the editors attention by Hans-Wolfgang Ackermann, and Ovid Medline, EMBASE, and BIOSIS (Biological Abstracts) online literature searches for respectively 1950–2006, 1980–2006, and 1969–2006. In the case of Medline and EMBASE the keyword "bacteriophage" was mapped to subject heading; all subheadings were included; and, the search strategy was focused rather than exploded. With BIOSIS the presence of "bacteriophage or phage" in the title was used to screen scientific articles. NB Although this gives a good indication of phage publications a more detailed analysis is required to tease apart the true number of phage publications, separating those on phage biology from, for example, those on phage typing.

such as carbon cycling *(38)*. Furthermore, recent genomic and transcriptomic studies have illustrated the extent of interlinked metabolisms of phage and host during a lytic infection for example with bacteriophages which infect cyanobacteria encoding and expressing key photosynthesis gene *(39–42)*.

Finally, but very importantly in terms of phage research are the concerns of the public, governmental healthcare agencies, and physicians that something must be done about the growing problem of antimicrobial resistance. This awareness is accompanied by the belated realization that phage therapy, which has been kept alive by the efforts of Eastern European scientists, offers a viable alternative to antibiotic therapy. In Canada, for example, it is now realized that much of the expertise in phage biology has disappeared as a result of retirements and the death of members of the phage community of scientists. The lack of "capacity issues" (i.e., knowledgeable young scientists) has resulted in the Canadian Institutes for Health Research issuing a call for research grants which will address the potential for using phage as a therapeutic agents. Similarly the same awareness in Europe and the United States has resulted in the number of new bacteriophage research groups increasing and the interest and attendance at bacteriophage conferences is increasing annually. What is apparent however is that for bacteriophages to be used therapeutically in countries such as Europe, Canada, and the US, we must properly understand the biology of the interaction between the phage and the bacterial pathogen. We are fortunate to be practicing phage biology in this exciting time where such experimentation is possible.

These volumes are designed to provide the amateur or professional with a step-by-step approach to many of the standard protocols in working with bacteriophages. We include both classical protocols which have been collected before they are forgotten and have to be re-invented, and also state-of-the-art protocols which use many of the latest molecular tools with which to study bacteriophages. It should take the new comer to bacteriophages from isolating these organisms to characterizing them at every level. It should also equip the experienced phage practitioner wishing to branch out into a new area of phage biology.

Unfortunately with time and space restrictions, it is not possible to be fully comprehensive. When we approached one scientist to contribute to this book he/she replied, "That's microbial archeology. I no longer have access to those laboratory research manuals." For similar reasons, phage immunoelectron microscopy is not covered, nor are the uses of maxi- *(43, 44)* or minicells *(45–49)* for studying phage gene expression.

We are indebted to our authors who have kindly shared their years of experience to make these volumes possible. They represent a truly multidisciplinary assemblage of scientists with a huge combined skill set. *Bacteriophages: Methods and Protocols* is a complete piece of biology, laid out in seven sections. Volume 1, Section 1 deals with methods of isolating bacteriophage (and archeophage) from a range of soil or aquatic environments using direct isolation and enrichment approaches. Volume 1, Section 2 covers the characterization of bacteriophages based upon their ability to form plaques, and their direct enumeration by fluorescent microscopy or flow cytometry. There is also a chapter on electron microscopy and a further one on classical phage taxonomy. The characterization of host range, adsorption and receptor interaction, and models of plaque development are also considered here. Finally there is a chapter on how to maintain phage stocks once you have them.

Bacteriophage–host interactions (Volume 1, Section 3) includes the construction of mutants using chemical mutagenesis or by recombineering, studies on lysogens, and transduction by temperate and lytic phages. A full scope of genomics is covered in Volume 2, Section 1, from DNA isolation and characterization (PFGE, base composition), through library construction, sequencing, annotation (termini, genes, promoters, terminators) and phylogenetics.

Volume 2, Section 2 concentrates on transcriptomics and proteomic approaches. These include: mRNA extraction during host infection, quantification of mRNA using real time PCR, and microarray construction. Isolation-independent methods of characterizing phage communities are described in Volume 2, Section 3. Volume 2, Section 4 describes the applied aspects of bacteriophage biology including phage typing, the isolation of lysins, and general and antibody phage display. There is also a final chapter to describe some online resources for phage workers.

To conclude, we hope that you find this book useful and inspiring, and we look forward to the next golden age of phage research.

References

1. Twort, F. W. (1915) *Lancet* **189**, 1241–1243.
2. d'Herelle, F. (1917) *Comptes rendus Académie Sciences* **165**, 373–375.
3. d'Herelle, F. (1926) *The bacteriophage and its behavior* (The Williams & Wilkins Company, Baltimore, MD).

4. Cairns, J., Stent, G. S., & Watson, J. D. (1966) *Phage and the Origins of Molecular Biology.* (Cold Spring Harbor Laboratory Press, Cold Spring Harbor, N.Y.).
5. Smith, G. P. (1988) in *Vectors: A survey of molecular cloning vectors and their uses*, eds. Rodrigues, L. C. & Denhardt, D. T. (Butterworths, Boston), pp. 61–83.
6. Sorge, J. A. (1988) in *Vectors: A survey of molecular cloning vectors and their uses*, eds. Rodriguez, R. L. & Denhardt, D. T. (Butterworths, Boston), pp. 43–60.
7. Denhardt, D. T. & Colasanti, J. (1988) in *Vectors: A survey of molecular cloning vectors and their uses*, eds. Rodrigues, L. C. & Denhardt, D. T. (Butterworths, Boston), pp. 179–203.
8. Brosius, J. (1988) in *Vectors: A survey of molecular cloning vectors and their uses*, eds. Rodrigues, L. C. & Denhardt, D. T. (Butterworths, Boston), pp. 205–225.
9. Mackie, G. A. (1988) in *Vectors: A survey of molecular cloning vectors and their uses*, eds. Rodrigues, L. C. & Denhardt, D. T. (Butterworths, Boston), pp. 253–267.
10. Collins, J. & Bruning, H. J. (1978) *Gene* **4**, 85–107.
11. Hohn, B., Koukolíková-Nicola, Z., Lindenmaier, W., & Collins, J. (1988) in *Vectors: A survey of molecular cloning vectors and their uses*, eds. Rodrigues, L. C. & Denhardt, D. T. (Butterworths, Boston), pp. 113–127.
12. Melnikov, A. A., Tchernov, A. P., Fodor, I., & Bayev, A. A. (1984) *Gene* **28**, 29–35.
13. Hermesz, E., Olasz, F., Dorgai, L., & Orosz, L. (1992) *Gene* **119**, 9–15.
14. Kirsanov, N. B., Mar'ina, O. V., & Ianenko, A. S. (1993) *Genetika* **29**, 1806–1810.
15. Soldatova, L. I., Sladkova, I. A., & Orekhov, A. V. (1994) *Antibiotiki i Khimioterapiia* **39**, 3–7.
16. Mead, D. A. & Kemper, B. (1988) in *Vectors: A survey of molecular cloning vectors and their uses*, eds. Rodrigues, L. C. & Denhardt, D. T. (Butterworths, Boston), pp. 85–102.
17. Cesareni, G. (1988) in *Vectors: A survey of molecular cloning vectors and their uses*, eds. Rodrigues, L. C. & Denhardt, D. T. (Butterworths, Boston), pp. 103–111.
18. Roberts, R. J., Vincze, T., Posfai, J., & Macelis, D. (2003) *Nucleic Acids Research* **31**, 418–420.
19. Brabban, A. D., Hite, E., & Callaway, T. R. (2005) *Foodborne Pathogens and Disease* **2**, 287–303.
20. Boyd, E. F., Davis, B. M., & Hochhut, B. (2001) *Trends in Microbiology* **9**, 137–144.
21. Brussow, H., Canchaya, C., & Hardt, W. D. (2004) *Microbiology & Molecular Biology Reviews* **68**, 560–602.
22. Canchaya, C., Fournous, G., & Brussow, H. (2004) *Molecular Microbiology* **53**, 9–18.
23. Canchaya, C., Fournous, G., Chibani-Chennoufi, S., Dillmann, M. L., Brussow, H., Canchaya, C., Fournous, G., Chibani-Chennoufi, S., Dillmann, M. L., & Brussow, H. (2003) *Current Opinion in Microbiology* **6**, 417–424.
24. Canchaya, C., Proux, C., Fournous, G., Bruttin, A., & Brüssow, H. (2003) *Microbiology & Molecular Biology Reviews* **67**, 238–276.
25. Fleischmann, R. D., Adams, M. D., White, O., Clayton, R. A., Kirkness, E. F., Kerlavage, A. R., Bult, C. J., Tomb, J. F., Dougherty, B. A., Merrick, J. M. et al. (1995) *Science* **269**, 496–512.
26. Banks, D. J., Beres, S. B., & Musser, J. (2005) in *Phages: Their Role in Bacterial Pathogenesis and Biotechnology*, eds. Waldor, M. K., Friedman, D. I., & Adhya, S. L. (ASM Press, Washington, D.C.), pp. 319–334.
27. Wagner, P. L., Waldor, M. K., Wagner, P. L., & Waldor, M. K. (2002) *Infection & Immunity* **70**, 3985–3993.
28. Moran, N. A., Degnan, P. H., Santos, S. R., Dunbar, H. E., Ochman, H., Moran, N. A., Degnan, P. H., Santos, S. R., Dunbar, H. E., & Ochman, H. (2005) *Proceedings of the National Academy of Sciences of the United States of America* **102**, 16919–16926.
29. Rohwer, F. (2003) *Cell* **113**, 141.
30. Lee, S., Kriakov, J., Vilcheze, C., Dai, Z., Hatfull, G. F., & Jacobs, W. R., Jr. (2004) *FEMS Microbiology Letters* **241**, 271–276.
31. Hatfull, G. F. & Sarkis, G. J. (1993) *Molecular Microbiology* 7, 395–405.
32. Ford, M. E., Sardis, G. J., Belanger, A. E., Hendrix, R. W., & Hatfull, G. F. (1998) *Journal of Molecular Biology* **279**, 143–164.
33. Ford, M. E., Stenstrom, C., Hendrix, R. W., & Hatfull, G. F. (1998) *Tubercle & Lung Disease* **79**, 63–73.
34. Hatfull, G. F., Pedulla, M. L., Jacobs-Sera, D., Cichon, P. M., Foley, A., Ford, M. E., Gonda, R. M., Houtz, J. M., Hryckowian, A. J., Kelchner, V. A. et al. (2006) *PLoS Genetics* **2**, e92.
35. Pedulla, M. L., Ford, M. E., Houtz, J. M., Karthikeyan, T., Wadsworth, C., Lewis, J. A., Jacobs-Sera, D., Falbo, J., Gross, J., Pannunzio, N. R. et al. (2003) *Cell* **113**, 171–182.
36. Breitbart, M., Felts, B., Kelley, S., Mahaffy, J. M., Nulton, J., Salamon, P., & Rohwer, F. (2004) *Proceedings of the Royal Society of*

London – Series B: Biological Sciences **271**, 565–574.
37. Breitbart, M., Salamon, P., Andresen, B., Mahaffy, J. M., Segall, A. M., Mead, D., Azam, F., & Rohwer, F. (2002) *Proceedings of the National Academy of Sciences of the United States of America* **99**, 14250–14255.
38. Suttle, C. A. & Suttle, C. A. (2005) *Nature* **437**, 356–361.
39. Mann, N. H., Clokie, M. R., Millard, A., Cook, A., Wilson, W. H., Wheatley, P. J., Letarov, A., & Krisch, H. M. (2005) *Journal of Bacteriology* **187**, 3188–3200.
40. Lindell, D., Jaffe, J. D., Johnson, Z. I., Church, G. M., Chisholm, S. W., Lindell, D., Jaffe, J. D., Johnson, Z. I., Church, G. M., & Chisholm, S. W. (2005) *Nature* **438**, 86–89.
41. Sullivan, M. B., Coleman, M. L., Weigele, P., Rohwer, F., Chisholm, S. W., Sullivan, M. B., Coleman, M. L., Weigele, P., Rohwer, F., & Chisholm, S. W. (2005) *Plos Biology* **3**, e144.
42. Clokie, M. R., Shan, J., Bailey, S., Jia, Y., Krisch, H. M., West, S., Mann, N. H., Clokie, M. R. J., Shan, J., Bailey, S. et al. (2006) *Environmental Microbiology* **8**, 827–835.
43. Mayo, O., Hernandez-Chico, C., & Moreno, F. (1988) *Journal of Bacteriology* **170**, 2414–2417.
44. East, A. K. & Errington, J. (1989) *Gene* **81**, 35–43.
45. Reeve, J. N. (1977) *Molecular & General Genetics* **158**, 73–79.
46. Reeve, J. N. & Cornett, J. B. (1975) *Journal of Virology* **15**, 1308–1316.
47. Mertens, G., Amann, E., Reeve, J. N., Mertens, G., Amann, E., & Reeve, J. N. (1979) *Molecular & General Genetics* **172**, 271–279.
48. Garcia, J. A. & Salas, M. (1980) *Molecular & General Genetics* **180**, 539–545.
49. Reeve, J. (1979) *Methods in Enzymology* **68**, 493–503.
50. Gratia, J.-P. (2000) *Genetics* **156**, 471–476.
51. Gratia, A. (1936) *Annales de l'Institut Pasteur* **57**, 652–676.
52. Adams, M. D. (1959) *Bacteriophages* (Interscience Publishers, Inc., New York).
53. Luria, S. E., Delbrück, M., & Anderson, T. F. (1943) *Journal of Bacteriology* **46**, 57–67.
54. Jiang, W., Chang, J., Jakana, J., Weigele, P., King, J., Chiu, W., Jiang, W., Chang, J., Jakana, J., Weigele, P. et al. (2006) *Nature* **439**, 612–616.
55. Wikoff, W. R., Conway, J. F., Tang, J., Lee, K. K., Gan, L., Cheng, N., Duda, R. L., Hendrix, R. W., Steven, A. C., & Johnson, J. E. (2006) *Journal of Structural Biology* **153**, 300–306.
56. Dokland, T., Lindqvist, B. H., & Fuller, S. D. (1992) *EMBO Journal* **11**, 839–846.
57. Bradley, D. E. & Bradley, D. E. (1966) *Journal of General Microbiology* **44**, 383–391.
58. Bradley, D. E. (1967) *Journal of Bacteriology* **31**, 230–314.
59. Lwoff, A., Horne, R. W., & Tournier, P. (1962) *Cold Spring Harbor Symposia on Quantitative Biology* **27**, 51–62.
60. Putman, F. W., Kozloff, L. M., & Evans, E. A. J. (1948) *Federation Proceedings* **7**, 179.
61. Espejo, R. T. & Canelo, E. S. (1968) *Virology* **34**, 738–747.
62. Hershey, A. D. & Chase, M. (1952) *Journal of General Physiology* **1**, 39–56.
63. Chu, F. K., Maley, G. F., Maley, F., & Belfort, M. (1984) *Proceedings of the National Academy of Sciences of the United States of America* **81**, 3049–3053.
64. Chu, F. K., Maley, G. F., Maley, F., & Belfort, M. (1984) *Federation Proceedings* **43**.
65. Chu, F. K., Maley, G. F., Belfort, M., & Maley, F. (1985) *Federation Proceedings* **44**.
66. Lazarevic, V., Soldo, B., Düsterhöft, A., Hilbert, H., Mauël, C., & Karamata, D. (1998) *Proceedings of the National Academy of Sciences of the United States of America* **95**, 1692–1697.
67. Lazarevic, V. (2001) *Nucleic Acids Research* **29**, 3212–3218.
68. Perler, F. B. (2006) *Nucleic Acids Research* **30**, 383–384.
69. Derbyshire, V. & Belfort, M. (1998) *Proceedings of the National Academy of Sciences of the United States of America.* **95**, 17.
70. Crick, F. H. C., Barnett, L., Brenner, S., & Watts-Tobin, R. J. (1961) *Nature* **192**, 1227–1232.
71. Wyatt, G. R. & Cohen, S. S. (1953) *Annales de l'Institut Pasteur* **84**, 143–146.
72. Guthrie, C., McClain, W. H., Guthrie, C., & McClain, W. H. (1973) *Journal of Molecular Biology* **81**, 137–155.
73. McClain, W. H., Guthrie, C., Barrell, B. G., McClain, W. H., Guthrie, C., & Barrell, B. G. (1972) *Proceedings of the National Academy of Sciences of the United States of America* **69**, 3703–3707.
74. Guthrie, C., Seidman, J. G., Altman, S., Barrell, B. G., Smith, J. D., McClain, W. H., Guthrie, C., Seidman, J. G., Altman, S., Barrell, B. G. et al. (1973) *Nature - New Biology* **246**, 6–11.
75. McClain, W. H., Guthrie, C., Barrell, B. G., McClain, W. H., Guthrie, C., & Barrell, B. G. (1973) *Journal of Molecular Biology* **81**, 157–171.
76. Bertani, G. & Weigle, J. J. (1953) *Journal of Bacteriology* **65**, 113–121.

77. Arber, W. & Dussoix, D. (1962) *Journal of Molecular Biology* **5**, 18–36.
78. Sinsheimer, R. L. (1959) *Brookhaven Symposia in Biology* **No 12**, 27–34.
79. Loeb, T. & Zinder, N. D. (1961) *Proceedings of the National Academy of Sciences of the United States of America* **47**, 282–289.
80. Semancik, J. S., Vidaver, A. K., & Van Etten, J. L. (1973) *Journal of Molecular Biology* **78**, 617–625.
81. Vidaver, A. K., Koski, R. K., & Van Etten, J. L. (1973) *Journal of Virology* **11**, 799–805.
82. Ortin, J., Viñuela, E., Salas, M., & Vásquez, C. (1971) *Nature New Biology* **234**, 275–277.
83. Fiers, W., Contreras, R., Duerinck, F., Haegeman, G., Iserentant, D., Merregaert, J., Min, J. W., Molemans, F., Raeymaekers, A., Van den, B. A. et al. (1976) *Nature* **260**, 500–507.
84. Sanger, F., Air, G. M., Barrell, B. G., Brown, N. L., Coulson, A. R., Fiddes, C. A., Hutchison, C. A., Slocombe, P. M., & Smith, M. (1977) *Nature* **265**, 687–695.
85. Benzer, S. (1955) *Proceedings of the National Academy of Sciences of the United States of America* **41**, 344–354.
86. Benzer, S. & Benzer, S. (1956) *Brookhaven Symposia in Biology* 3–5.
87. Luria, S. & Delbrück, M. (1943) *Genetics* **2**, 491–511.
88. Wollman, E. (1934) *Bulletin de L'Institut Pasteur* **32**, 945–955.
89. Bertani, G. (2004) *Journal of Bacteriology* **186**, 595–600.
90. Burnet, F. M. (1934) *Biological Review* **9**, 332–350.
91. Lederberg, E. M. & Lederberg, J. (1953) *Genetics* **38**, 51–64.
92. Lwoff, A., Siminovitch, L., & Kjeldgaard, N. (1950) *Annales de l'Institut Pasteur* **79**, 815–859.
93. Campbell, A. (1962) *Advances in Genetics* **11**, 101–145.
94. Signer, E. R. & Weil, J. (1968) *Cold Spring Harbor Symposia on Quantitative Biology* **33**, 715–719.
95. Echols, H., Gingery, R., & Moore, L. (1968) *Journal of Molecular Biology* **34**, 251–260.
96. Jacob, F. & Monod, J. (1961) *Journal of Molecular Biology* **3**, 318–356.
97. Monod, J. & Jacob, F. (1961) *Cold Spring Harbor Symposia on Quantitative Biology* **26**, 389–401.
98. Ptashne, M. (1967) *Nature* **214**, 232–234.
99. Ptashne, M. (1967) *Proceedings of the National Academy of Sciences of the United States of America* **57**, 306–313.
100. Taylor, A. L. (1963) *Proceedings of the National Academy of Sciences of the United States of America* **50**, 1043–1051.
101. Bertani, L. E. (1951) *Journal of Bacteriology* **62**, 293–300.
102. Ikeda, H. & Tomizawa, J. I. (1968) *Cold Spring Harbor Symposia on Quantitative Biology* **33**, 791–798.
103. Ravin, N. V. (2006) in *The Bacteriophages*, ed. Calendar, R. (Oxford University Press, New York, NY), pp. 448–456.
104. Freeman, V. J. & Morse, I. U. (1952) *Journal of Bacteriology* **63**, 407–414.
105. Freeman, V. J. (1951) *Journal of Bacteriology* **61**, 675–688.
106. Uetake, H., Nakagawa, T., & Akiba, T. (1955) *Journal of Bacteriology* **69**, 571–579.
107. Uetake, H., Luria, S. E., & Burrous, J. W. (1958) *Virology* **5**, 68–91.
108. Zinder, N. D. & Lederberg, J. (1952) *Journal of Bacteriology* **64**, 679.
109. Lennox, E. S. (1955) *Virology* **1**, 190–206.
110. Morse, M. L., Lederberg, E. M., & Lederberg, J. (2007) *Genetics* **41**, 142–156.
111. Weigle, J. (1957) *Virology* **4**, 14–25.
112. Ebel-Tsipis, J., Botstein, D., & Fox, M. S. (1972) *Journal of Molecular Biology* **71**, 433–448.
113. Bayer, M. E., Thurow, H., & Bayer, M. H. (1979) *Virology* **94**, 95–118.
114. Roessner, C. A., Struck, D. K., & Ihler, G. M. (1983) *Journal of Biological Chemistry* **258**, 643–648.
115. Furukawa, H., Kuroiwa, T., & Mizushima, S. (1983) *Journal of Bacteriology* **154**, 938–945.
116. Molineux, I. J. (2001) *Molecular Microbiology* **40**, 1–8.
117. Ellis, E. L. & Delbrück, M. (2007) *Journal of General Physiology* **22**, 365–384.
118. Delbrück, M. (1945) *Journal of Bacteriology* **50**, 131–135.
119. Doermann, A. H. & Dissosway, C. (1948) *Year Book Carnegie Institute of Washington* **48**, 170–176.
120. Doermann, A. H. (1952) *Journal of General Physiology* **35**, 645–656.
121. Gellert, M. (1967) *Proceedings of the National Academy of Sciences of the United States of America* **57**, 148–155.
122. Dressler, D. (1970) *Proceedings of the National Academy of Sciences of the United States of America* **67**, 1934–1942.
123. Gilbert, W. & Dressler, D. (1968) *Cold Spring Harbor Symposia on Quantitative Biology* **33**, 473–484.
124. Okazaki, T. & Okazaki, R. (1969) *Proceedings of the National Academy of Sciences of the United States of America* **64**, 1242–1248.

125. Wickner, W., Brutlag, D., Schekman, R., & Kornberg, A. (1972) *Proceedings of the National Academy of Sciences of the United States of America* **69**, 965–969.
126. Meselson, M. & Weigle, J. J. (1961) *Proceedings of the National Academy of Sciences of the United States of America* **47**, 857–868.
127. Wolfson, J., Dressler, D., & Magazin, M. (1972) *Proceedings of the National Academy of Sciences of the United States of America* **69**, 499–504.
128. Wolfson, J., Dressler, D., & Magazin, M. (1972) *Proceedings of the National Academy of Sciences of the United States of America* **69**, 499–504.
129. Watson, J. D. (1972) *Nature New Biology* **239**, 197–201.
130. Brenner, S., Jacob, F., & Meselson, M. (1961) *Nature* **190**, 576–581.
131. Astrachan, L. & Volkin, E. (1959) *Biochimica et Biophysica Acta* **32**, 449–456.
132. Volkin, E. & Astrachan, L. (1956) *Virology* **2**, 433–437.
133. Volkin, E., Astrachan, L., & Countryman, J. L. (1958) *Virology* **6**, 545–555.
134. Roberts, J. W. (1969) *Nature* **224**, 1168–1174.
135. Weber, K. & Osborn, M. (1969) *Journal of Biological Chemistry* **244**, 4406–4412.
136. Laemmli, U. K. (1970) *Nature* **227**, 680–685.
137. Dunn, J. J. & Studier, F. W. (1973) *Proceedings of the National Academy of Sciences of the United States of America* **79**, 237–248.
138. Levin, M. E., Hendrix, R. W., & Casjens, S. R. (1993) *Journal of Molecular Biology* **234**, 124–139.
139. Coppo, A., Manzi, A., Pulitzer, J. F., & Takahashi, H. (1973) *Journal of Molecular Biology* **76**, 61–87.
140. Georgopoulos, C. P., Hendrix, R. W., Casjens, S. R., & Kaiser, A. D. (1973) *Journal of Molecular Biology* **76**, 45–60.
141. Revel, R. H., Stitt, B. L., Lielausis, I., & Wood, W. B. (1980) *Journal of Virology* **33**, 366–376.
142. Sternberg, N. (1973) *Journal of Molecular Biology* **76**, 25–44.
143. Takano, T. & Kakefuda, T. (1972) *Nature New Biology* **239**, 34–37.
144. Duda, R. L., Martincic, K., Xie, Z., & Hendrix, R. W. (1995) *FEMS Microbiology Reviews* **17**, 41–46.
145. Duda, R. L. (1998) *Cell* **94**, 55–60.
146. Guo, P., Erickson, S., & Anderson, D. (1987) *Science* **236**, 690–694.
147. Dubos, R. J., Straus, J. H., & Pierce, C. (1943) *Journal of Experimental Medicine* **20**, 161–168.
148. Raettig, H. (1958) *Bacteriophagie, 1917 bis 1956; Zugleich en Vorschlag zur Dokumentation Wissenschaftlichen Literatur.* (G. Fisher, Stuttgart).
149. Raettig, H. (1967) *Bacteriophagie 1957–1965* (G. Fisher, Stuttgart).

Section 1

Bacteriophage Isolation

Chapter 1

Methods for the Isolation of Viruses from Environmental Samples

K. Eric Wommack, Kurt E. Williamson, Rebekah R. Helton, Shellie R. Bench, and Danielle M. Winget

Abstract

Viruses are omnipresent and extraordinarily abundant in the microbial ecosystems of water, soil, and sediment. In nearly every reported case for aquatic and porous media environments (soils and sediments) viral abundance exceeds that of co-occurring host populations by 10–100-fold. If current estimates based on metagenome DNA sequence data are correct, then viruses represent the largest reservoir of unknown genetic diversity on Earth. Microscopy and molecular genetic tools have been critical in demonstrating that viruses are a dynamic component of microbial ecosystems capable of significantly influencing the productivity and population biology of their host communities. Moreover, these approaches have begun to describe and constrain the immense genetic diversity of viral communities. A critical first step in the application of many cultivation-independent approaches to virus ecology is obtaining a concentrate of viruses from an environmental sample. Culture-dependent methods also rely on viruses being present at a high enough abundance to detect. Here, methodological details for the isolation and concentration of viruses from water, soil, and aquatic sediment samples are covered in detail.

Key words: Virioplankton, viral concentrate, elution, ultrafiltration, tangential flow, microporous filtration membrane.

1 Introduction

Starting with Anton van Leeuwenhoek and the first microscope *(1)* it can be argued that some of the most dramatic changes in our understanding of the biological world have come through direct observation of microorganisms within environmental samples. An appreciation of the immense diversity and biogeochemical importance of bacteria in the sea was only realized about 30 years ago with the first reports of accurate direct counts in

Martha R. J. Clokie, Andrew M. Kropinski (eds.), *Bacteriophages: Methods and Protocols, Volume 1: Isolation, Characterization, and Interactions, vol. 501*, © 2009 Humana Press, a part of Springer Science+Business Media
DOI 10.1007/978-1-60327-164-6_1 Springerprotocols.com

1977 *(2)*. This discovery also highlighted the inadequacies of cultivation-based approaches as direct counts of bacteria from an environmental sample (water, soil or sediment) typically exceed corresponding cultivation counts (colony-forming units or most probable number) by 100–1000-fold. Now known as the "great plate-count anomaly" *(3)* this phenomenon has been an important driving force in the application of sensitive molecular genetic tools to the study of bacterial diversity and ecology.

Likely owing to their small size, viruses have been the last microbial group to be examined using direct observation. The first accurate direct counts of viruses in water samples revealed that viruses are the most abundant microorganisms on Earth *(4)*. With abundances ranging from 10^4 to 10^8 viruses ml^{-1}, viruses typically outnumber co-occurring bacterial cells by 10-fold in most aquatic environments [*see (5, 6)* for review]. Although studies in aquatic environments were the first indication of the pervasiveness of viruses within microbial ecosystems; aquatic viral abundance is dwarfed by abundances in soils and aquatic sediments. The first broad report of viral direct counts in soils appeared only 6 years ago *(7, 13)*. Within the small sample set of soils examined to date, viral abundance ranges from high 10^8 to mid 10^9 viruses per gram dry weight of soil with highest abundance occurring in wetter forested soils and lower abundances in intensively managed agricultural soils *(7)*. Another 10–100-fold increase over the range of viral abundance in soil can be seen in direct counts of viruses extracted from aquatic sediments. The lowest observed viral abundances in aquatic sediments roughly equal the highest viral abundances seen in soils, $\sim 10^9$ viruses/ml^{-1} sediment. In the organic rich estuarine sediments extracted viral abundance can approach $\sim 10^{11}$ viruses ml^{-1} sediment with corresponding viral abundance in sediment pore waters comprising 1% or less of total abundance *(14, 16)*.

Despite the high abundance of viruses within aquatic and porous media environments, application of microscopy and molecular genetic tools to the study of viral ecology often requires that virus particles be extracted and concentrated within small volumes. Because the ideal objective of synecological (ecological relationship) studies is to assay all members of a community with equal efficiency, methods are chosen which minimize both artifactual bias in the concentration of viruses and provide a sample devoid of contaminants that can interfere with microscopy or molecular genetic assays. As such, the adsorption–elution methods traditionally used in cultivation-based studies of viruses in environmental samples *(8)* have been largely inappropriate for cultivation-independent studies of viral ecology. Here, we present methodological approaches for obtaining concentrates of viruses from water, soil, and sediment samples which can subsequently be applied to transmission electron and epifluorescence microscopy as well as a variety of molecular genetic analyses.

2 Materials

2.1 Large-Scale Tangential Flow Filtration of Water Samples

1. Large 20 l and 50 l polypropylene carboys with caps (Nalgene)
2. 9.5 mm diameter PharMed peristaltic pump tubing (Cole-Parmer Instrument Co.; Vernon Hills, IL; http://www.coleparmer.com/)
3. 9.5 mm diameter Tygon tubing for plumbing of water sample between filters, pumps, and reservoirs (Nalgene; Nalge Nunc International; Rochester, NY; http://www.nalgenunc.com/)
4. Cartridge sediment filter made of string wound polypropylene fiber (Kenmore, Sears, Roebuck & Co. Hoffman Estates, Il) capable of removing particles > 25 μm.
5. Filter housing for cartridge sediment filter.
6. Peristaltic pump capable of providing sufficient cross-flow for a $0.85\,m^2$ ultrafiltration membrane (e.g., M12, Millipore, Corp. Bedford, MA).
7. $0.5\ m^2 0.22\,\mu m$ Pellicon, microporous tangential flow membrane (Millipore, Corp.; Billerica, MA; http://www.millipore.com/)
8. $0.85\ m^2$ Helicon S10 30 kD molecular weight cut-off (MWCO) ultrafiltration cartridge (Millipore, Corp.)

2.2 Small-Scale Tangential Flow Ultrafiltration of Water Samples

1. Small 2 l and 500 ml polycarbonate bottles with caps (Nalgene)
2. 6.4 mm diameter PharMed peristaltic pump tubing (Cole-Parmer)
3. 6.4 mm diameter Tygon tubing for plumbing of water sample between filters, pumps, and reservoirs (Nalgene)
4. Peristaltic pump capable of providing sufficient cross-flow for a $0.1\ m^2$ ultrafiltration membrane (Cole Parmer).
5. $0.1 m^2$ Prep-Scale molecular weight cut-off (MWCO) ultrafiltration cartridge (Millipore, Corp.)
6. Stopwatch timer
7. 1 and 2 l graduated cylinders

2.3 Post-Processing of Viral Concentrates from Water Samples

1. Sterile 60 ml luer lock syringe (Becton Dickenson; Franklin Lakes, NJ; http://www.bd.com/)
2. Sterile $0.22\,\mu m$ luer lock cartridge filter, e.g., Sterivex (Millipore Corp.)
3. 50 ml conical sterile polypropylene centrifuge tubes
4. Centri-prep 80 spin filter cartridge with 30 kD MWCO ultrafiltration membrane (Millipore, Corp.)

5. Low speed benchtop centrifuge with swinging bucket rotor capable of holding 250 ml bottles.

2.4 Extraction: Soils

1. Potassium citrate (1%), per liter: 10 g $K_3C_6H_5O_7 \bullet H_2O$ (Fisher); 1.44 g $Na_2HPO_4 \bullet 7H_2O$ (Fisher); 0.24 g KH_2PO_4(Fisher); pH 7
2. Branson 185 Sonifier (Branson Ultrasonics Corp.; Danbury, CT; http://www.bransonultrasonics.com/)
3. Low speed swing-out benchtop centrifuge
4. Ice bucket
5. 50 ml conical sterile polypropylene centrifuge tubes

2.5 Extraction: Aquatic Sediments

1. Sodium pyrophosphate 10 mM and EDTA 5 mM
2. Plastic zip-close bags
3. Multi-tube vortex mixer capable of holding 50 ml centrifuge tubes
4. 0.22 μm Sterivex (Millipore, Corp.) filters
5. 15 and 50 ml conical sterile polypropylene centrifuge tubes
6. 10 ml luer lock syringes
7. Low speed swing-out benchtop centrifuge

3 Methods

Owing to the physical differences between the samples, methodological approaches for the concentration of virus particles from water, soil, and sediment samples are fundamentally different. While viral abundance in many aquatic environments exceeds 10^6 viruses ml^{-1}, the frequency of even the most abundant strain within the assemblage is <1% *(9)*. Therefore, greater than 100-fold concentration of virus particles is necessary to obtain sufficient material for downstream molecular genetic analyses (e.g., community profiling and metagenomic analyses) and cultivation. Efficient recovery and concentration of viruses from water samples requires use of ultrafiltration membranes with a retention cut-off of at least 100 kD or smaller. Typically a 30 kD molecular weight cut-off (MWCO) filter is used in conjunction with tangential flow filtration (TFF), a filtration approach that minimizes filter clogging. In TFF, the flow of the process fluid (i.e., the water sample) is directed tangentially across the filter while a slight cross-membrane pressure is applied to push water and smaller solutes through the filter membrane pores. The tangential flow of the recirculating water sample prevents filter clogging despite the >100-fold concentration of particles in the retained portion of the water sample. An excellent review of the technical details of TFF can be found in technical brief #32 from Millipore Corporation *(10)*. Several manufacturers supply TFF membranes which are appropriate for the concentration of viruses from water samples. The specific combination of pre-filtration and virus concentration filters and apparatus detailed here have been found to be 50–80% efficient in the recovery and concentration of viruses from marine water samples *(11)*. Application of these methods to filters and ultrafiltration membranes from other manufacturers will require testing to determine proper running conditions for maximal concentration efficiency. Concentration of virus particles from large volume water samples (i.e., >20 l) occurs in six steps: (1) pre-filtration to remove all particles and cells > 25 μm in size; (2) tangential flow filtration through a 0.22 μm microporous filtration membrane to remove bacterial cells; (3) virus concentration into a 2 l volume using a large (~0.85 m^2) 30 kD MWCO tangential flow filtration membrane; (4) reduction of sample volume to ~250 ml using a small (~0.1 m^2) 30 kD MWCO TFF membrane; (5) filter sterilization of viral concentrate using a 0.22 μm syringe filter; and (6) post concentration of viruses into a small, ~2 ml volume using ultrafiltration spin columns.

Obtaining a sample of virus particles from soil or aquatic sediment samples requires the extraction of virus particles from the porous media into a buffer solution. Ideally, the buffer solution should compete and displace viruses from sediment particles by disrupting the electrostatic and hydrophobic interactions between the virus and porous media *(12)*. For this reason, a proteinaceous buffer such as 10% Beef Extract has been favored for the recovery of viruses from soils and aquatic sediments. Unfortunately, samples of extracted virus particles within a beef extract solution are incompatible with many downstream analyses, especially epifluorescence microscopy *(13)*. For this reason, it is preferable to use buffered salt solutions such as potassium citrate or sodium pyrophosphate at neutral pH along with physical disruption methods such as sonication or vortex mixing to elute virus particles from the porous media sample. With elution and physical disruption methods around 60–70% of extractible viruses can be removed from a soil *(7)* or aquatic sediment *(16)* sample with one extraction. After extraction, the virus suspension is filter sterilized to remove any co-extracted bacterial cells. If an ultra-pure sample is needed, viruses within the suspension can be further concentrated and separated from contaminants on a CsCl step-gradient in an ultracentrifuge following well- established protocols [*(15)* also *see* **Chapters 21, 33,** and **34**]. Because of the extraordinary viral abundance within soils and sediments, further concentration of virus particles after extraction is typically not necessary prior to microscopy or application of molecular genetic analyses.

3.1 Concentration of viruses from large (> 20 l) volume water samples

1. Pre-filter 20–50 l water sample through a sediment cartridge filter of string wound polypropylene fiber capable of removing particles > 25 μm (**Note 1**). These filters are commercially available as home water filtration devices. The water sample should be pumped through the filter using a diaphragm or peristaltic pump and collected into an appropriately sized carboy.
2. Prior to filling, rinse the carboy 3× with a few liters of the 25 μm pre-filtered water. Rinse all carboys 3× with appropriate filtrate prior to filling.
3. Plumb the 0.22 μm 0.5 m^2 Pellicon TFF filter by placing Tygon tubing lines to the retentate and permeate ports. The tube from the carboy to the peristaltic pump head and one of the retentate ports is the feed line, while the tube from the other retentate port back to the carboy is the return line. The section of the feed line within the peristaltic pump head should be replaced by equally sized PharMed tubing, which can withstand long-term peristaltic use. The retentate ports on the Pellicon housing can serve as either the feed or return as there is no set direction for the tangential flow in the Pellicon cartridge. The permeate line is a section of tubing from the permeate port of the Pellicon filter housing to a different carboy that will collect the 0.22 μm filtered water. A schematic diagram of the plumbing for the large large-scale TFF system is shown in **Fig. 1.1**.
4. Because TFF is a pressure filtration, the direction of the peristaltic flow should be from the water sample towards the filter
5. Turn the permeate control valve off and start pumping the 25μm filtered water through the Pellicon at a low 10% pump speed and an occlusion setting of 5 on the pump head. Continue pumping at low speed until the lines and filter is completely full and all air has been completely expelled. It may be necessary to elevate the feed and return lines to assist in air expulsion, if necessary. Once the system is fully primed, adjust the occlusion setting back to 3 to minimize wear on the PharMed tubing.
6. Slowly raise the pump speed to 45% and open the permeate valve so that the ratio of retentate to permeate flow is 2:1. Relative flow rates can be determined by collecting permeate and retentate for a given amount of time with a graduated cylinder. A 2:1 flow ratio is achieved when the volume of water collected from the retentate is 2× the volume collected from the permeate for a given same amount of time.
7. As tangential flow filtration proceeds, bacterial cells and particles between 0.22 and 25μm in size will be concentrated in the retentate of the Pellicon filter. The Pellicon filter permeate will include viruses and dissolved solutes < 0.22 μm.

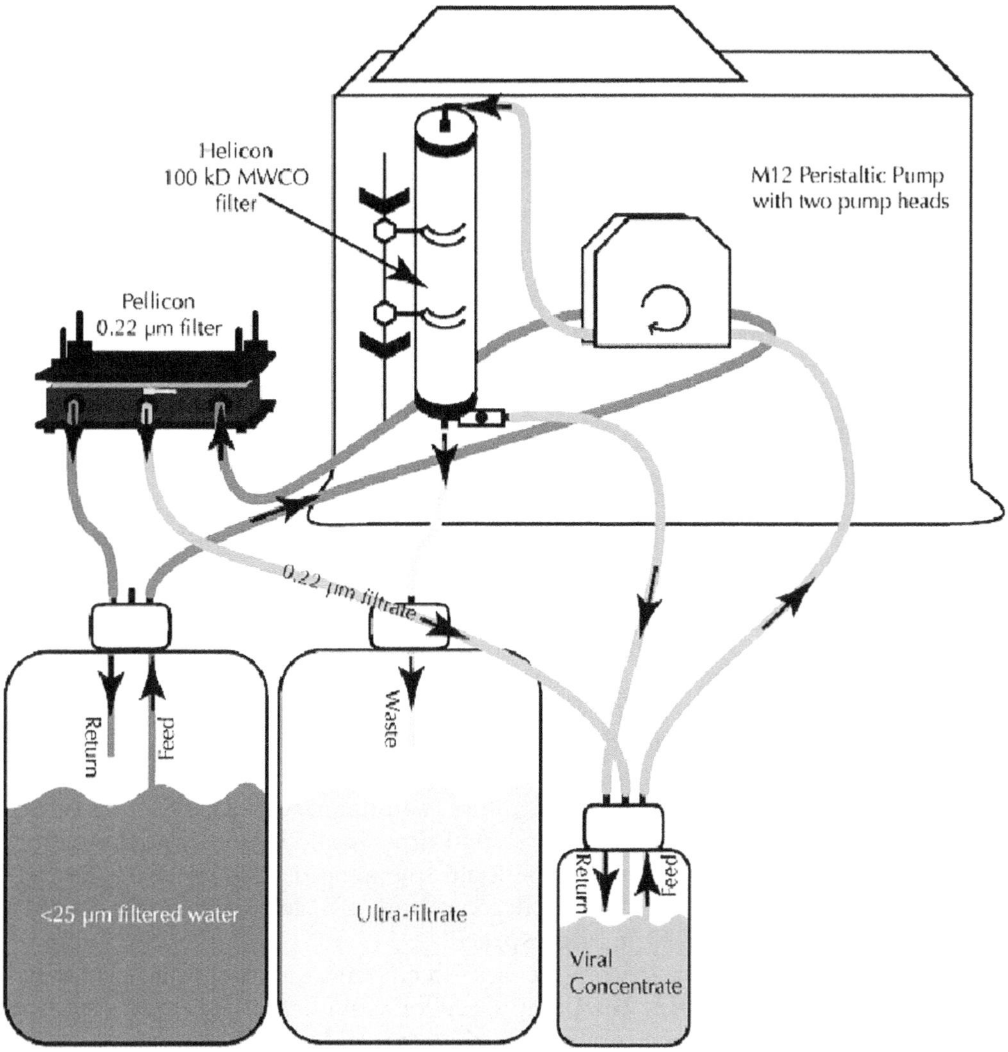

Fig. 1.1. Schematic diagram of the TFF filtration apparatus used for the concentration of viruses from water samples.

Do not allow the volume of the $> 0.22\,\mu\mathrm{m}$ retentate to fall below 1/20 of the original volume (e.g., 50 l–2.5 l; **Note 2**).

8. Once at least 20 l of $0.22\mu\mathrm{m}$ filtered water has been obtained, TFF to concentrate virus particles can begin. Use of two peristaltic pump heads on one pump allows for simultaneous operation of the Pellicon filter and $0.85\ \mathrm{m}^2$ Helicon S10 30 kD MWCO ultrafiltration cartridge.
9. After reducing the pump speed to 10% and stopping the pump, plumb the feed and return lines of the Helicon filter cartridge in a similar fashion to that of the Pellicon in step 3. Use the $< 0.22\,\mu\mathrm{m}$ permeate from the Pellicon filter as the feed water for the Helicon cartridge (**Note 3**).

10. Fully open the permeate control valve and set the occlusion knob on the pump head to 5 to allow for priming of the lines and Helicon cartridge. Start the pump at a low setting of 10%, then raise to 20% to fill the cartridge and expel air. Once the cartridge is full and all air is expelled adjust the occlusion knob back to 3.
11. Slowly raise the pump speed to 45%, then slowly close the permeate control valve until a retentate : permeate flow of 3:1 is achieved.
12. Viruses and dissolved solutes between 30 kD and 0.22 μm will be concentrated within the retentate of the Helicon filter. The permeate of the Helicon filter is effectively free of viruses.
13. Collect 1 l of the virus-free permeate in a 2 l plastic bottle for subsequent rinsing of the Helicon.
14. Collect the rest of the virus-free permeate in a 50 l carboy. This water will be used for washing the TFF filters.
15. Continue TFF until the level of the retentate of the Helicon is near the minimum hold-up volume of the filtration system, ~1 l. Do not allow the filer to run dry or bubble excessively. Open the permeate valve, reduce the pump speed to 10% and then stop the pump. Release the feed line from the peristaltic pump head and allow the Helicon filter and line to drain completely. Store the 1 l viral concentrate at 4°C.
16. Place the feed, return, and permeate lines into a 1 l volume of previously collected virus-free permeate. Place the feed line back into the peristaltic pump head, then prime the lines and filter at an occlusion setting of 5 and a pump speed of 10% until all air is expelled.
17. Recirculate the virus-free permeate through the Helicon filter at 30% pump speed for about 5 min. Avoid excessive bubbling of air during this step.
18. Reduce the pump speed to 10% stop the pump, open the pump head and drain all lines into the 2 l virus-free permeate rinse bottle.
19. Add the 1 l rinse water to the 1 l viral concentrate. Store the viral concentrate at 4°C until further TFF concentration.
20. Wash the Pellicon filter by first flushing with 25 l of virus-free permeate at a pump setting of at least 45%. Follow this initial wash with ~25 l of reverse osmosis or distilled water. Flushing is achieved by not recirculating, but rather directing the return and permeate lines to waste. Reverse the direction of tangential flow 2× during washing to assist in rinsing. Do not drain the filter. Instead, clamp the lines to keep the cartridge filled with ultrapure water.
21. Wash the Helicon in a similar fashion; however, the direction of flow cannot be reversed in this filter. Keep the cartridge filled with ultrapure water for up to 2 days (**Note 4**).

3.2 Concentration of Viruses from Small (< 2 l) Volume Water Samples

1. In processing a small water sample, remove bacteria and small particulates by filtration through a $0.22\,\mu m$ filter. Pre-filtration can be avoided with viral concentrates from the Helicon filter, as this sample is already free of bacteria and particulates.
2. Because the flow direction of the Prep-Scale filter is fixed, plumb the filter similarly to that of the Helicon filter. Fully open the permeate valve and slowly increase the pump speed to expel air from the filter. Once the pump speed has reached $1\,l\ min^{-1}$ slowly close the permeate valve until a retentate : permeate ratio of 3:1 is achieved with a steady drip out the permeate line.
3. Collect 100 ml of virus-free permeate in a 250 or 500 ml polycarbonate bottle.
4. Once the retentate volume is close to the minimum hold up volume of the Prep-Scale filter and lines ($\sim$100 ml), open the permeate control valve, and slow the pump down to a full stop. Unclamp the feed line from the peristaltic pump head and drain the system. This is the viral concentrate sample.
5. Run the feed, retentate, and permeate lines from the polycarbonate bottle containing virus-free permeate. Prime and then rinse the Prep-Scale filter by recirculating virus-free permeate for about 5 min. Avoid excessive bubbling of air through the sample.
6. Drain the filter and pool the rinse water with the viral concentrate from step #4. The final volume should be $\sim$300 ml.
7. To insure sterility of the viral concentrate, use a $0.22\,\mu m$ syringe filter to remove any unintended bacterial contamination
8. Wash the Prep-Scale filter by flushing with $\sim$2 l virus-free permeate followed by 4 l ultrapure water.
9. If the cartridges are not to be used for 2 days or longer, filter membranes should be cleaned following manufacturer's instructions. Care must be taken to ensure that cleaning and storage solutions are compatible with the filtration membrane. Test filter for integrity according to manufacturer's directions after every 5–10 uses.

3.3 Extraction and Elution of Viruses from Aquatic Sediment Samples

1. Sediment samples can be processed immediately or stored frozen at –20°C if they are to be used for viral analysis only.
2. Although it is not necessary prior to the extraction of viruses from sediment, pore water can be selectively removed from the sample by low speed centrifugation. Pore water removal can be useful in cases where specific characterization of this portion of the sediment viral community is desired. In the case of estuarine sediments, pore water viral populations comprise 5% or less of total sediment viral abundance *(16)*. If the sediment is

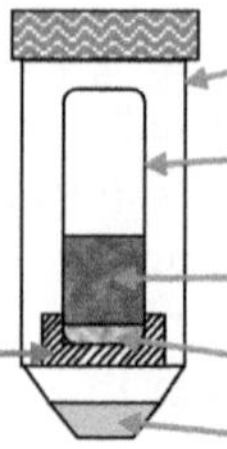

Fig. 1.2. Device for extraction of pore water from sediment samples. Use an 18 gauge needle to poke five holes in a cross pattern (i.e., X) into the cap of the Falcon tube. Place 2–5 g of sediment inside Falcon tube and recap with a GFF filter between the perforated cap and the tube opening. Invert the Falcon tube into a 50 ml centrifuge tube and centrifuge the device in a swinging-bucket rotor at $1000 \times g$, 20°C, for 20 min. Pore water will collect at the bottom of the 50 ml tube.

too thick to separate pore water by direct centrifugation, then the apparatus shown in **Fig. 1.2** can be used to remove it.

3. Place a 2 cc sediment sample in a 50 ml conical centrifuge tube and add 8 ml of 0.02 μm–filtered 10 mM sodium pyrophosphate and 8 μl of 5 mM EDTA to each tube. Samples that do or do not contain pore water can also be used for viral extraction.
4. Place tube(s) inside two plastic zip-style bags, place horizontally on a vortex mixer and vortex on the highest setting for 20 min.
5. Centrifuge tubes at $2000 \times g$, 20°C, for 25 min. Remove tubes from centrifuge carefully to prevent resuspension of the pellet.
6. Carefully decant the supernatant into a 10 ml luer lock syringe to which a 0.22 μm Sterivex (Millipore Corp.) filter has been attached. Replace plunger and gently push sample through the filter into a clean sterile 15 ml tube. Sample storage conditions or addition of biological fixatives depend on subsequent use of the sample. Typically, snap freezing in liquid N_2 followed by storage at –80°C provides the best preservation of virus particles *(16)*.

3.4 Extraction and Elution of Viruses from Soil Samples

1. Determine, using standard techniques, the dry weight and moisture content of the soil sample *(17)*.
2. Place a 5 gm dry weight soil sample into a 50 ml conical centrifuge tube and add 15 ml 1% potassium citrate extraction buffer and mix with brief vortexing.
3. Place sample on ice for 20 to 30 min.
4. Sonicate samples on ice in 3 one minute intervals with one minute rest on ice in between sonications. Use a Branson model S-450a sonicator with a 1/2" externally threaded disrupter + 3/4" and a 1/8" tapered microtip at a duty cycle setting of 30% and an output control setting of 3.
5. Centrifuge soil buffer mixture at $3000 \times g$ for 30 min. at 4°C in a low speed table top centrifuge.

6. Carefully remove the supernatant and filter through a 0.22 μm Sterivex (Millipore, Corp.) filter. The filtered supernatant contains the extracted virus particles.
7. If the sample is to be used for microscopic observation of virus particles then the sample should be placed into a cryovial, snap frozen in liquid N_2 and stored at –80°C until preparation for microscopy.

4 Notes

1. Pre-filtration is necessary as subsequent filtration of particulates > 100 μm can result in clogging of passages within the tangential flow filtration cartridges.
2. Excessive concentration of bacteria and particles between 0.22 and 25 μm makes cleaning difficult and may result in filter clogging.
3. The direction of tangential flow is set for the Helicon cartridge and should not be reversed.
4. If the cartridges are not to be used for 2 days or longer, filter membranes should be cleaned following manufacturer's instructions. Care must be taken to ensure that cleaning and storage solutions are compatible with the filtration membrane. Test filter for integrity according to manufacturer's directions after every 5–10 uses.

Acknowledgments

The authors gratefully acknowledge support from the following agencies: National Science Foundation Microbial Observatories program (grant number MCB-0132070 awarded to K.E. Wommack); the National Research Initiative of the USDA Cooperative State Research, Education, and Extension Service (grant 2005-35107-15214 to K.E. Wommack); a USDA National Needs Graduate Fellowship (awarded to S.R. Bench); an NSF predoctoral fellowship (awarded to D.M. Winget); and an EPA STAR Graduate Fellowship (U916129 to K.E. Williamson).

References

1. Ford, B.J., *The Royal Society and the microscope*. Notes Rec. R. Soc. Lond., 2001. **55**(1): 29–49.
2. Hobbie, J.E., R.J. Daley, and S. Jasper, *Use of nucleopore filters for counting bacteria by fluorescence microscopy*. Appl. Environ. Microbiol., 1977. **33**: 1225–1228.
3. Staley, J.T. and A. Konopka, *Measurement of in situ activities of nonphotosynthetic microorganisms in aquatic and terrestrial habitats*. Annu. Rev. Microbiol., 1985. **39**: 321–46.
4. Bergh, O., et al., *High abundance of viruses found in aquatic environments*. Nature (London), 1989. **340**: 467–468.

5. Weinbauer, M.G., *Ecology of prokaryotic viruses.* FEMS Microbiol. Rev., 2004. **28**(2): 127–81.
6. Wommack, K.E. and R.R. Colwell, *Virioplankton: Viruses in aquatic ecosystems.* Microbiol. Molec. Biol. Rev., 2000. **64**: 69–114.
7. Williamson, K.E., M. Radosevich, and K.E. Wommack, *Abundance and diversity of viruses in six Delaware soils.* Appl. Environ. Microbiol., 2005. **71**(6): 3119–25.
8. Sobsey, M.D., et al., *Development and evaluation of methods to detect coliphages in large volumes of water.* Water Sci. Technol., 2004. **50**(1): 211–7.
9. Edwards, R.A. and F. Rohwer, *Viral metagenomics.* Nat. Rev. Microbiol., 2005. **3**(6): 504–10.
10. Millipore. *Protein concentration and diafiltration by tangential flow filtration.* 2003 6 (cited; Available from: http://www.millipore.com/publications.nsf/docs/tb032.
11. Suttle, C.A., A.M. Chan, and M.T. Cottrell, *Use of ultrafiltration to isolate viruses from seawater which are pathogens of marine phytoplankton.* Appl. Environ. Microbiol., 1991. **57**: 721–726.
12. Gerba, C.P., *Applied and theoretical aspects of virus adsorption to surfaces.* Adv. Appl. Microbiol., 1984. **30**: 133–68.
13. Williamson, K.E., K.E. Wommack, and M. Radosevich, *Sampling natural viral communities from soil for culture-independent analyses.* Appl. Environ. Microbiol., 2003. **69**(11): 6628–33.
14. Danovaro, R., et al., *Determination of virus abundance in marine sediments.* Appl. Environ. Microbiol., 2001. **67**(3): 1384–1387.
15. Sambrook, J. and D.W. Russell, *Molecular cloning: a laboratory manual.* 2001, Cold Spring Harbor: Cold Spring Harbor Laboratory.
16. Helton, R.R., L. Liu, and K.E. Wommack, *Assessment of factors influencing direct enumeration of viruses within estuarine sediments.* Appl. Environ. Microbiol., 2006. **72**(7): 4767–74.
17. Dane, J.H. and G.C. Topp, eds. *Methods of Soil Analysis, Part 4, Physical Methods.* Soil Science Society of America Book Series. Vol. 5. 2002, Soil Science Society of America: Madison, WI. 1692.

Chapter 2

Bacteriophage Enrichment from Water and Soil

Rohan Van Twest and Andrew M. Kropinski

Abstract

Classical bacterial enrichment devised by Sergius Winogradsky (1856–1953) and Martinus Beijerinck (1851–1931) can be modified to enrich for bacteria-specific viruses. In this chapter simple protocols are presented for the enrichment of phages from water samples, such as sewage, and soil.

Key words: Enrichment, soil, sewage, *Bacillus*, *Pseudomonas*, enterics, enrichment bias, membrane filtration, water.

1 Introduction

Bacteriophages are the most abundant life forms on earth. They play major roles in bacterial ecology, adaptation to novel environments, and in bacterial evolution *(1–5)* and pathogenesis *(6)*. Bacteriophages are common in soils and direct counting methods have found 10^7–10^9 per gram of soil *(7, 8)*. Viruses are also highly abundant in freshwater and ocean waters and may approach 10^7 per milliliter *(9–13)*. Global numbers have been estimated as 10^{31}. However, specific phages are sometimes difficult to isolate directly so enrichment procedures are often required. Unfortunately, this technique will limit the diversity of phages that are isolated since any enrichment protocol induces a bias in that the phage which can propagate most efficiently under the experimental conditions predominate *(14)*.

In this chapter we will describe simple enrichment techniques which have been used successfully in both postgraduate research and undergraduate laboratory exercises. For example it has been effectively used at the University of British Columbia

Martha R. J. Clokie, Andrew M. Kropinski (eds.), *Bacteriophages: Methods and Protocols, Volume 1: Isolation, Characterization, and Interactions, vol. 501*, © 2009 Humana Press, a part of Springer Science+Business Media
DOI 10.1007/978-1-60327-164-6_2 Springerprotocols.com

(Vancouver, Canada), Queen's University (Kingston, Ontario, Canada), and the University of Guelph (Ontario, Canada). In addition to the basic enrichment procedure, we will describe a new technique developed by Francisco Lucena's group at Universidad de Barcelona (Spain) which they have extensively utilized for studying phage ecology *(15–17)*.

Sewage treatment plants are ideal sources from which bacteriophages are isolated. Due to their different ecosystems including raw input sewage, and aerobic and anaerobic digestion tanks, they contain high numbers of diverse bacteria. Alternative sources include pond, lake, river or ocean waters.

2 Materials

2.1 Collecting Materials

1. Soil samples collected in sterilized screw-capped jars
2. 1 l samples of pond water, pond sediment, raw sewage etc., collected in sterilized Erlenmeyer flasks or screw-capped bottles (**Note 1**). This can best be accomplished by using a can or wide-necked plastic jar nailed to a long pole. It is recommended that one obtain water samples at least several days after a heavy rain since the latter will dilute the sewage.

2.2 Enrichment of Phage from Aqueous Materials

1. Centrifuge (such as Beckmann Avanti series centrifuge with rotors for 50–250 ml samples)
2. Sterile double-strength Difco Luria broth (LB) or Tryptic Soy broth (TSB) supplemented with 2 mM $CaCl_2$, stored in 100 or 500 ml sterile Pyrex screw-capped bottles.
3. Sterile 125 ml Erlenmeyer flasks.
4. Overnight 5–10 ml broth cultures of the host bacterium grown in single-strength LB or TSB
5. Agar plates containing LB or TSB supplemented with 1 mM $CaCl_2$.

2.3 Phage Concentration from Aqueous Samples Using Membrane Filter Adsorption and Elution

1. Centrifuge (such as Beckmann Avanti series centrifuge with rotors for 50–250 ml samples)
2. Solid $MgSO_4$
3. 0.22 μm (47 mm diameter) mixed cellulose ester GSWP filter (Millipore Corp.; Billerica, MA; http://www.millipore.com/; catalogue number: GSWP04700).
4. Eluting solution; 1–3% (w/v) Bacto beef extract – 3% (v/v) Tween 80 (polyoxyethylene sorbitan monooleate; Sigma-Aldrich St. Louis, MO; http://www.sigmaaldrich.com) – 50 mM NaCl.
5. Ultrasonic cleaning bath
6. Overnight 5–10 ml broth cultures of the host bacterium grown in single-strength LB or TSB

2.4 Phage Enrichment from Soil Environments

1. Sterile Difco Tryptic soy broth
2. Sterile 15 ml plastic Falcon tubes
3. Benchtop centrifuge
4. Overnight 5–10 ml broth cultures of your host bacterium grown in TSB

2.5 The Underlay Protocol for Rapid Purification of Phages

1. LB or TSB agar plate (supplemented with 1 mM $CaCl_2$)
2. Overnight 5–10 ml broth cultures of the host bacterium grown in LB or TSB
3. Tube containing 3 ml of overlay medium (TSB or LB, containing 1 mM $CaCl_2$, and 0.6% wt/vol agar)

3 Methods

3.1 Enrichment

1. Centrifuge the sewage suspensions at $10,000 \times g$ for 10 min to remove particulates. The supernatant can be used directly in enrichments involving rapidly dividing bacterial cultures (enterics, *Pseudomonas, Bacillus*; alternatively, it can be filter sterilized by passage through $0.2\,\mu m$ low protein binding membrane filters: Millipore Express (PES) or Durapore (PVDF), Pall-Gelman Supor membrane (PES), Whatman inorganic Anopore membranes. Membrane chemistries with reported low protein binding are polytetrafluoroethylene, polypropylene, polycarbonate, polysulfone, and cellulose acetate.
2. The supernatant can be stored at 4°C in the presence of several ml of chloroform in a glass or solvent-resistant plastic bottle.
3. For aerobic hosts, pipette 10 ml of sterile double-strength broth containing 2 mM $CaCl_2$ into a 125 ml Erlenmeyer flask, and add 10 ml of clarified (and filtered) sewage (**Notes 2, 3** and **4**).
4. Inoculate the flask with 0.1 ml of an overnight broth culture of the desired host bacterium and incubate at the appropriate growth temperature with gentle mixing (50 rpm).
5. After 24–48 h incubation, centrifuge the contents of the flask at $10,000 \times g$ for 10 min. Decant the supernatant into a small screw-capped bottle or a series of capped test tubes. N.B. If you do not use filtered sewage, the bacterial spores which are present in the clarified sewage will germinate and after 48 h incubation the smell will elicit very negative remarks from your coworkers.
6. As a general protocol add approximately 0.5 ml of chloroform to the clarified crude lysate, shake and store at 4°C (**Note 5**).

3.2 Testing Enrichment for Phage

At this stage of the enrichment it is recommended that you spot-test your lysate on the host cells to see whether it contains any phage active on your host bacterium. This can be accomplished in the following manner:

1. Spread loopfuls of host bacteria down the surface of agar plates prepared with the same medium supplemented with Ca^{2+}.
2. Allow the fluid to dry and then place 5 μl of the clarified enrichments on the streaks, incubate overnight and observe for zones of cell killing (**Fig. 2.1**).

3.3 Phage Concentration from Aqueous Samples Using Membrane Filter Adsorption and Elution (16)

1. Clarify the water sample by centrifugation at $10,000 \times g$ for 10 min to remove large particulates and bacteria.
2. Add solid $MgSO_4$ to 50 mM.
3. Filter the water sample slowly through the 0. 22 μm (47 mm diameter) mixed cellulose ester GSWP filter.
4. Cut the filter into a number of pieces which are placed in a flask with 5 ml of Eluting solution.
5. Place in an ultrasonic cleaning bath for 4 min to aid in the elution of the viruses.

This physically enriched preparation can be directly plated or processed using the "Underlay Procedure" described in **Section 3.5** of this chapter.

3.4 Phage Enrichment from Soil Using Bacillus as the Model Bacterium

1. Collect at least 5 g of soil into a sterile container and note the source and nature of the soil (moist or dry, texture [coarse, medium, fine], and type [clay, loam, sand, compost]; **Notes 6, 7** and **8**).
2. Weigh out 0.5 g of soil and add to a sterile 15 ml centrifuge tube. Add 4.5 ml of TSB to sample (i.e., a 1:10 dilution) and mix thoroughly by inversion (**Notes 9** and **10**).

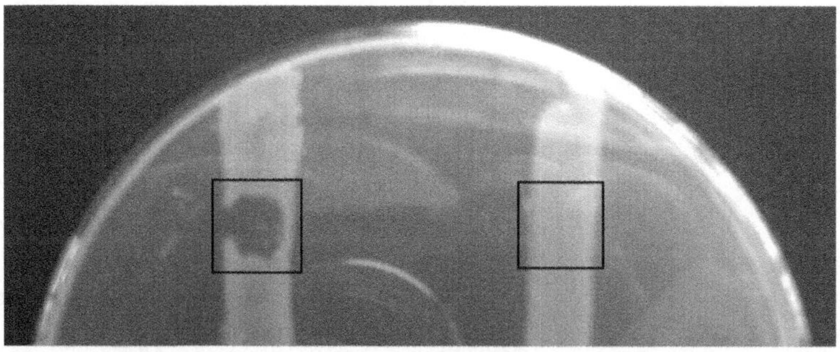

Fig. 2.1. Spot test on phage on a bacterial streak (*left boxed area*). A control, in which broth was spotted on the streak, is shown on the *right*. Photograph kindly provided by Erika J. Lingohr (PHAC-LFZ).

3. Allow the sample to incubate at room temperature for at least 1 h to allow free phage to be suspended in the liquid component. Invert frequently during the incubation period to increase the disruption of particulate material and to distribute the phage throughout the solution.
4. Centrifuge the sample in a benchtop centrifuge at full speed for 5 min. Remember to use a balance tube before starting the centrifuge.
5. One can then proceed as outlined in the previous protocol starting from step 3. Alternatively, we have found the following protocol effective for isolating *Bacillus* phages:
6. Pipette 3 ml of the supernatant into a sterile 15 ml centrifuge tube and inoculate with 0.1 ml of an overnight (ON) culture of *Bacillus*.
7. Incubate overnight (ON) with shaking at 30°C to allow the growth of the bacterium and specific phage enrichment.
8. After ON enrichment, centrifuge at 10,000 × *g* for 10 min and carefully decant the supernatant to another sterile tube containing several drops of chloroform. Label the tube appropriately and store at 4°C.

After either enrichment, the phage suspension should be titered using one of the protocols described in **Chapter 7**. Individual plaques can then be picked, resuspended in fresh medium, diluted, and plated. As an alternative to the techniques offered in **Chapter 7** the underlay technique provides rapid single plaque purification.

3.5 Underlay Procedure for Phage Purification

1. Resuspended phage particles from a well-isolated plaque in 1 ml of sterile broth.
2. Streak an LB or TSB agar plate (supplemented with 1 mM $CaCl_2$) as though one were attempting to obtain single colony isolates from a bacterial culture. Allow the plate to dry at room temperature.
3. Prepare an overlay containing host cells as described in **Chapter 7** and pour it over the surface of the plate starting from the region containing the most dilute phage particles and allowing it to diffuse towards the primary inoculation area.
4. After the overlay sets (15 min) incubate the plate. Individual well-separated plaques will be visible after overnight incubation.

4 Notes

1. It is recommended that careful attention should be paid to collecting the samples in such a way as not to contaminate the outside of the storage container or the person collecting

the samples. Please note that raw sewage will most definitely contain human pathogens.

2. For the bacteria listed above, Difco Luria broth or Tryptic Soy Broth work well. For bacteria not mentioned above we would suggest initially using the medium recommended for propagating the host bacterium (see for example the American Type Culture Collection (ATCC at http://www.atcc.org/Home.cfm) or the Deutsche Sammlung von Mikroorganismen und Zellkulturen (DSMZ at http://www.dsmz.de/).
3. Many phages require 1–10 mM divalent ions such as Ca^{2+} or Mg^{2+} for attachment or intracellular growth *(18–20)*. It is a good idea when working with phage mixtures or uncharacterized phages to include 1–2 mM Ca^{2+} in the all media.
4. Please note that lipid-containing phages (members of the *Cystoviridae* [φ6], *Corticoviridae* [PM2], *Plasmaviridae* [L2], and *Tectiviridae* [PRD1]) will be inactivated by this treatment. In addition, certain members of the *Inoviridae*, while lacking lipids, are also solvent sensitive. In the latter cases the primary lysate could be filter sterilized.
5. As with the sewage sample you do not want to collect a soil sample immediately after a heavy rain. Our lab experience had been that moist soils are much less successful as a source for *Bacillus* phages than dry soil; in fact the class soil that has been sitting out in the open for more than 5 years is our fallback source. We find if the freshly collected soil is dried overnight at 37°C the success rate for enrichment improves significantly.
6. For publication purposes you might want to consult a soil expert to critically evaluate the type of soil sample that you have used.
7. The type of soil can make about a fivefold difference in the absolute numbers of viral particles with forest soils having higher numbers than agricultural soil *(7)*.
8. Williamson et al. *(7, 8)* tested a number of solutions for their efficacy in releasing phage from soil. These include 10% beef extract in 0.05 M disodium phosphate – 5.7 mM citric acid monohydrate (pH 9.0); 1% potassium citrate – 5.4 mM $Na_2HPO_4.7H_20$–1.8 mM KH_2PO_4 (pH 7). Other eluants included water *(21)*; 5–10 mM sodium pyrophosphate (pH 7.0) *(8, 22, 23)*; and, trypticase soy broth (pH 9.0) *(13)*. The organic-based solutions appear to function better than the inorganic-based solutions.
9. While many of the protocols call for vigorous mixing using a Vortex-type mixer or even sonication, we think that this approach may damage many of the larger viruses. We therefore recommend mixing by inversion.

References

1. Daubin, V. and H. Ochman. 2004. Bacterial genomes as new gene homes: the genealogy of ORFans in *E. coli*. Genome Research *14*:1036–1042.
2. Miao, E.A. and S.I. Miller. 1999. Bacteriophages in the evolution of pathogen-host interactions. Proceedings of the National Academy of Sciences of the United States of America *96*:9452–9454.
3. Ai, Y., F. Meng, and Y. Zeng. 2000. The evolution of pathogen-host interactions mediated by bacteriophages. Wei Sheng Wu Hsueh Pao – Acta Microbiologica Sinica *40*: 657–660.
4. Dobrindt, U. and J. Reidl. 2000. Pathogenicity islands and phage conversion: evolutionary aspects of bacterial pathogenesis. Ijmm International Journal of Medical Microbiology *290*:519–527.
5. Canchaya, C., G. Fournous, S. Chibani-Chennoufi, M.L. Dillmann, and H. Brussow. 2003. Phage as agents of lateral gene transfer. Current Opinion in Microbiology *6*: 417–424.
6. Sakaguchi, Y., T. Hayashi, K. Kurokawa, K. Nakayama, K. Oshima, Y. Fujinaga, M. Ohnishi, E. Ohtsubo, et al. 2005. The genome sequence of *Clostridium botulinum* type C neurotoxin-converting phage and the molecular mechanisms of unstable lysogeny. Proceedings of the National Academy of Sciences of the United States of America *102*: 17472–17477.
7. Williamson, K.E., M. Radosevich, and K.E. Wommack. 2005. Abundance and diversity of viruses in six Delaware soils. Applied & Environmental Microbiology *71*:3119–3125.
8. Williamson, K.E., K.E. Wommack, M. Radosevich, K.E. Williamson, K.E. Wommack, and M. Radosevich. 2003. Sampling natural viral communities from soil for culture-independent analyses. Applied & Environmental Microbiology *69*:6628–6633.
9. Rohwer, F. 2003. Global phage diversity. Cell *113*:141.
10. Breitbart, M., L. Wegley, S. Leeds, T. Schoenfeld, and F. Rohwer. 2004. Phage community dynamics in hot springs. Applied & Environmental Microbiology *70*: 1633–1640.
11. Filee, J., F. Tetart, C.A. Suttle, H.M. Krisch, J. Filee, F. Tetart, C.A. Suttle, and H.M. Krisch. 2005. Marine T4-type bacteriophages, a ubiquitous component of the dark matter of the biosphere. Proceedings of the National Academy of Sciences of the United States of America *102*:12471–12476.
12. Kepner, R.L., Jr., R.A. Wharton, Jr., and C.A. Suttle. 1998. Viruses in Antarctic lakes. Limnology & Oceanography *43*:1754–1761.
13. Paul, J.H., J.B. Rose, S.C. Jiang, C.A. Kellogg, L. Dickson, J.H. Paul, J.B. Rose, S.C. Jiang, et al. 1993. Distribution of viral abundance in the reef environment of Key Largo, Florida. Applied & Environmental Microbiology *59*:718–724.
14. Dunbar, J., S. White, and L. Forney. 1997. Genetic diversity through the looking glass: Effect of enrichment bias. Applied & Environmental Microbiology *63*:1326–1331.
15. Mendez, J., A. Audicana, M. Cancer, A. Isern, J. Llaneza, B. Moreno, M. Navarro, M.L. Tarancon, et al. 2004. Assessment of drinking water quality using indicator bacteria and bacteriophages. Journal of Water & Health *2*:201–214.
16. Mendez, J., A. Audicana, A. Isern, J. Llaneza, B. Moreno, M.L. Tarancon, J. Jofre, F. Lucena, et al. 2004. Standardised evaluation of the performance of a simple membrane filtration-elution method to concentrate bacteriophages from drinking water. Journal of Virological Methods *117*:19–25.
17. Lucena, F., F. Ribas, A.E. Duran, S. Skraber, C. Gantzer, C. Campos, A. Moron, E. Calderon, et al. 2006. Occurrence of bacterial indicators and bacteriophages infecting enteric bacteria in groundwater in different geographical areas. Journal of Applied Microbiology *101*:96–102.
18. Haberer, K. and J. Maniloff. 1982. Adsorption of the tailed mycoplasma virus L3 to cell membranes. Journal of Virology *41*:501–507.
19. Landry, E.F. and R.M. Zsigray. 1980. Effects of calcium on the lytic cycle of *Bacillus subtilis* phage 41c. Journal of General Virology *51*:125–135.
20. Mahony, D.E., P.D. Bell, and K.B. Easterbrook. 1985. Two bacteriophages of *Clostridium difficile*. Journal of Clinical Microbiology *21*:251–254.
21. Ashelford, K.E., M.J. Day, and J.C. Fry. 2003. Elevated abundance of bacteriophage infecting bacteria in soil. Applied & Environmental Microbiology *69*:285–289.
22. Danovaro, R., E. Manini, and A. Dell'Anno. 2002. Higher abundance of bacteria than viruses in deep Mediterranean sediment. Applied & Environmental Microbiology *68*:1468–1472.
23. Danovaro, R., A. Dell'Anno, M. Serresi, and S. Vanucci. 2001. Determination of virus adundance in marine sediments. Applied & Environmental Microbiology *67*:1384–1387.

Chapter 3

Isolation of Phage via Induction of Lysogens

Raúl R. Raya and Elvira M. Hébert

Abstract

Most bacterial cells carry prophage genomes either integrated into the host DNA or present as repressed plasmids. Methods are described for the induction of prophages using Mitomycin C, and for the isolation of prophage-cured bacterial cell lines.

Key words: Lysogen, induction, lytic cycle, lysogenic cycle, prophage, integration, SOS response, mitomycin, RecA, repressor.

1 Introduction

After infection, a temperate bacteriophage can follow two alternative cycles of replication, lytic or lysogenic. In the lytic cycle, the phage replicates its genome and assembles hundreds of new progeny which are then released after cell lysis. In the lysogenic cycle, the temperate phage genome interacts reversibly with the host genetic apparatus in such a way which does not lead to multiplication, but allows the viral genome to replicate in synchrony with the host DNA replication and cell division. The phage genome remaining in the host, either as a plasmid or integrated into the host chromosome is termed the prophage. Bacteria carrying a prophage are termed lysogens since, under certain conditions the prophage is induced into the lytic cycle of phage replication.

Lysogeny is widespread in bacteria *(1)*. Comparative analysis of whole bacterial genomes *(2)* has shown that, with few exceptions, prophages are quantitatively very important components of bacterial genomes (e.g., the polylysogenic Sakai strain

Martha R. J. Clokie, Andrew M. Kropinski (eds.), *Bacteriophages: Methods and Protocols, Volume 1: Isolation, Characterization, and Interactions, vol. 501*, © 2009 Humana Press, a part of Springer Science+Business Media
DOI 10.1007/978-1-60327-164-6_3 Springerprotocols.com

of *Escherichia coli* O157:H7 contains 18 prophages; about 12% of the complete bacterial genome corresponds to prophage DNA sequences). The temperate phages include dsDNA tailed members of the viral families *Myoviridae, Siphoviridae,* and*Podoviridae*, as well as members of the *Inoviridae* family (ssDNA; filamentous). Most prophage genomes share the same genetic organization: they are organized in two clusters of genes each controlling related functions, and which are transcribed divergently. One of these clusters comprises genes participating in the integration and maintenance of lysogeny; the other includes genes involved in the lytic life cycle. These genomes have a mosaic structure of conserved sequences interspersed by non-homologous regions *(3, 4)*. These data suggest that prophages evolve by horizontal DNA transfer through the exchange of modules via homologous recombination *(4, 5)*.

Prophages and phage-like elements are among the major contributors to bacterial diversity and bacterial evolution *(4–7)*. They provide the infected bacteria with immunity against superinfection by other related phages, modify the genome structure, and can be passive vectors for the transfer of virulence genes (via transduction or lysogenic conversion) or active components in the regulation of bacterial pathogenesis. Some prophages encode a wide variety of putative and established virulence factors which may alter the phenotype of lysogenized cells resulting in the production of toxins, virulence factors or expression of modified cell surface antigens [*see* also *(2, 4–6, 8–10)*].

In the bacteriophage lambda genetic switch, the paradigm for lysogeny, the choice between lysis and lysogeny depends on the relative expression rates of the regulatory proteins CI , which promotes lysogeny, and Cro, which favors the lytic cycle *(11)*. During lysogeny, the key regulator is the CII protein, which represses transcription from the lytic promoters as well as positively regulating its own synthesis.

The stability of CII is determined by factors that measure the energy level of the cell including the levels of intracellular signaling molecule cyclic AMP (cAMP). When cells are starved, the cAMP concentration is high, which promotes CII stabilization and thus lysogeny. However when cells have sufficient energy they have low intracellular levels of cAMP and a reduced level of lysogenization, (due to enhanced proteolysis of CII). The effect of cAMP levels can be exploited during experimentation as the starvation of cells prior to phage adsorption enhances lysogeny. A further way to promote lysogenization is to infect cells at high multiplicities of infection.

It is clearly an advantage for a phage entering a new cell to be able to sense the cell density and determine whether there is sufficient energy to enter the lytic cycle and make a large burst size or whether the energy level is low, so the best strategy for survival is to enter the prophage state. The transition from lysogeny to lytic development, called lysogenic induction or prophage induction, occurs on activation of the SOS system in response to DNA damage (i.e., after treatment with mitomycin C). At this point the phage makes a last ditch opportunity for propagation.

Upon induction of the SOS response, the activated RecA protein degrades the CI repressor. The prophage is then excised from the bacterial genome and the lytic pathway resumes. As stated, the energetic state of the cell has significant effects on the lytic phage infection process, so phage induction is highly affected by what the host was experiencing shortly prior to induction, as well as by nutrients and other conditions present at the moment of induction. Very poor growth conditions favor the lysogenic pathway while good growth conditions favor the lytic response *(11)*.

In this chapter, we present the most important facts concerning temperate bacteriophages and their induction conditions with special reference to phages of lactic acid bacteria. We must point out that although the general features we have described are applicable to a wide range of phages, nearly all strains of phage differ from one another in their development and thus the protocols will have to be modified for the specific phages under study.

2 Materials

2.1 Culture Media

1. **MRS (for lactobacilli; 12)**: 10 g/l peptone, 10 g/l meat extract, 5 g/l yeast extract, 20 g/l glucose, 5 g/l sodium acetate, 1 g/l Tween 80, 2 g/l ammonium citrate, 2 g/l K_2HPO_4, 0.2 g/l $MgSO_4.7H_2O$, and 0.05 g/l $MnSO_4.4H_2O$; pH 6.5. Autoclave at 15 psi [103.4 kPa] for 15 min at 121°C. Add agar at 1.5% (w/v) to make media "bottom agar," and at 0.65% (w/v) to prepare media "soft agar." If required, add $CaCl_2$ to a final concentration of 10 mM.
2. **M17-glu (for lactococci; 13):** 5 g/l phytone peptone, 5 g/l polypeptone, 2.5 g/l yeast extract, 5 g/l beef extract, 0.5 g/l ascorbic acid, 5 g/l glucose, 19.0 g/l beta-disodium glycerophosphate, 1 ml 1 *M* $MgSO_4.7H_2O$. Autoclave for 15 min at 121°C. Add agar at 1.5% (w/v) to make media "bottom agar," and at 0.65% (w/v) to prepare media "soft agar." If required, add $CaCl_2$ to a final concentration of 10 mM.

Media can be stored in bottles at room temperature for 4–6 months.

2.2 Solutions and Buffers

Mitomycin C 0.5 μg/μl (MitC stock solution)

Mitomycin	2 mg
Add sterile 0.1 M $MgSO_4$ to	4 ml

Mitomycin C 50 ng/μl (working solution)
Dilute 1/10 MitC stock solution in sterile 0.1 *M* $MgSO_4$

1 *M* $CaCl_2$

Calcium chloride anhydrous	11.1 g
Add distilled water to	100 ml
Autoclave for 20 min at 121°C	

0.1 *M* $MgSO_4$

Magnesium sulfate anhydrous	12.0 g
Add distilled water to	100 ml
Autoclave for 20 min at 121°C	

0.1 *N* NaOH

Sodium hydroxide	0.2 g
Add distilled water to	50 ml

1 *M* Tris–HCl, pH 7.6

Tris(hydroxymethyl) aminomethane	12.1 g
Add distilled water to	90 ml
Adjust pH to 7.6	
Bring final volume with water to	100 ml
Autoclave for 20 min at 121°C	

Phage buffer

10 mM Tris–HCl, pH 7.6	1 ml 1 *M* of Tris–HCl, pH 7.6
NaCl (sodium chloride) 0.4%	0.4 g
gelatin 0.1%	0.1 g
Autoclave for 20 min at 121°C	

2.3 Equipment and Materials

Water bath set at the optimal growth temperature of the bacterium in study (i.e., 30°C for lactococci and mesophilic lactobacilli and 37°C for thermophilic lactobacilli).
Water bath set at 45°C
Incubators set at 30°C or 37°C
Spectrophotometer (absorbance at 600 nm)
Refrigerated centrifuge
pH meter
Shaker
UV box with General Electric G1578 15 watt germicidal short wave lamp
Membrane filter of 0.45 μm pore size
Sterile tubes
Sterile swabs
Drigalski spatula (aka glass hockey stick)
Sterile 1- and 5-ml pipettes

Sterile centrifuge tube with cap
Sterile glass Petri dish
Micropipettes and sterile tips
A 24 h culture of the strain to be tested for lysogeny
MRS or M17 broth tube (5 ml) in screw-capped tube
MRS or M17 broth with 0.05*M* $MgSO_4$ (5 ml)
Double strength MRS or M17 broth (5 ml)
1/10 strength MRS or M17 broth
Petri dishes containing bottom agar plus 10 mM $CaCl_2.6H_2O$
A bottle containing 50 ml soft agar plus 10 mM $CaCl_2.6H_2O$
Petri dishes with bottom agar media
Ethanol

3 Methods

Unless otherwise stated, media MRS and M17-glu are used, respectively, for lactobacilli and lactococci. Mesophilic lactobacilli and lactococci are incubated at 30°C, while thermophilic lactobacilli are incubated at 37°C.

3.1 Bacteriophage Induction with Mitomycin C (Note 1–3)

1. Inoculate 5 ml of fresh broth growth media with 50 μl of an overnight culture of the strain to be tested for lysogeny. Measure the initial absorbance at 600 nm. Incubate in a water bath for 30 min at the optimal bacterial growth temperature (**Note 2**).
2. Add mitomycin C to a final concentration of 0.1–0.5 μg/ml (**Notes 4–5**). Measure the absorbance at 600 nm each hour for 6–8 h (or until a decrease of the optical density is observed).
3. Centrifuge the culture at 3000 × g for 12 min at 4°C.
4. Neutralize the supernatant to pH 7.0 with 0.1 *N* NaOH.
5. Filter the supernatants through a membrane filter of 0.45 μm of pore size.
6. Store the sterile supernatant at 4°C (**Note 12**).

3.2 Bacteriophage Induction by Ultraviolet Light (Notes 1–3)

1. Transfer 100 μl of an overnight bacterial culture of the strain to be tested for lysogeny to 5 ml fresh broth (MRS or M17).
2. Incubate at 30 or 37°C for 3 h.
3. Transfer cells to sterile centrifuge tubes and centrifuge at 6000 × g for 10 min at room temperature.
4. Resuspend the cells in 5 ml sterile 0.1*M* $MgSO_4$.
5. Transfer to a sterile glass Petri dish.
6. Irradiate with constant swirling (120 rpm) for 20–30 s. Petri dish should be placed 16 cm from the germicidal short wave lamp.

7. Transfer cells to a screw-capped tube containing 5 ml double strength (MRS or M17) broth.
8. Incubate at 30°C (lactococci) or 37°C (lactobacilli) in a constant temperature water bath. Protect cells from light exposure.
9. Read absorbance at 600 nm every 45 min for 6 h or until there is a marked decrease in the absorbance reading (use MRS or M17 broth with 0.05*M* $MgSO_4$ as control zero of absorbance). Plot time (h) versus absorbance (600 nm).
10. Centrifuge lysate at 3000 × g for 12 min at 4°C.
11. Check the pH and neutralize the supernatant to pH 7.0 with 0.1 *N* NaOH.
12. Store the sterile supernatant at 4°C.

3.3 Host Range Determination by Spot Test (Note 6)

1. Mix in a sterile tube 100 μl of an overnight culture of the potential indicator strain with 3 ml of soft agar and pour the content in Petri dishes containing 15 ml of bottom agar. Swirl the plate in circles to spread the mixture evenly over the plate.
2. Leave the plate on the bench until the soft agar has solidified (approximately 10 min). An alternative to steps 1 and 2 is to moisten a sterile swab with an overnight culture of indicator cells, and generously swab the surface of a bottom agar plate. Then follow procedure as stated in steps 3 and 4.
3. Pour 5 μl of the sterile supernatant (obtained in **3.1**) over the solidified soft agar. Let it absorb and incubate the Petri dishes upside up overnight at the optimal growth temperature of the putative indicator strain.
4. The following day, check for zones of clearing.

3.4 Isolation of Prophage-Cured Derivatives (Note 7)

1. Follow procedure **3.2** from step 1 through step 6.
2. With irradiated cells, prepare decimal dilutions in 1/10 strength broth and plate 100 μl of dilutions 10^0, 10^{-1}, and 10^{-2} on proper agar media.
3. Incubate at optimal bacterial growth temperature for 36 h.
4. Transfer several colonies to fresh media broth. Incubate overnight at optimal bacterial growth temperature.
5. Continue steps 1–4, as stated in **3.3** Host Range Determination by Spot test.
6. Keep cells that are positive for phage attack (clearing zone) for their further characterization and potential use as indicator cells.

3.5 Double-Layer Plaque Assay to Enumerate Phages (Note 8 and 9). N.B. (see also Chapters 14–16)

1. In a sterile test tube, mix 100 μl of an overnight culture of the indicator strain with 100 μl of the decimal dilution (generally until 10^{-7}) of the phage lysate suspension. Use sterile 1/10 media broth to prepare the phage dilutions. Add $CaCl_2$ to a final concentration of 10 mM.

2. Preincubate the tubes 8 min at 30°C or 37°C to allow adsorption of the phages.
3. Add 3 ml of molten soft agar at 45°C containing 10 mM $CaCl_2$.
4. Gently mix the tube by rolling it between the hands until the cells are evenly distributed (care must be taken to avoid the formation of bubbles into the molten agar) and pour its content on Petri dishes containing 25 ml of bottom agar. Swirl the plate in circles to spread the mixture evenly over the plates (**Note 10**).
5. Leave the plate on the bench until the soft agar has solidified (about 10 min), and incubate the Petri dishes upside down overnight at the optimal growth temperature.
6. The following day, count the number of plaques. Calculate the titer in pfu/ml (plaque-forming units per milliliter which is equal to the number of plaques × 10 × inverse of the dilution factor).

3.6 Propagation of Phages in Solid Medium (Note 11)

1. Mix 100 μl of an overnight culture of the indicator or prophage-cured derivative strain grown in MRS or M17 broth with 100 μl of the phage lysate suspension containing 10^4–10^5 pfu/ml (phage dilution should be done in phage buffer). Add $CaCl_2$ to a final concentration equal to 10 mM.
2. Incubate the mixture at 30°C or 37°C for 8 min to allow adsorption of the phages.
3. Add the bacteria–phage mixture to a tube containing 3 ml to molten MRS–Ca^{2+} soft agar at 45°C. Gently mix the tube and pour the content in Petri dishes containing 25 ml of solidified MRS – Ca^{2+} bottom agar. Swirl the plate on circles on the bench immediately after pouring to spread evenly the soft agar over the plate.
4. Leave plates on the bench until the soft agar has solidified (10 min).
5. Incubate the plate without inversion overnight at the appropriate temperature.
6. Add 4 ml of phage buffer over the plates that show semi-confluent lysis (phage plaques should touch one another with bacterial growth visible only in the junction between adjacent plaques), and shake it slightly for 30 min.
7. With a micropipette harvest as much of the phage buffer as possible and transfer it to a sterile centrifuge tube.
8. Repeat steps 6 and 7 (using fresh buffer).
9. Centrifuge the combined phage containing buffers at 4000 × g for 15 min at 4°C
10. Filter the supernatants through a membrane filter of 0.45 μm pore size.
11. Store the sterile supernatant at 4°C.

3.7 Propagation of Phages in Liquid Medium (Note 11)

1. Inoculate 100 μl of an overnight host bacteria culture to 5 ml of fresh MRS or M17 broth.
2. Add $CaCl_2$ to a final concentration of 10 mM.
3. Incubate at proper temperature for 30 min in water bath.
4. Add 100 μl phage lysate with at least 5×10^6 pfu/ml – multiplicity of infection (MOI) of about 0.1.
5. Continue incubating for 6–8 h or until lysis occurred.
6. Centrifuge at 3000 × g for 10 min at 4°C.
7. Neutralize the supernatant to pH 7.0 with 0.1 *N* NaOH.
8. Filter the supernatants containing the phage lysates through a membrane filter of 0.45 μm pore size.
9. Store sterile lysate at 4°C.

4 Notes (*see* also ref. *(14)*)

1. The lytic cycle of temperate phages can either be induced spontaneously or be induced by physiological stimuli and treatments that insult DNA integrity. Mitomycin C, which causes damage to the bacterial DNA, is the standard chemical used to activate the lytic life cycle of temperate phages. Other physical or chemical treatments that induce prophages are agents that damage DNA (i.e., UV light, antitumor drugs); agents that disrupt DNA replication (i.e., antigyrase drugs, antifolates, and inhibitors of DNA topoisomerase II, such as fluoroquinolone antibiotics); hydrogen peroxide; etc. Secondary phage infection can also promote prophage induction. However, many prophages are not inducible whilst many others are inducible with different efficiencies with alternative inducers.
2. The concentration of mitomycin C, age of the culture, and temperature of incubation are all important factors for proper phage induction. Optimal growth conditions for each strain to be tested should be used. High nutritional media are commonly used for inducing and growing phages. However complex media is easier to prepare and generally yields high titers of phages. Some phages are most stable in the presence of divalent cations; however, the level of such ions present in commonly used complex media is sufficient, so the addition of divalent cations to the medium is not generally necessary at this point.
3. In cases of motile microorganisms, it should be keep in mind that some phages attaches to the bacterial flagellum and will only infect highly motile cultures. In this case, cell motility should be examined by microscopy.
4. Mitomycin C should be used in the range of 0.1–2 μg/ml; higher concentrations of mitomycin C can be toxic for the

cells. We recommend using 0. 1–0. 2 μg/ml for thermophilic lactobacilli; 0. 5–1. 0 μg/ml for mesophilic lactobacilli; and 1. 0 μg/ml for*Lactococcus*. It is recommended that one add the inducing agent 30 min after the fresh inoculum is made or at an optical density of 0.1.

5. Wear disposable gloves and suitable protecting clothes when handling mutagenic solutions of mitomycin C. Store mitomycin solution at 4°C and protect the drug from the light and high temperatures.
6. Temperate bacteriophages often exhibit narrow host ranges. For most lysogens, suitable indicator strains or conditions for lytic propagation are not easily defined and evidence of these phage-like particles is limited to their visualization under the transmission electron microscope (TEM) or, after staining the phage genomes with SYBR Gold or SYBR Green I, by epifluorescence microscopy (EFM) [*(15)* and **Chapter 8**]. The narrow host range that most temperate phages exhibit suggests the presence of homoimmunity or restriction modification systems. Plasmid DNA transduction has been used as an alternative to lytic plaque assays to define the host range of a *Lactobacillus* phage, φadh: high-frequency transducing particles of φadh transduced plasmid DNA to several *L. gasseri* cells which, otherwise, did not support lytic growth of phage φadh *(16)*.
7. Prophage-cured derivative strains can be used to demonstrate formal proof of classical lytic and lysogenic cycles of phage replication. The cured strain, sensitive to the phage lytic growth and not inducible with mitomycin C, serves as host where lysogeny can be reestablished. The original host and new lysogens confer immunity to superinfecting phage
8. Some bacteriophages have revealed divalent cation dependency for optimal infectivity (mainly Ca^{2+} and in some cases Mg^{2+}). The initial pH of the agar can also be important; i.e., *L. gasseri* phage φadh, although induced broth lysis and form clear zones on its indicator cells, does not form plaques on MRS agar plus $CaCl_2$ at pH 6.5. Plaques were detected only when the initial pH of the medium was 5.5.
9. If the bacteria do not form lawns on solid culture media, plaque assays cannot be used to estimate viable phage titers. In such cases, most probable number can be used instead.
10. Vigorous mixing of the tube containing the phage–host mixture may damage the phage particles and introduce air bubbles into the soft agar that could look like plaques, especially to the inexperienced eye.
11. Some temperate phages replicate poorly in liquid medium, resulting in incomplete lysis. In general, propagation of phages in solid media is more efficient than in liquid culture.

12. Lysates of phages that do not contain lipids in their structures can be stored at 4°C for several years with the addition of a few drops of chloroform. Some phages can be stored for several months and even years without loosing titer. However, most phages are unstable and have to be amplified frequently to keep them active.

References

1. Weinbauer, M.G. (2004). Ecology of prokaryotic viruses. *FEMS Microbiol. Rev.* **28,** 127–181.
2. Canchaya, C., G. Fournous and H. Brüssow. (2004) The impact of prophages on bacterial chromosomes. *Mol. Microbiol.* **53,** 9–18.
3. Botstein, D. (1980). A theory of modular evolution in bacteriophages. *Ann. NY Acad. Sci.* **354,** 484–491.
4. Brüssow, H., C. Canchaya and W.-D. Hardt. (2004). Phages and the evolution of bacterial pathogens: from genomic rearrangements to lysogenic conversion. *Microbiol. Mol. Biol. Rev.* **68,** 560–602.
5. Casjens, S. (2003). Prophages and bacterial genomics: what we have learned so far? *Mol. Microbiol.* **49,** 277–300.
6. Hendrix, R.W., M.C. Smith, R.N. Burns, M.E. Ford and G.F. Hatfull. (1999). Evolutionary relationships among diverse bacteriophages and prophages: all the world's a phage. *Proc. Natl. Acad. Sci. USA* **96,** 2192–2197.
7. Moreira, D. (2000). Multiple independent horizontal transfers of informational genes from bacteria to plasmids and phages: implications for the origin of bacterial replication machinery. *Mol. Microbiol.* **35:** 1–5.
8. Boyd, E.F. (2005). Bacteriophages and bacterial virulence, in *Bacteriophages: Biology and Applications*, (Kutter E. and Sulakvelidze A., ed.), CRP Press, FL, pp. 223–265.
9. Chopin, A., A. Bolotin, A. Sorokin, S.D. Ehrlich and M.-C. Chopin. (2001). Analysis of six prophages in *Lactococcus lactis* IL1403: different genetic structure of temperate and virulent phage populations. *Nucleic Acids Res.* **29,** 644–651.
10. Waldor, M. K. and J. J. Mekalanos. (1996). Lysogenic conversion by a filamentous phage encoding cholera toxin. *Science* **272,** 1910–1914.
11. Guttman, B., R. Raya and E. Kutter. (2005). Basic phage biology, in *Bacteriophages: Biology and Applications*, (Kutter E. and Sulakvelidze A., ed.), CRP Press, FL, pp. 29–66.
12. De Man, J., M. Rogosa and M. Sharpe. (1960) A medium for the cultivation of lactobacilli. *J. Appl. Bacteriol.* **23,** 130–135.
13. Terzaghi, B.E. and W.E. Sandine. (1975). Improved medium for lactic streptococci and their bacteriophages. *Appl. Microbiol.* **29,** 807–813.
14. Carlson, K. (2005) Appendix: working with bacteriophages: Common techniques and methodological approaches, in *Bacteriophages: Biology and Applications*, (Kutter E. and Sulakvelidze A., ed.), CRP Press, FL, pp. 437–494.
15. Wen, K., A.C. Ortmann and C.A. Suttle. (2004). Accurate estimation of viral abundance by epifluorescence microscopy. *Appl. Environ. Microbiol.* **70,** 3862–3867.
16. Raya, R.R., E.G. Kleeman, J.B. Luchansky and T.R. Klaenhammer. (1989). Characterization of the temperate bacteriophage phi-adh and plasmid transduction in *Lactobacillus acidophilus* ADH. *Appl. Environ. Microbiol.* **55,** 2206–2213.

Chapter 4

Isolation of Cyanophages from Aquatic Environments

Andrew D. Millard

Abstract

Cyanophages are a group of viruses which specifically infect cyanobacteria. The cyanobacteria are predominantly aquatic phototrophic bacteria and the two dominant genera *Synechococcus* and *Prochlorococcus* contribute significantly to primary production in the oceans. Cyanophages that infect marine cyanobacteria were first isolated in the early 1990s and it is now known that by lysing their host cells they play an important role in the microbial loop and alter biogeochemical cycles. They are also thought to influence the community structure and evolution of the cyanobacteria that they infect. Most recently cyanophages have been shown to carry host photosynthetic genes. It was only by the isolation and purification of cyanophage isolates have these important functions become known. Although this chapter illustrates cyanophage isolation with *Synechococcus*, the same techniques will apply to isolation of phage from other cyanobacteria. With the continued isolation of cyanophages a greater understanding of their biological importance will be gained.

Key words: Cyanophages, *Synechococcus*, purification, isolation, plaque assays, ASW, artificial seawater.

1 Introduction

Cyanophages are defined as viruses that specifically infect cyanobacteria. They were first reported in the freshwater environment over 40 years ago *(1)*. It was nearly 25 years ago that cyanophages capable of infecting both unicellular and filamentous marine cyanobacteria were first reported, *(2)*. However, it was not until the early 1990s that cyanophages infecting the marine cyanobacteria *Synechococcus* were isolated *(3,4)*. Since then marine cyanophages have been isolated from a number of different geographical locations including the English Channel, Sargasso Sea, Gulf of Mexico, Red Sea and Bermuda costal waters *(3–7)*. The abundance of cyanophages is variable in different regions with

Martha R. J. Clokie, Andrew M. Kropinski (eds.), *Bacteriophages: Methods and Protocols, Volume 1: Isolation, Characterization, and Interactions, vol. 501*, © 2009 Humana Press, a part of Springer Science+Business Media
DOI 10.1007/978-1-60327-164-6_4 Springerprotocols.com

numbers typically in the range of 10^6–10^8 infective particles per litre *(4, 7–9)*, but are sometimes found in excess of 10^9 per litre *(3, 4)*. All cyanophages isolated to date that infect marine *Synechococcus* MC-A are lytic and the vast majority are members of the *Myoviridae* family of viruses. The exception to this is phages which infect *Procholorococcus* where a large proportion of them belong to the family *Podoviridae (10)*. Through the isolation of cyanophages a number of genetic probes have been developed to determine the diversity of natural cyanophage populations; these are specific to cyanomyoviruses and based on the major capsid and portal protein, gp23 and gp20 respectively *(6, 9)*. The isolation of clonal cyanophages has also allowed the genome of a number of cyanophages to be completely sequenced *(11, 12)*; this has lead to the discovery that core photosynthetic genes and other host genes of importance, e.g., those involved in nutrient acquisition, are carried by cyanophages. It is only by the continued isolation of cyanophages followed by appropriate studies of their physiology and observation of their behaviour under laboratory and simulated environmental conditions that a greater understanding of their diversity and interaction with their hosts will be gained.

2 Materials

2.1 *Synechococcus* Growth Medium

1. Artificial seawater medium (ASW) as devised by Wyman *et al.*, (1985) was used to culture *Synechococcus*. Stock solutions of individual components were made as follows:

Stock solutions	g per 500 ml
$MgCl_2$	100
$CaCl_2$	50
KCl	50
$NaNO_3$	37.5
$MgSO_4$	175
Tris	100
K_2HPO_4	6

2. ASW (1 l): 10 ml of $NaNO_3$, $MgCl_2.6H_20$ and $MgSO_4.7H_20$, with 5 ml of $CaCl_2H_20$ and KCl, with 5.5 ml of Tris and 2.5 ml of K_2HPO_4. This is combined with 25 g of NaCl and 1 ml of trace metals and the volume is first made up to 800 ml with water, the pH is then adjusted to 8.0 with HCl before the volume is topped up to 1 l (**Note 1**). Once sterilised by autoclaving ASW can be kept at room temperature for several months.

3. ASW trace metal stock: $H_3BO_3$2.86 g/l, $MnCl_2.4H_20$ 1.81 g/l, $ZnSO_4.7H_20$ 0.222 g/l, Na2Mo0$_4$.2H_20 0.39 g/l, $CuSO_4.5H_20$ 0.008 g/l, $Co(N0_3)_2.6H_20$ 0.0049 g/l, $FeCl_3.6H_20$ 3.0 g/l and EDTA 0.5 g/l. Once sterilised by filtering through a filter with a pore size of 0. 22 μm, the trace metal stock can be stored at 4°C for several years.

2.2 Cleaning of Agar

(1) 27 cm diameter porcelain Buchner funnel
(2) 2 l borosilicate Buchner funnel flask with side arm for 9 mm tubing
(3) 2 l borosilicate conical flask with side arm
(4) 9 mm tubing
(5) 5 l plastic beaker
(6) Magnetic stirrer
(7) Vacuum pump
(8) Bacto Agar (Difco)

2.3 Water Samples

Water samples that contain cyanophages will be needed to be collected in advance. Normally only a small volume of water is required to isolate cyanophages. If host cell numbers are known by either flow cytometry or haemocytometer counts, the phage numbers can be estimated at one order of magnitude higher. It must be remembered however that as there will be a wide variety of cyanobacterial hosts not all phage will infect the model organism being used in isolation. If hosts are present at 10^4/ml and phage at 10^5/ml, in theory they should be detected in 1 μl of sea water. If host and thus phage numbers are low, phage can be concentrated prior to isolation using techniques described in **Chapter 2**. Once collected water must be filtered through a 0. 2 μm filter to remove any bacteria and can be kept at 4°C in the dark.

3 Methods

The isolation of cyanophages is not fundamentally any different from the isolation of other phages, however, it is sometimes more problematic as the host cyanobacteria can be difficult to culture and timescales are longer as the doubling time of the cells is much slower. Some host strains appear to be significantly better for isolating phages than others, so it is worth establishing this for your own cyanobacterial phages. The basic principle is the attachment of a single phage particle to a susceptible host resulting in the lysis of that host and the release in a number of progeny. This can either occur on agar plates where subsequent rounds of infection and lysis cause the localised death of host cells and formation of a discrete plaque as progeny phage slowly diffuse through the agar.

This technique is known as a plaque assay (N.B. for details on the theory of this procedure *see* **Chapters 7** and **14–16**).

The alternative is for this to occur in liquid medium where the resulting progeny phage are not inhibited by the agar and the phage can absorb to all remaining hosts, eventually resulting in lysis of the entire culture. This procedure can be carried out in volumes from microlitres up to hundreds of millilitres of media. The method described in this chapter is for 200 μl that can be carried out in the wells of microtitre plate. Cyanophages can be isolated directly from seawater by the use of plaque assays or well assays. Both these methods have different advantages. The main advantage of using well assays with cyanobacteria is that it does not require the growth of cyanobacteria on agar plates. However, samples must be serially diluted so that only one phage particle is present in a water sample. With plaque assays it is not necessary for such dilution series as a single discrete plaque is the end product of a single successful infection. The procedures described in this chapter are for the isolation and purification of cyanophages that will infect the marine cyanobacterium *Synechococcus* sp. WH7803 but they are more widely applicable (**Note 2**).

3.1 Preparation of Agar

Synechococcus WH7803 is sensitive to trace amounts of elements and to detergents. Therefore, it is necessary to wash agar to remove any impurities that may inhibit *Synechococcus* growth (**Note 3**). The method is a modification of the method developed by Waterbury and Willey *(13)*.

1. Agar must be cleaned several days in advance of when it is to be used. Set up the Buchner funnel as shown in **Fig. 4.1**. Make sure the Buchner funnel fits into the top of the Buchner flask and forms a tight seal. Attach a rubber hose from side arm of the Buchner conical flask into the top glass tube in

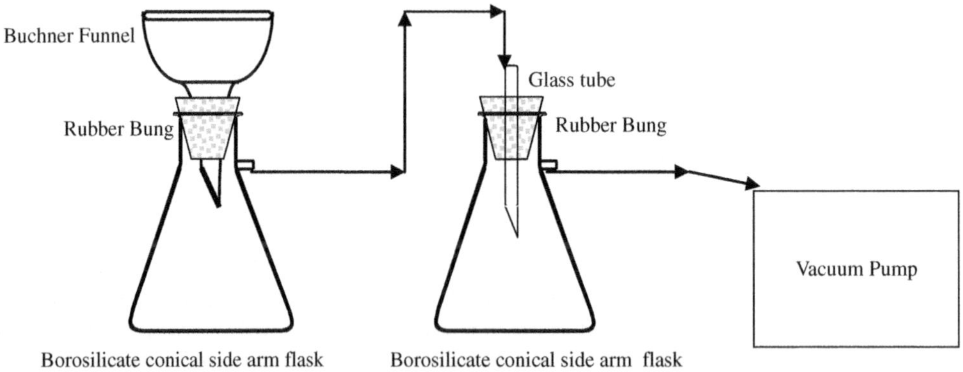

Fig. 4.1. Water is removed from the agar by filtration. A Buchner funnel is fixed to a conical sidearm arm flask which is in turn attached to a second flask, which is joined to a vacuum pump.

the second conical flask. Attach a second rubber hose from the side arm of the conical flask to the vacuum pump.

2. Draw around the top of the Buchner funnel on 3MM Whatman filter paper to make a template. Cut out several circles of filter paper that will be used for filtering.
3. Mix 250 g of Bacto agar (Difco) with 5 l of water in a 5 l plastic beaker (**Note 4**).
4. Drop the magnetic stirbar into the bottom of the beaker and place the beaker on the magnetic stirrer and stir for 30 min.
5. Allow the agar to settle to the bottom of the beaker. Straw coloured water will now form a distinct layer on top. Pour off as much water as possible, whilst trying to minimise the amount of agar that is poured away.
6. Place a pre-cut filter paper into the top of the Buchner funnel and lightly dampen with water.
7. Pour the water-saturated agar into the Buchner funnel.
8. Turn on the vacuum pump. The water should be sucked from the agar. Maintain a vacuum of approximately 40–50 cm Hg. The agar should start to turn white in colour as the water is removed. Keep the pump on until there is no longer a steady drip of water from the Buchner funnel (**Note 5**).
9. Use a spatula to loosen the agar from the edges of Buchner funnel and tip the agar back into the 5 l beaker. Add 5 l of water and repeat the washing procedure (steps 3–9) until the water being removed runs clear. This normally requires three washes with water.
10. After washing with water, mix the agar with 5 l of ethanol and repeat the washing procedure (steps 3–9). It is only necessary to wash once with ethanol (**Note 6**).
11. After one wash with ethanol, repeat the washing procedure (steps 3–9) with acetone.
12. This should result in a Buchner funnel full of compacted agar that is white in colour. This is then removed with a spatula and placed on tray lined with sheets of 3 MM Whatman paper.
13. Ensure that there are no large lumps of agar stuck together and place the trays into a fume hood. Do not turn the fume hood on, leave to dry out completely. The dried agar should appear lighter in colour than unwashed Bacto agar.
14. Store the dried agar in a plastic airtight container.

3.2 Growth of *Synechococcus*

Synechococcus WH7803 is cultured routinely in 100 ml and 1 l volumes. For 40 plaque assays 1 l of exponentially growing cells is required. Initially, 1 l of 1× ASW is inoculated with 100 ml of *Synechococcus* WH7803 that has an OD of > 0.5 at 750 nm. An exponentially growing culture has an OD of 0.35–0.40 at 750 nm, this normally occurs 7–10 days after the initial inoculation. The culture is grown under constant aeration and illumination of 2–25 $\mu Em^{-2}s^{-1}$, at 25°C.

3.3 Preparation of 1% ASW/Agar Plates

1. Mix 500 ml of water with 10 g of cleaned agar in a 1 l bottle and autoclave.
2. Mix this with 500 ml of 2× ASW (**Note 7**).
3. 1 l should provide enough agar to pour 40 plates using 18 cm Ø Petri dishes
4. Plates can be stored inverted in sealed bags for several weeks at 4°C.

3.4 Preparation of Soft Agar

1. Fresh 'sloppy' 0.4% (w/v) agar is required every time plaque assays are carried out.
2. For each plaque assay 2.5 ml of sloppy agar is needed, sloppy agar is always prepared in excess.
3. Add 60 ml of 2× ASW to a 150 ml bottle.
4. In a second bottle add 60 ml of water to 0.48 g of cleaned agar.
5. Autoclave and allow to cool in a water bath at 50°C.

3.5 Plaque Assays

1. For 40 plaque assays a litre of culture is required. The culture is centrifuged at 6000×*g* for 15 min at 25°C to pellet the *Synechococcus* cells.
2. Pour off the clear supernatant without disturbing the pellets of *Synechococcus* cells.
3. Resuspend the cell pellets with ASW by pipetting ASW up and down the side of the centrifuge bottle with a 25 ml pipette. Resuspend the first pellet in 10 ml of 1× ASW and then use this to resuspend the remaining three pellets.
4. Transfer to a 25 ml clear glass screw top vial and note the volume transferred. With 5 ml of fresh 1× ASW wash any remaining cells from the sides of the centrifuge tubes and transfer to the same glass vial.
5. Add 1× ASW to the universal so that the total volume is equal to 20 ml.
6. Transfer 0.5 ml of concentrated *Synechococcus* cells to a 5 ml glass screw top vial; 40 screw top vials will be needed for 20 ml of concentrated cells.
7. Add the diluted seawater samples to each vial, one sample per glass vial (**Note 8**).
8. Gently shake the mixture of *Synechococcus* cells and water. Then leave at 25°C in constant light of 2–25 $\mu E\ m^{-2}s^{-1}$ for 1 h.
9. Add 2.5 ml of cooled molten 0.4% (w/v) agar to each vial containing 0.5 ml of cells. Gently mix the agar and cells with a gentle swirling action. Pour this mixture onto the top of a solid 1% ASW agar plate. Allow the agar to set for 1 h before inverting the plate and placing in a suitable incubator with 2–25 $\mu E\ m^{-2}s^{-1}$ of light at 25°C (**Note 9**).

10. Monitor the plates daily for the appearance of plaques. The appearance of plaques can vary from 3 to 10 days.
11. Remove a single plaque by the use of Pasteur pipette. Place the end of the pipette over a plaque and gently press into the top layer of agar, rotate the pipette to gently remove a plug of agar.
12. Resuspend the plug of agar in 1 ml of fresh ASW and leave for at least 1 h at 4°C in the dark, to allow the phage to diffuse into the fresh medium to form a lysate.
13. Make a dilution series of the lysate and use this in a second round of plaque assays (steps 1–12). This process has to be repeated 3 times to ensure that a clonal cyanophage has been isolated (**Note 10**).
14. The lysate from the final purification can then be used in a further plaque assay to produce confluent lysis of the *Synechococcus* lawn. The top layer of sloppy agar can then be removed and resuspended in 2 ml of fresh ASW and left overnight at 4°C.
15. The lysate can then be centrifuged at 13,000×g in a benchtop centrifuge to pellet the agar. The supernatant can then be removed and passed through a 0.22 μm filter to remove any bacterial contaminants and produce a phage stock (**Note 11**). Stocks can be stored at 4°C for several years in the dark at 4°C (**Chapter 14**).

Figure 4.2 A lawn of *Synechococcus* sp. WH7803 on a 1% ASW/agar plate. It is possible to see discrete plaques where the *Synechococcus* cells have been lysed by the cyanophages found in

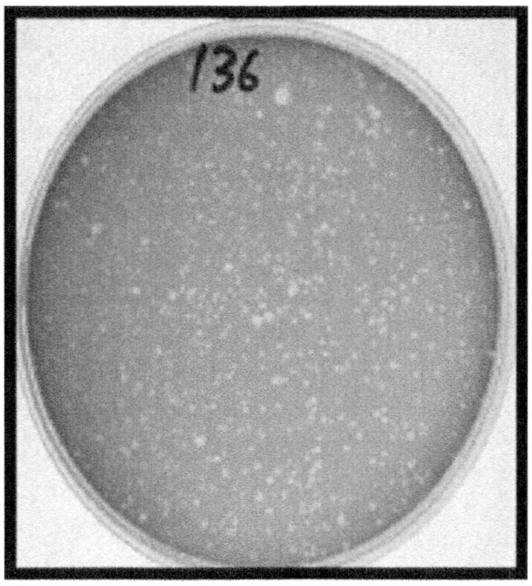

Fig. 4.2. A lawn of *Synechococcus* sp. WH7803 on a 1% ASW/agar plate. It is possible to see discrete plaques where the *Synechococcus* cells have been lysed by the cyanophages found in water sample collected from the Red Sea. A 50 μl sample of seawater was used and plaques had formed after 10 days of growth.

water sample collected from the Red Sea. A 50 μl sample of seawater was used and plaques had formed after 10 days of growth.

3.6 Well Assays

The use of well assays avoids the need to culture cyanobacteria on a solid medium, smaller volumes of culture are needed and if desired phage may be suitably enriched for observation by plaque assay.

1. Add 160 μl of ASW to each well of a 96-well microtitre tray (**Note 12**). Then add 40 μl of *Synechococcus* cells from an exponentially growing culture (OD 0.35–0.40 at 750 nm).
2. Add 10 μl of diluted water sample to individual wells.
3. Control wells with no seawater added serve as negative controls.
4. The plate is then incubated in a suitable incubator with 2–25 μE $m^{-2}\,s^{-1}$ of light at 25°C for 7–21 days. The wells are monitored for the lysis of *Synechococcus* cells compared to the negative control.
5. The lysate from a well which has completely lysed is removed and centrifuged at 13,000×*g* for 5 min in a benchtop centrifuge to pellet any cells debris. The supernatant is then serially diluted and used for further round of well assays (**Note 13**). This process is repeated 3 times to ensure a clonal isolate.

4 Notes

1. All water used is of milliQ grade, i.e., it has a resistance of 18.2 Ω.
2. This method can be easily adapted to isolate phages that infect other strains of *Synechococcus* and*Procholorococcus* or cyanophages infecting freshwater cyanobacteria can also be isolated using this method by simply changing the medium and culture conditions used to grow the cyanobacterial host.
3. It is possible to culture some strains such as *Synechococcus* WH7803 on commercially cleaned agar such as Plant Cell agar A8678 (Sigma-Aldrich St. Louis, MO; http://www.sigmaaldrich.com). However, several other strains of *Synechococcus* can be cultured on cleaned agar from the method described in this chapter, but will not grow on commercially cleaned agar.
4. It is often necessary to stir the agar first with a glass rod to ensure it is thoroughly mixed with the water, otherwise the agar quickly settles to the bottom forming a viscous 'gloop' that cannot be stirred by a magnetic stirrer.
5. Prolonged filtering can result in the filter becoming plugged, which can result in tears appearing and a loss of vacuum. The

agar has to be removed, the filter paper replaced and the process started again.

6. It may take significantly longer to remove the ethanol than it does to remove water from the agar. If the removal of ethanol is slow, yet the agar is still saturated then changing the filter can speed up the process.
7. It is necessary to autoclave ASW separately from agar as the combination of the two can result in agar plates that are sub-optimal for the growth of *Synechococcus.*
8. It is not normally known how many cyanophages are present in a seawater sample, unless the area has been sampled before. It is therefore necessary to use a range of dilutions of the neat seawater sample to obtain single plaques as well as range of volumes to obtain single plaques. It is also necessary to have a negative control where no water is added to the *Synechococcus* cells and a positive control (if available) of a cyanophage that will cause lysis.
9. It is often necessary to cover the plates with a neutral grey film to reduce the amount of light that reaches the *Synechococcus* cells during the first few days of incubation – this prevents photo-bleaching of the cells.
10. It is necessary to obtain single plaques, therefore a dilution series needs to be carried. A dilution series of 10^{-1}, 10^{-2} and 10^{-3} is normally sufficient. However, it is dependent on individual phages and plaque size on the number of dilutions needed to obtain single plaques.
11. By filtering there may be a reduction in titre. Alternatively 5 μl of chloroform can be added to the lysate to kill any bacterial contaminants. However, some phages are sensitive to chloroform so care should be taken not to try this on the only stock of an isolate.
12. If the plate is stored in incubator that has a constant draft caused by cooling fans, put medium only in the wells around the edge of the plate. This will help to prevent the evaporation of samples.
13. It is necessary to produce a dilution series to extinction. This should ensure that one at least one well in a dilution series does not lyse. The dilution before the sample that does not lyse should therefore contain only a single phage.

References

1. Safferman, R.S. and M.E. Morris, *Algal virus: isolation*. Science, 1963. **140**: 679–80.
2. Moisa, I., E. Sotropa, and V. Velehorschi, *Investigations on the presence of cyanophages in fresh and sea waters of Romania*. Virologie, 1981. **32**(2): 127–32.
3. Wilson, W., et al., *Isolation and Molecular Characterization of Five Marine Cyanophages Propagated on Synechococcus sp. Strain WH7803*. Appl. Envir. Microbiol., 1993. **59**(11): 3736–3743.

4. Suttle, C.A. and A.M. Chan, *Marine Cyanophages Infecting Oceanic and Coastal Strains of Synechococcus – Abundance, Morphology, Cross-Infectivity and Growth-Characteristics.* Marine Ecology-Progress Series, 1993. **92**(1–2): 99–109.
5. Zhong, Y., et al., *Phylogenetic diversity of marine cyanophage isolates and natural virus communities as revealed by sequences of viral capsid assembly protein gene g20.* Appl Environ Microbiol, 2002. **68**(4): 1576–84.
6. Fuller, N.J., et al., *Occurrence of a sequence in marine cyanophages similar to that of T4 g20 and its application to PCR-based detection and quantification techniques.* Appl Environ Microbiol, 1998. **64**(6): 2051–60.
7. Lu, J., F. Chen, and R.E. Hodson, *Distribution, isolation, host specificity, and diversity of cyanophages infecting marine Synechococcus spp. in river estuaries.* Appl Environ Microbiol, 2001. **67**(7): 3285–90.
8. Waterbury, J. and F. Valois, *Resistance to Co-Occurring Phages Enables Marine Synechococcus Communities To Coexist with Cyanophages Abundant in Seawater.* Appl. Envir. Microbiol., 1993. **59**(10): 3393–3399.
9. Muhling, M., et al., *Genetic diversity of marine Synechococcus and co-occurring cyanophage communities: evidence for viral control of phytoplankton.* Environ Microbiol, 2005. 7(4): 499–508.
10. Sullivan, M.B., J.B. Waterbury, and S.W. Chisholm, *Cyanophages infecting the oceanic cyanobacterium Prochlorococcus.* Nature, 2003. **424**(6952): 1047–51.
11. Mann, N.H., et al., *The genome of S-PM2, a "photosynthetic" T4-type bacteriophage that infects marine Synechococcus strains.* J Bacteriol, 2005. **187**(9): 3188–200.
12. Sullivan, M.B., et al., *Three Prochlorococcus cyanophage genomes: signature features and ecological interpretations.* PLoS Biol, 2005. **3**(5): e144.
13. Waterbury, J.B. and M. Willey, *Isolation and Growth of Marine Planktonic Cyanobacteria.*, in *Methods Enzymol.* 1989.

Chapter 5

Isolation of Viruses from High Temperature Environments

Jennifer Fulton, Trevor Douglas, and Mark Young

Abstract

The detection and isolation of viruses directly from high temperature (>80°C) acidic (pH<4) hot springs, fumaroles, and soils has long been challenging. These extreme environments tend to have a low biomass (typically $<10^6$ cells/ml) and low free viral abundance (10^3–10^6particles/ml) compared to eutrophic freshwater lakes *(1)* or near- shore marine environments *(2)*. Establishing laboratory cultures from these environments pose challenges due to the extreme and poorly defined growth conditions that must be mimicked in the laboratory. Likewise, culture-independent approaches for isolation of viruses are problematic because of our rudimentary knowledge of viral diversity in these environments and the lack of universally conserved signatures that can be used for virus detection and isolation. Here we discuss both culture-based and culture-independent techniques for detection and isolation of viruses from acidic thermal features.

Key words: Archaea, virus, thermophiles, detection, isolation, extremophile, *Sulfolobus*, *Acidianus*, *Pyrobaculum*, *Thermoproteus*.

1 Introduction

The dominant viral hosts present in high temperature acidic environments are members of the Archaea. Unfortunately, we have only a limited understanding of the Archaea and their viruses in these environments. Compared to Bacteria and Eukarya, only a relatively small number of viruses have been identified and characterized for members of the Archaea. Of over 5000 viruses known, only a few are from archaeal hosts. Little more than a decade ago there were only four characterized viruses of this domain *(3)*. Today there are approximately 25 viruses replicating in only four host genera (*Sulfolobus, Acidianus, Pyrobaculum and Thermoproteus*) have been isolated from high temperature terrestrial

Martha R. J. Clokie, Andrew M. Kropinski (eds.), *Bacteriophages: Methods and Protocols, Volume 1: Isolation, Characterization, and Interactions, vol. 501*, © 2009 Humana Press, a part of Springer Science+Business Media
DOI 10.1007/978-1-60327-164-6_5 Springerprotocols.com

environments *(4)* and a number of other viruses from the Euryarchaeota.

There appears to be a trend emerging relating to the viruses that have been isolated from members of the two largest phyla of Archaea. Those isolated from hosts belonging to the phylum Euryarchaeota (predominantly from methanogens and halophiles) tend to have head and tail morphologies characteristic of many bacteriophages. In contrast, no head and tail virus morphologies have been isolated in the phylum Crenarchaeota *(5)*. In fact, many of the viruses of the Crenarchaeota bear little resemblance to viruses of the other domains of life. For example, the 60×90 nm spindle-shaped virus particle morphology observed in *Sulfolobus* spindle-shaped viruses (SSVs) is unique and is the most common virus morphotype for entities infecting crenarchaeal hosts. In addition to the unique morphologies, the proteins encoded by many of the viruses isolated from archaeal hyperthermophiles show little to no sequence similarity to other proteins in the databases *(4)*.

Traditionally, most viruses have been isolated from the environment by first culturing their hosts in the laboratory and subsequently isolating the viruses by propagating them on the established bacterial/archeal culture. The obvious advantage of this approach is that the isolated virus is already associated with its host. The disadvantage of culture-dependent approaches is that one is limited to only isolating viruses from hosts that can be successfully cultured. This may severely limit the likelihood of isolating the dominant virus in a particular environment. It also introduces culturing bias in the population structures of the viruses themselves *(2)*, and limits the ability to assess the total diversity of viruses present in a particular environment. The ability to forgo enrichment cultures in favor of direct detection of viruses and subsequent isolation of genetic material directly from environmental samples will greatly expand our understanding of the diversity of viruses present in high temperature environments.

Despite the inherent difficulties of culturing Archaeal hosts from high temperature acidic environments, this methodology has provided the majority of viruses that have been isolated to date. These viruses have predominantly been isolated from *Sulfolobus* spp. partly because of its relative ease of culturing and ubiquitous presence in high temperature acidic environments.

High temperature acidic hot springs are typically lower both in cell and virus abundance as compared to lower temperature neutral environments. Our estimates of host abundance in acidic hot springs range from 10^5 to $10^7\,ml^{-1}$ depending on the hot spring. Our estimates of free virus particles in acidic hot springs based on epifluorescent microscopic (EFM) and transmission electron microscopic (TEM) counts range from 10^4 to 10^6 free viruses per milliliter. The low ratio of free virus particles to cells (estimated to be 1:10) may reflect the instability of virus particles free of their hosts in near boiling acid conditions. This is in contrast to viral counts as high as $10^8 - 10^9$ per milliliter *(2)*, and a 3:1 to 10:1 ratio of virus to host *(6)*, in marine coastal waters and sediments.

A number of culture-dependent and culture-independent methods have been utilized to detect or isolate viruses from a wide range of environments. These include the screening of primary enrichment cultures by TEM *(7)* and EFM methods *(8)* (**Chapter 7**), precipitation of virus directly from environmental samples utilizing polyethylene glycol (PEG) *(3)*, end point filtration *(9)* and tangential flow filtration *(10)* (**Chapter 1**). Here we present methods that have been proven successful in both detecting and isolating viruses from high temperature acidic thermal features in Yellowstone National Park (YNP) and other thermal sites worldwide. Most of the viruses characterized from these environments have been isolated from aerobic and anaerobic enrichment cultures. Although we concentrate on viruses from *Sulfolobus*, the methodology can be modified for the organism of interest. Recent efforts to expand our understanding of total viral diversity utilizing culture-independent methods are also presented. The continued and expanded discovery of new viruses of hyperthermophilic Archaeans will likely depend on both culture-dependent and culture-independent methods.

2 Materials

2.1 Culturing

1. Telescoping concrete or window washing pole fitted with an appropriate hose clamp.
2. Long-neck 125 ml flasks with metal caps (custom, Chemglass, Vineland, NJ; http://www.chemglass.com/).
3. Oil bath shaking incubators at the appropriate growth temperature (Innova 3100 water shaker bath, New Brunswick Scientific, Edison, NJ; http:www.nbsc.com/Main.asp).
4. *Sulfolobus* medium, e.g., DSMZ Media 182 for *Sulfolobus solfataricus (7)* or other medium as appropriate.

2.2 Culture-Independent Detection

1. Pre-filter of glass wool to remove large materials/sediments. One can construct these filters from 0.5 m of 25 cm diameter PVC pipe with hose fittings on each end and densely packed with glass wool.
2. 100,000 Da or smaller molecular weight cut-off tangential flow filter units (MaxCell Process-scale Hollow fiber cartridge UFP-100-E65; GE Healthcare, Piscataway, NJ; http://www.gehealthcare.com/).

3. 10,000 Da molecular weight cut-off tangential flow filter (MaxCell Process-scale Hollow fiber cartridge, UFP-10-C45, GE Healthcare, Piscataway, NJ; http://www.gehealthcare.com/).
4. Power source (i.e., gasoline generator)
5. Peristaltic Pump – preferably with two pump heads for efficient concentration of sample (i.e., Masterflex I/P double headed long shaft peristaltic pump, Vernon Hills, IL; http://www.masterflex.com/).
6. 10 l or larger carboy container
7. Peristaltic tubing

2.3 Virus Purification

1. 0.45 μm filter (i.e., Pall FP-450 0.45 μm 47 mm filter #66480)
2. Vacuum filter flask
3. Ultracentrifuge (i.e., Beckman model L8-M with T30 rotor, Beckman Coulter, Inc., Fullterton, CA; http://www.beckmancoulter.com/)
4. Centrifuge tubes appropriate to rotor type

2.4 Nucleic Acid Extraction

1. TEN (0.1 M NaCl, 10 mM Tris Cl (pH 8), 1 mM EDTA (pH 8)
2. TENST (TEN, 0.12% Triton X-100, 1.6% N-lauryl sarcosine)
3. TE (10 mM Tris–HCl, 1 mM EDTA at pH 8)
4. 20 mg/ml Proteinase K (Sigma-Aldrich, St. Louis, MO; http://www.sigmaaldrich.com)
5. Phenol (Fisher)
6. Phenol: chloroform: isoamyl alcohol (25:24:1, v/v)
7. 100% Ethanol
8. 70% Ethanol
9. 3 M Sodium acetate, pH 5.2 (Fisher)

2.5 Genome Amplification

1. GenomiPhi DNA Amplification Kit (GE Healthcare, Piscataway, NJ; http://www.gehealthcare.com/).

2.6 Long-Term Storage

1. 80°C freezer
2. 80% (w/v) autoclaved glycerol

3 Methods

3.1 Culture-Dependent Approaches

Isolating viruses from established enrichment cultures has the practical advantages of allowing relatively large numbers of environmental samples to be taken and surveyed for viruses. Little specialized equipment in the field is required for sampling. It has the disadvantage in that most laboratories are restricted to testing a limited number of culturing media and conditions. It also requires specialized culturing conditions, equipment, and dedicated personnel to establish and maintain cultures.

We will use a specific example of aerobic culturing of *Sulfolobus* as an illustration of culture-dependent approach to the isolation of archaeal viruses. The procedure is schematically presented in **Fig. 5.1**. The reader is directed to the literature for examples of both anaerobic and aerobic culturing techniques to the isolation of archaeal viruses from various hosts found in high temperature acidic environments *(2, 9–14)*.

Locate a thermal feature that fits within the desired range of temperature (>75°C) and pH (pH<4.0) for *Sulfolobus.* If a pH meter is being used, ensure that it has been calibrated to operate in the temperature range of the thermal feature you are sampling. Alternatively, pH paper and thermometers can provide a reliable measure of *in situ* temperature and pH. Place a sterile 50 ml tube in the clamp of the telescoping pole and collect sample from the thermal feature. After sample collection, tightly cap the tube for transport back to the laboratory. We have found that if the environmental sample will be used to inoculate media within 24 h, there is little need to take special precautions to preserve the sample or maintain it at its field temperature.

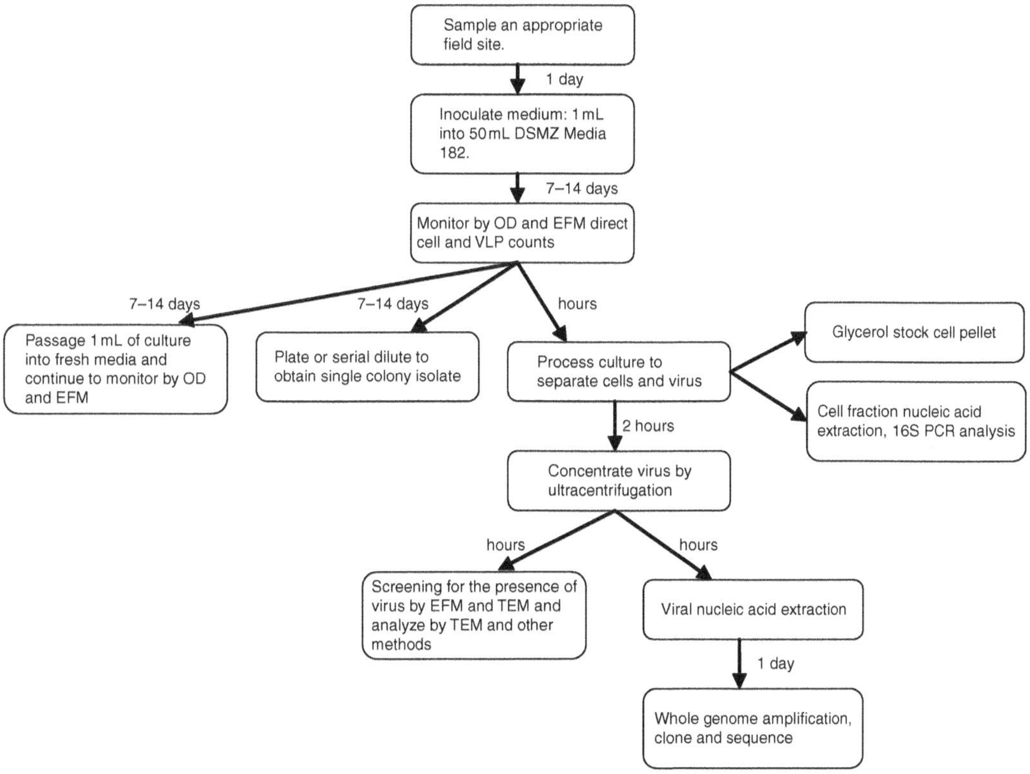

Fig. 5.1. Schematic illustration of the culture-dependent method of isolating viruses from thermal sites.

Once in the laboratory, inoculate 0.5–1 ml into 50 ml of sterile medium in a 125 ml long-neck flasks. Appropriate media can be found through culture collection centers such as the German Resource Center for Biological Materials (DSMZ; http://www.dsmz.de/) and the American Type Culture Collection (ATCC; http://www.atcc.org/) or by using filter sterilized hot spring water supplemented with minimal carbon and nutrient sources. Incubate in an oil bath incubator at 78 – 80°C. Monitor culture growth indirectly by optical density at 600 nm or by direct cell and VLP counts (**Section 3.3**). Typically, enrichment cultures need to be monitored for 7–14 days before cell growth and or VLP production is evident. When the maximum or desired VLP levels are obtained, as estimated by EFM (typically $10^5 - 10^9$ VLPs/ml), the primary enrichment can be further processed for VLP characterization and purification.

If the culture is to be maintained, 1 ml of the original enrichment culture can be inoculated into fresh media. If obtaining a single isolate is the objective the sample can be plated on solid media to obtain individual colonies or subjected to end point dilution procedures *(3)*.

The use of culture-dependent approaches to the isolation of virus can be a long and tedious process requiring months to years to perform. However, once optimized, the subsequent isolation and characterization of the same or similar virus types from different thermal features is relatively rapid. Development of a protocol to maintain and produce a constant supply of virus is a requirement if the detailed characterization of the virus and its interaction with its host is a long-term objective.

3.2 Culture-Independent Approach to Virus Isolation

Culture-independent approaches to virus isolation from high temperature environments are based on taking advantage of the relative abundance and the unique size and density characteristics of virus particles. The general approach is to filter a relatively large volume of environmental sample (for hot springs this may range from liters to hundreds of liters), while simultaneously concentrating the cells and the viruses present in the sample. In the second step, the VLPs are separated from the cells. In the final step, the virus is further purified and concentrated. Overall, the object is to capture as much of the total virus diversity present in the environment at the time of sampling.

We use the example of a tangential flow filtration protocol as an approach for culture-independent isolation of virus from acidic hot spring environments. Tangential flow filtration allows the volume of an environmental sample to be vastly reduced. For example, if the direct environmental sample contains 10^4 VLPs ml^{-1} you can theoretically increase this to 10^8 VLPs ml^{-1} simply by concentrating 100 l to 10 ml. Further concentration to 10^{11} VLPs ml^{-1} can be accomplished by ultracentrifugation and resuspension of the virus pellet in 10 µl. Of course, one seldom achieves theoretical VLP yields (**Note 1**).

An outline of the basic protocol follows and is schematically presented in **Fig. 5.2**. Position the intake end of collection tubing at an appropriate location within the hot spring. Suspend the intake line in water in such a way that minimizes the uptake of sediments. In pools with large particulate matter or other debris, place a metal screen fitted over the intake hose. Place the peristaltic pump in-line and pump through the filters at an appropriate rate (typically 2 l/min). Pump from the source through a glass wool filter to remove as much sediment as possible. The water is pumped through the 100,000 MWCO tangential flow filter unit placed in-line with the glass wool filter. We generally flush the filter for several minutes prior to collecting sample. To begin collecting sample, direct the retentate hose from the tangential flow unit into the 10 l carboy and begin to apply backpressure on the outlet of the filter in order to force water through the filter tangentially. The retentate flow rate should approximately equal the filtrate flow rate (approximately 2 l/min). In order to maximize surface area of the filter being used, ensure that the entire cartridge is wetted. It may be necessary to support the filter on an

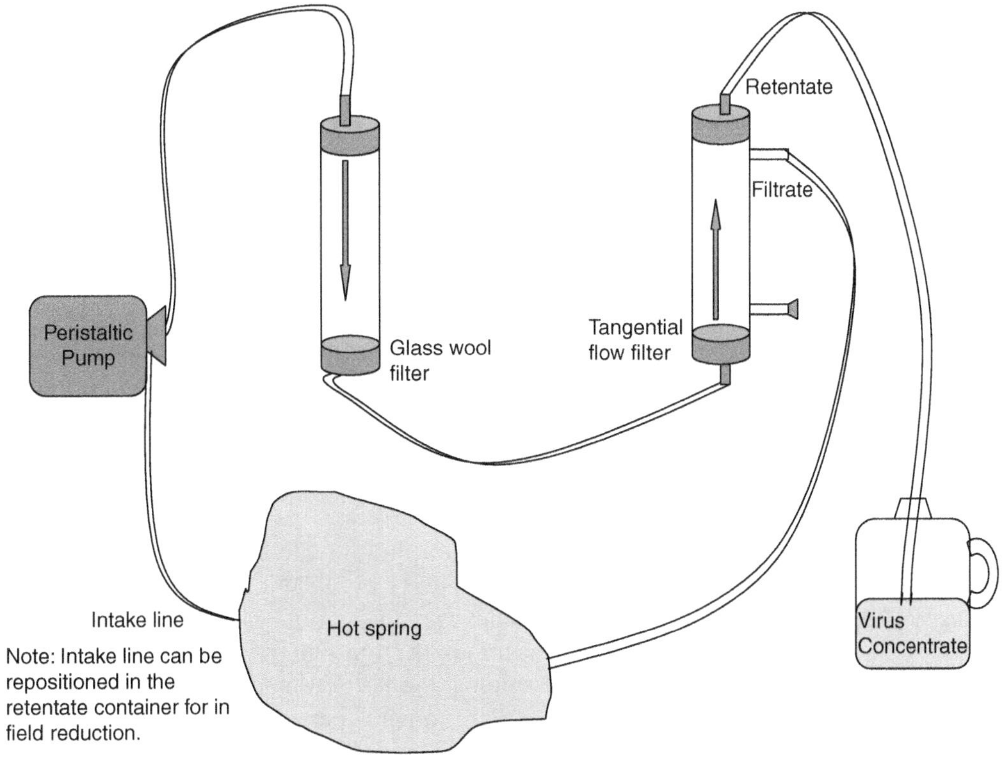

Fig. 5.2. Schematic representation of tangential flow filtration at a thermal feature.

incline so that water is being pumped up the filter and the filtrate output is at the top of the filter with the lower filtrate output valve closed.

By utilizing two tangential filters units simultaneously (which requires a double headed peristaltic pump) one can continue to filter directly from the spring while you concentrate with a second filter with both intake and retentate lines in the carboy and the filtrate being directed as waste. Again, ensure that the retentate and filtrate flow rates are approximately equal and that the entire surface area of the filter is wetted. With only one filter and pump head fill the carboy completely, and then place both intake and retentate output into the carboy while reducing by forcing half the volume through filter. Continue to filter and concentrate until a desired volume has been sampled. Reduction of 200 l to approximately 2 l within 3 h has been successful in YNP hot acidic pools. The concentrated sample contains VLPs, cells and other hot spring components greater than 100 kDa.

Upon returning to the lab, the 2 l concentrated sample is centrifuged ($6000 \times g$ for 20 min) to remove sediments and cells. The resulting supernatant is filtered through a double 0.45 μm filter to remove any remaining cells and sediments. The sample is further concentrated 10-fold (to approximately 200 ml) using the 10,000 Da MWCO tangential flow filter unit in a manner similar to that described for the in-field 100,000 MWCO tangential flow unit. The supernatant is now ready for further virus concentration if necessary (**Section 3.5**). We have averaged 1.5 l/min pumping from the hot spring, requiring approximately 2 h 15 min to concentrate 200 l to 2 l (**Note 2**). The advantage of this type of culture-independent method is that it allows for sampling of large volumes in order to increase virus biomass. However, tangential flow based filtration and concentration methods can introduce sampling bias. In addition, this methodology requires a considerable time commitment at the site and generally only one sample per day can be colleted.

3.3 Cell and Virus Enumeration by Epifluorescent Microscopy

Direct cell and virus counts can be determined by Epifluorescent Microscopy. This can serve as a method for the survey of sampling sites and directing future filtering efforts. Please refer to **Chapter 8** in this publication for detailed protocols (**Note 3**).

3.4 Virus Detection by Transmission Electron Microscopy (TEM)

Detailed analysis of viral morphology can be performed by TEM. Evidence of virus-like particles by EFM can also be verified by TEM. TEM allows direct visualization of viruses, but is not a quantitative measure as the amount of sample loaded onto grids varies between preparations. TEM analysis is best performed after all cells and other nonviral particulate matter have been removed by centrifugation and filtration. Additionally, it may require further concentration by ultracentrifugation (see below) for virus like particles to be detectable.

3.5 Further Virus Concentration by Ultra Centrifugation

Viruses can be further concentrated from enrichment cultures or environmental samples. First, any remaining cells and debris should be removed by low speed centrifugation and filtration. Briefly outlined, the sample should be centrifuged at $6000 \times g$ for 20 min, followed by end point filtration of the resulting supernatant through 0.45 μm filters. The virus in the resulting filtrate can then be concentrated by high speed centrifugation (2 h at 25,000 rpm in Beckman Type 30 rotor works well). After ultracentrifugation, the supernatant should be carefully removed and the pellet, which may only be visible as an area of hydrophobicity on the wall of the tube, resuspended in a small volume (generally 1/1000 the original supernatant volume) of distilled water or an appropriate buffer (typically a buffer near the same pH as in the sample site). At this point, the sample is suitable for further analysis by TEM (**Chapter 10**), viral nucleic acid extraction, or further purification by ion or size exchange chromatography, by sedimentation velocity on sucrose gradients, and/or isopycnic banding on density gradients (**Chapters 21, 33, and 34**).

3.6 Viral Nucleic Acid Isolation

Viral nucleic acids can be extracted from the total cell pellets or from virus enriched concentrates. Extraction of total nucleic acids directly from cell pellets can allow for the detection of high abundance extrachromosomal elements including viruses *(10)*. Briefly outlined, cell pellets directly from an environmental sample or from an enrichment culture are collected by low speed centrifugation ($6000 \times g$ for 10 min). The resulting cell pellet is resuspended in 250 μl TEN followed by addition of 250 μl TENST. Incubate for 30 min at room temperature prior to extraction by phenol chloroform method described below.

For isolation of viral nucleic acids from virus concentrates bring the sample up to a total volume of 200 μl in TE buffer with a final concentration of 0.1% SDS (i.e., 2 μl of 10% SDS), and with 400 μg Proteinase K. Incubate the sample at 37°C ramping to 55°C over 30 min and then continue with phenol extraction. If the virus DNA is protected by capsids then to reduce contamination from bacterial DNA and RNA, the sample can be nuclease treated prior to nucleic acid extraction. To do this add an appropriate amount of RNase A and DNase I (typically 0.1–0.5 units/μl solution) in an appropriate buffer (**Note 4**) to the sample and incubate at 37°C for 30 min. This should remove any nucleic acid not protected by a viral capsid. The DNase and RNase will be eliminated by the Proteinase K during the nucleic acid extraction. However, avoid adding excess enzyme to ensure that they are degraded by the Proteinase K before virus disassembly and the viral nucleic acid is exposed.

3.6.1 Phenol Extraction

Add an equal volume of sample to warm TE saturated phenol (55–60°C) and vigorously vortex. Centrifuge at 14000×*g* for 5 min. Remove the upper aqueous layer to a new tube. Repeat the phenol extraction. Extract a third time with an equal volume of phenol:CH_3Cl:isoamyl alcohol. Remove the upper aqueous layer to a new tube and add 2. 5× the sample volume of 100% ethanol and 1/10 volume 3M sodium acetate, pH 5.2. Incubate on ice for > 20 min to precipitate the nucleic acids. Nucleic acid samples in ethanol can be safely stored for months at –20°C. To collect the nucleic acids, centrifuge at 14000×*g* for 20 min. Immediately and carefully remove supernatant. Add 1 ml 70% ethanol to wash the pellet. Centrifuge at 14000×*g* for 2 min. Carefully remove the ethanol and let air dry, but do not overdry the pellet. Resuspend the pellet in appropriate volume of distilled water or TE (generally 50 μl). For chromosomal DNA resuspend the pellet in TE and 1 μl RNase A *(11)*.

3.7 Genome Amplification

Whether one takes a culture-dependent or independent approach to virus isolation, sufficient genetic material for cloning and sequencing is often limiting. Multiple displacement amplification (MDA) can be effectively used to overcome this limitation. MDA uses random hexamer primers and the highly processive strand displacing φ29 viral polymerase. This procedure can amplify 1–10 ng of template DNA, or significantly less with reduction of the total reaction volume, to detectable levels *(11)*. We have had success amplifying viral genomes both from low abundance anaerobic enrichment cultures and from concentrated environmental samples using the GenomiPhi DNA Amplification kit from GE Healthcare. This method, with very high processivity along with random priming, allows for initial amplification of all DNA in a given sample. As a result care must be taken to reduce any contaminating DNA which will be amplified efficiently along with viral DNA. A first round of amplification using MDA, followed by sequencing, can provide sufficient information to design specific primers and PCR sequence gaps in the genome (**Chapter 25**).

3.8 Storage of Samples

Long-term storage of the enrichment culture can be achieved by adding autoclaved 80% (w/v) glycerol to the cell pellet or viral pellet and storing at −80°C.

4 Notes

1. Anecdotal evidence suggests that there is an upper limit of approximately 200 l beyond which yields drop off quickly. Other limitations to this system are the nature of the VLPs

themselves. For example, highly charged particles are more likely to adhere to the negatively charged filters. Additionally, the clarity of the sample plays a role; the more sediment present in the environmental sample the less successful tangential flow filtration will be. Despite these limitations, tangential filtration-based methods can be a successful method to capture a broad diversity of viruses present in high temperature acidic environments.

2. Site selection is very important with this method. Hot springs with high clay content are not ideal as they quickly clog the filter and decrease yields.
3. Performing the appropriate controls is particularly important when counting VLPs by EFM as they can easily be mistaken for other particulate matter in the media or the environmental sample. For example, remaining clay particles in samples can give erroneously high virus counts which are not verifiable by other methods. We generally count viruses after the sample has been processed to remove all sediments and cells, allowing us to reduce background from media precipitates or environmental sediments.
4. The buffer generally comes with the RNAse/DNAse enzymes, if not suitable buffer can be made from 100 mM Tris–HCl pH7.5, 25 mM $MgCl_2$, and 5 mM $CaCl_2$.

References

1. Bettarel, Y., Sime-Ngando, T., Amblard, C., Dolan, J. (2004) Viral activity in two contrasting lake ecosystems. *Applied and Environmental Microbiology,* **70**, 2941–2951.
2. Snyder, J.C., Spuhler, J., Wiedenheft, B., Roberto, F.F., Douglas, T., Young, M.J. (2004) Effects of culturing on the population structure of a hyperthermophilic virus. *Microbial Ecology,* **48**, 561–566.
3. Zillig, W., Kletzin, A., Schleper, C., Holz, I., Janekovic, D., Hain, J., Lanzendörfer, Kristjansson, J.K. (1994) Screening for *Sulfolobales,* their plasmids and their viruses in Icelandic solfataras. *Systematic Applied Microbiology.* **16**, 609–628.
4. Prangishvili, D., Garrett, R. (2005) Viruses of hyperthermophilic Crenarchaea. *Trends in Microbiology,* **13**, 535–542.
5. Suttle, C.A. (2005) Viruses in the sea. *Nature,* **437**, 356–361.
6. Wommack, K.E., Colwell, R.R. (2000) Virioplankton: Viruses in aquatic ecosystems, *Microbiology and Molecular Biology Reviews,* **64,** 69–114.
7. Borsheim, H.Y., Bratbak, G., Heldal, M. (1990) Enumeration and biomass estimation of planktonic bacteria and viruses by transmission electron microscopy. *Applied Environmental Microbiology,* **56**, 352–356.
8. Wen, K., Ortmann, A., Suttle. C.A. (2004) Accurate estimation of viral abundance by epifluorescence microscopy. *Applied and Environmental Microbiology,* **70**, 3862–3867.
9. Vaidya, S.R., Kharul, H.K., Chitambar, S.D., Wanjale, S.D., Bhole, Y.S. (2004) Removal of hepatitis A virus from water by polyacrylonitrile-based ultrafiltration membranes. *Journal of Virological Methods,* **119**, 7–9.
10. Venter, J.C., Remington, K., Heidelberg, J.F., Halpern, A.L., Rusch, D., Eisen, J.A., Wu, D., Paulsen, I., Nelson, K.E., Nelson, W., Fouts, D.E., Levy, S., Knap, A.H., Lomas, M.W., Nealson, K., White, O., Peterson, J., Hoffman, J., Parsons, R., Baden-Tillson, H., Pfanndock, C., Rogers, Y-H., Smith, H.O. (2004) Environmental genome shotgun sequencing of the Sargasso Sea. *Science,* **304**, 66–74.

11. Zhang, K., Martiny, A.C., Reppas, N.B., Barry, K.W. Malek, J., Chisholm, S.W., Church, G. (2006) Sequencing genomes from single cells by polymerase cloning. *Nature Biotechnology*, **24**, 680–686.
12. Snyder, J.C., Stedman, K., Rice, G., Wiedenheft, B., Spuhler, J., Young, M. (2003) Viruses of hyperthermophilic Archaea. *Research in Microbiology*, **154**, 474–482.
13. Bettarel, Y., Sime-Ngando, T., Amblard, C., Dolan, J. (2004) Viral activity in two contrasting lake ecosystems. *Applied and Environmental Microbiology*, **70**, 2941–2951.
14. German National Resource Centre for Biological Material (DSMZ) website: www.dsmz.de.
15. Itoh, T., Suzuki, K., Nakase, T. (1998) *Thermocladium modestius* gen. nov., sp. nov., a new genus of rod-shaped, extremely thermophilic crenarchaeote, *International Journal of Systematic Bacteriology*, **48**, 879–88.
16. Segerer, A.H., Trincone, A., Gahrz, M., Stetter, K. (1991) *Stygiolobus azoricus* gen. nov., sp. nov. represents a novel genus of anaerobic, extremely thermoacidiophilic Archaebacteria of the order *Sulfolobales. International Journal of Systematic Bacteriology*, **41**, 495–501.
17. Huber, G., Spinnler, C., Gambacorta, A., Stetter, K.O. (1989) *Metallosphaera sedula* gen. and sp. nov. Represents a new genus of aerobic, metal-mobilizing, thermoacidophilic Archaebacteria. *Systematic Applied Microbiology*, **12**, 38–47.
18. Balch, W.E., Fox, G.E., Magrum, L.J., Woese, C.R., Wolfe, R.S. (1979) *Methanogens:* Reevaluation of a unique biological group. *Microbiological Reviews*, **43**, 260–296.
19. Itoh, T., Suzuki, K., Takashi, N., (2002) *Vulcanisaeta distributa* gen. nov., sp. nov., and *Vulcanisaeta souniana* sp. nov., novel hyperthermophilic, rod-shaped crenarchaeotes isolated from ot springs in Japan. *International Journal of Systematic and Evolutionary Microbiology*. **52**, 1097–1104.
20. Segerer, A., Neuner, A., Kristjansson, J.K., Stetter, K.O.(1986) *Acidianus infernus* gen. nov., sp. nov., and *Acidianus brierleyi* comb. nov.: Facultatively aerobic, extremely acidophilic thermophilic sulfur-metabolizing Archaebacteria. *International Journal of Systematic Bacteriology*, **36**, 559–564.

Chapter 6

Isolation of Novel Large and Aggregating Bacteriophages

Philip Serwer, Shirley J. Hayes, Julie A. Thomas, Borries Demeler, and Stephen C. Hardies

Abstract

Viruses are detected via either biological properties such as plaque formation or physical properties. The physical properties include appearance during microscopy and DNA sequence derived from community sequencing. The assumption is that these procedures will succeed for most, if not all, viruses. However, we have found that some bacteriophages are in a category of viruses that are not detected by any of these classical procedures. Given that the data already indicate viruses to be the "largest reservoir of unknown genetic diversity on earth," the implied expansion of this reservoir confirms the belief that the genome project has hardly begun. The first step is to fill gaps in our knowledge of the biological diversity of viruses, an enterprise that will also help to determine the ways in which (a) viruses have participated in evolution and ecology and (b) viruses can be made to participate in disease control and bioremediation. We present here the details of procedures that can be used to cultivate previously undetectable viruses that are either comparatively large or aggregation-prone.

Key words: Agarose gel, dilute, agarose gel, structure of, bacteriophage propagation, microbial diversity, microbial genomics.

1 Introduction

Bacteriophages are viruses that infect and often kill host bacteria. Bacteriophages are a major driving force in many aspects of ecology *(1)* and their killing of infected host cells is usually accompanied by bursting (lysis) of the infected cells. Lysis causes a decrease in the turbidity of a bacterial suspension. This decrease in turbidity both was and is used to assay the presence of a bacteriophage particle *(2, 3)*. During infection, virulent bacteriophages replicate in an infection/lysis cycle that does not stop until bacteriophage-resistant host mutants emerge. In

Martha R. J. Clokie, Andrew M. Kropinski (eds.), *Bacteriophages: Methods and Protocols, Volume 1: Isolation, Characterization, and Interactions, vol. 501*, © 2009 Humana Press, a part of Springer Science+Business Media
DOI 10.1007/978-1-60327-164-6_6 Springerprotocols.com

contrast, lysogenic bacteriophages only sometimes kill and cause lysis. Lysogenic bacteriophages also have an alternative life cycle in which they suppress their own host-toxic genes and do not kill the host cell. Instead, they co-replicate with the host, often by physically integrating the bacteriophage genome (prophage when integrated) with the genome of the host *(4)*.

Lysis-based assays for bacteriophages are typically either (a) single-plaque assays in which a supporting gel maintains a zone of host cell lysis next to a more turbid zone of intact host cells or (b) liquid clearing assays in which an entire liquid bacterial culture loses turbidity. The uses of these two assays for isolating new bacteriophages are discussed in the excellent reviews of Adams *(2)* and Carlson *(3)* and also in **Chapters 2, 4,** and **5** in this volume. We will focus here on some recent, nonclassical variations of previous procedures. These variations appear to have the potential for expanding the range of those viruses that can be characterized via propagation and purification. Much room exists for doing this given the "great plaque count anomaly" whereby the number of viruses detected by microscopy is minimally 2–3 orders of magnitude higher than the number culturable *(1, 5)*. The highest bacteriophage densities are in soil and aquatic sediments *(6, 7)*, sometimes in the absence of any bacteriophages that can be cultured with classical procedures *(7)*.

1.1 Nonclassical Bacteriophage and Host Isolation: Long-Genome Viruses

In the case of isolated eukaryotic double-stranded DNA viruses, long genomes have been found with complexity far beyond anything expected, based on what is needed for virus-specific events. These events include virus DNA replication, transcription, capsid assembly and DNA packaging. The longer genomes are 313–415 Kb long in the case of phycodnaviruses *(8, 9)* and 1,200 Kb in the case of mimivirus *(9, 10)*.

The second longest viral genome is the genome of bacteriophage G which infects *Bacillus megatherium* originally reported at 500 Kb *(11)*, but sometimes at ~670 Kb. Given that bacteriophage G was isolated over 40 years ago, one expects other comparable long-genome bacteriophages to have been isolated by now. However, that is not the case. The next longest bacteriophage genome is, to the authors' knowledge, the 316.675 Kb genome (unpublished data) of *Pseudomonas chlororaphis* bacteriophage 201ϕ2-1, isolated in **ref.** *(12)* with the procedures described here. The 201ϕ2-1 genome is followed by the 280 Kb genome of *Pseudomonas aeruginosa* bacteriophage ϕKZ *(13)* and then by several genomes in the 200–280 Kb range, including the genome of bacteriophage 0305ϕ8-36 (below) and the genomes of several bacteriophages related to *Escherichia coli* bacteriophage T4 *(14)*. The shortage of long-genome bacteriophage isolations suggests that (a) classical isolation procedures will have to be modified in order to detect and propagate the long-genome bacteriophages and (b) at least part of the reason for the great plaque count anomaly is the absence of suitable isolation procedures in the past.

In fact, bacteriophage G itself had a nonclassical isolation and also a nonclassical history of post-isolation propagation. Bacteriophage G originally was a contaminant discovered by electron microscopy of a preparation of bacteriophage α (15). The α host did not propagate G. A host for G was subsequently found by trial and error *(15)*. We originally received bacteriophage G via the Fangman laboratory and have always had difficulty in growing G in liquid culture. Plate stocks in 0.7% agar gels were used for our early studies of the physical properties of G and its genome *(16)*. Subsequently, at one point, we were unable to retrieve activity in agar gels from any of our stocks of bacteriophage G. We also could not find anyone else who had an active preparation. Had we not changed our procedures, even bacteriophage G might today have only historical significance. The key change in procedure was the use of dilute agarose gels for single-plaque propagation.

1.2 The Structure of Agarose Gels

The rationale for propagating G in dilute agarose gels was that increase in the radius of the gel's effective pore (P_E) was needed to permit greater diffusion during plaque formation. The evidence was that $2 \times P_E$ of the classical 0.4–0.7% agar gels was comparable to the size of bacteriophage G. Bacteriophage G is a myovirus with a DNA-containing outer shell that has a diameter of 140–160 nm and a tail that has a length of 400 nm *(15)*.

The estimate of the P_E of agar gels was based on the following background information. Agar is a mixture of related linear polysaccharides with a backbone of agarobiose. Agar polysaccharides are derivatized with negatively charged groups to an extent that varies among different polysaccharide chains both within one agar preparation and among the various sources of agar. Agar is a major component of the intercellular "connective tissue" of red algae. Agarose is derived from agar by purification of the agar components with the least derivatization and electrical charge [reviewed in **ref.** *(17)*]. Thus, the working assumption was that the P_E of agar gels was close to the P_E of agarose gels at any given gel concentration and conditions of gelation. According to one set of studies in which two independent measures of P_E were compared *(18)*, the $2 \times P_E$ of a 0.7% gel of the least purified agarose is ~260 nm [HEEO agarose in Table II of **ref.** *(17)*]. A $2 \times P_E$ of 260 nm is less than 1.8× the average diameter of the bacteriophage G DNA-containing outer shell and about 0.65× the length of the G tail. Thus, restriction of G diffusion by a 0.7% agar gel is a reasonable assumption.

This assumption was confirmed by determination of the dependence of plaque size on gel concentration for G. The plaques of G converted from < 1 mm to > 0.5 cm in diameter as the concentration of a supporting agarose gel changed from 0.4 to 0.1% *(19)*.

Potentially useful for the propagation of long-genome bacteriophages are the observations that the value of P_E increased as the temperature of gelation increased *(20)* and as the ionic strength during gelation increased *(21)*. That is to say, a very simple practical way (tested below) to increase P_E during single-plaque propagation is to increase the temperature of the gelation of the plaque-supporting gel, independent of the temperature of bacteriophage propagation in this gel.

1.3 Objective and Related Assumptions

The major objective of the procedures described here is to isolate and propagate large and aggregating bacteriophages so that they can be characterized in terms of the content and evolution of their protein structure, and in terms of their interactions in the complex systems in which they are found. Achieving this objective will also provide materials for (a) understanding microbial ecology and (b) managing bacterial growth by both phage therapy and phage prophylaxis of infectious disease *(22, 23)*, among other practical applications *(24–26)*. Virulent bacteriophages are preferred for therapeutic purposes and they do not produce prophages with which other bacteriophages can subsequently recombine. Therefore, virulent bacteriophages should undergo less horizontal gene transfer than lysogenic bacteriophages *(27)*. Horizontal gene transfer interferes with establishing homology relationships *(28, 29)*. The most immediate objective, therefore, is to improve procedures to find the long genome, virulent bacteriophages that we assume to be present in environmental samples, but, somehow, are missing in previous isolations.

The use of dilute agarose gels is the strategy presented here. In addition to being effective for isolating long-genome bacteriophages, this strategy should also be effective for isolating bacteriophages that aggregate during plaque formation and, therefore, behave during growth as though larger than they are. Aggregating bacteriophages (and aggregating eukaryotic viruses) are almost completely uninvestigated and may well have a major component of the earth's genome. The literature is curiously silent in the area of aggregating viruses in general, although other infectious agents (prions) are already known to typically be in an aggregated state *(30)*. Aggregating bacteriophages will be lost when procedures such as filtration and centrifugal pelleting are used to remove bacteria before assay. To the authors' knowledge, all currently used procedures remove bacteria before either microscopy-based or community sequencing-based assay for bacteriophages. Bias against aggregating bacteriophages can also be present during liquid enrichment culture, as illustrated by the observation that the aggregating bacteriophage described below (0305ϕ8-36) does not lyse liquid cultures, although it forms clear plaques.

Thus, at least some aggregating bacteriophages have been largely, if not entirely, invisible to previous studies.

2 Materials

1. Petri plates containing 1% gelled sterile bottom agar dissolved in a nutrient broth that contains the following: 10 g Bacto tryptone, 5 g NaCl in 1000 ml water.
2. Molten agarose that contains the following (growth medium): 10 g Bacto tryptone, 5 g KCl in 1000 ml water with 0.002 M $CaCl_2$ added post-autoclaving.
3. An incubator held at 50°C for the long-term storage of molten agarose.
4. A plating block with culture tubes (1.0 cm in inner diameter) held at 50°C.
5. An overnight culture of host cells in growth medium. This culture has been grown with aeration at room temperature (25 ± 3°C) and then transferred to a sterile dropper bottle.
6. A solution of the fluorescent dye, DAPI (4′, 6-diamidino-2-phenylindole) at 10 μg/ml.
7. Glass microscope slides and cover glasses.
8. Fluorescence microscope with filter set appropriate for DAPI (U-MNU filter cube [Olympus], for example).
9. Centrifuge tube (40 ml), sterile glass pipettes and sterile spatula.
10. Cryoprotecting solution: 10% Dextran (molecular weight = 10,000) in growth medium.
11. –70°C freezer.

3 Methods

Isolation of bacteriophages from environmental water typically requires that initial propagation be started either immediately in the field or, in any case, in a time that is less than the time for the sample to change its microbiology. However, we have found that isolation of bacteriophages from dry soil samples can be performed months to years after taking the sample and storing it in a laboratory cabinet. Our soil samples were dry and obtained in Southern Texas, including the King Ranch (Kingsville, Texas), where ground temperatures reach 49–60°C in the summer under intense irradiation from the sun. Thus, soil-associated bacteriophage stability in the laboratory is not a surprise. The procedure below produces new bacteriophages at an average rate of

1 bacteriophage per host, per 2–3 attempts made with 1–5 g of soil. The soil used was not precisely quantified.

Thus, the isolations are not part of a rigorous study of soil ecology. Considering the absence of aggregating bacteriophages from past studies of soil microbial ecology, no rigorous and complete study of soil microbial ecology has, apparently, yet been performed.

3.1 *De novo* Bacteriophage Isolation

1. Remove molten agarose solution from the oven and pipette 4 ml of molten agarose into each culture tube in the plating block at 50°C. The agarose concentration is typically either 0.1% or 0.15% to increase P_E and optimize the growth of both large and aggregating bacteriophages (**Notes 1** and **2**).
2. Place a soil sample on the gelled agar in a Petri plate. This is usually done with a sterile spatula, but can also be done by pouring.
3. Place 4 drops of overnight culture in the culture tube with 4 ml molten agarose and mix.
4. Pour the host cell/molten agarose mixture over the soil sample and rock the Petri plate gently.
5. Place the Petri plate in location for overnight incubation, usually on a laboratory bench. Once placed, Petri plates with dilute agarose gels should not be disturbed for about 10 h because of the fragility of the gels.
6. Incubate the Petri plate, typically for 16 h, though some bacteriophages have taken up to 48 h to make a visible plaque. Moving the Petri plate even after 16 h should be done with care if agarose gels more dilute than 0.25% are used. Otherwise the agarose gel may slide and distort. Petri plates should not be inverted if a gel this dilute is used.
7. Observe the Petri plate and scan for clearing within a turbid bacterial lawn, i.e., potential bacteriophage-induced clearing. Take a photograph. Examples are in **Figs. 1a** and **2a** of **ref.** *(12)*. The cleared areas will be tested for the presence of bacteriophages by the procedure in the next section. Bacteriophages are not the only source of clearing (**Notes 3** and **4**). The larger the P_E of the agarose gel, the larger the bacteriophage that can be detected by this procedure

3.2 Single-Plaque Propagation, Cloning, and Storage

1. Touch a sterile needle to the surface of a cleared zone from the de novo experiment of **Section 3.1**. Bacteriophages, if present, adhere to the needle.
2. Stab the needle 3–10 times into the center of the bottom agar in a new Petri plate.
3. Place 4 drops of overnight host culture in a culture tube with 4 ml of molten agarose at 50°C and mix. Pour the molten agarose near the edge of the Petri plate to minimize uncontrolled turbulence around the inoculated

bacteriophages. Allow the molten agarose to flow over the entire plate. The intent is to produce fluid flow that just partially spreads the inoculated bacteriophages. The targeted result is a single-plate dilution series. That is to say, one area of the incubated plate is confluent and other areas have progressively fewer plaques, eventually single isolated plaques that are used for single-plaque cloning and further propagation.

4. Gently rock the Petri plate to increase the spreading of bacteriophage particles. This aspect has some experience-based art in it. If rocking is too great, the spreading will be sufficient to make the plate confluent after incubation.
5. Incubate the plate overnight at room temperature.
6. Examine the plate and photograph. If the original clear area had bacteriophage particles, this second Petri plate should have areas of both confluent lysis and single plaques [**Figs. 1b** and **2b** in **ref.** *(12)*].
7. To further purify bacteriophages, steps 1–6 are repeated, but the inoculum now becomes a single plaque from the previous cloning. Cloning is continued until single-plaque cloning and propagation has been accomplished three times.
8. For freezing and storage, an entire plaque is removed and placed in a screw cap freezer vial that has cryoprotecting solution added in an approximately 10× volume. Then the vial is placed in the –70°C freezer. Storage is done in quadruplicate in at least two different freezers (For further details on storing *see* **Chapter 16**).
9. The procedures of 1–8 are slightly modified for preparative propagation and storage in liquid at 4°C (Note #5).

3.3 Screening for Aggregating Bacteriophages

1. Use a pipette to dissect a bacteriophage plaque. Place the dissected contents in a glass tube.
2. Add growth medium and dice/homogenize the contents of the plaque. Gels of 0.1–0.2% agarose behave as a quasi-liquid when this is done.
3. Add 0.05× amount of the DAPI solution and mix.
4. Place 2–5 ml on a glass microscope slide and place a cover glass on the stained extract.
5. View in the fluorescence microscope [*see*, for example, **refs.** *(31, 32)* and **Chapter 8**]. Aggregates are sometimes large enough (multi-μm dimensions) to be obvious. The larger aggregates sediment at 1 g and, therefore, are found at the surface of the slide, not near the cover glass. Smaller aggregates are also usually obvious based on size, fluorescence intensity, and resolution of more than one particle. Bacteriophage 0305ϕ8-36 was over 99% aggregated when observed with this procedure *(32)*. Further work on characterizing bacteriophage aggregates by fluorescence microscopy is in **ref.** *(32)*. Note #6

describes characterization by centrifugation of more purified 0305ϕ8-36 aggregates.

4 Notes

1. Dilution of agar gels (to 0.3%) has also been found to be one of several ways to increase plaque sizes *(33, 34)*. We used agarose, rather than agar, primarily because agarose gels were stronger at any given concentration. Thus, concentrations as low as 0.1% were readily accessible and we have had success with agarose concentrations as low as 0.075%.
2. To further increase the P_E of the upper layer agarose gel used for bacteriophage propagation, a possible procedure is raising the temperature of gelation without changing the temperature of bacteriophage propagation (**Section 1.2**). This procedure was tested for an aggregating bacteriophage, 0305ϕ8-36 (genome length = 218.948 Kb) by (a) pouring host cell-containing, molten 0.1% agarose in several different Petri plates (the host was *Bacillus thuringiensis*), (b) gelling the agarose at different temperatures, (c) inoculating bacteriophage particles with a stab (four separate stabs per plate) and (d) forming plaques with all plates at the same temperature, in thermal contact with each other. The result was that the plaque radius decreased as the gelation temperature decreased from 42°C (**Fig. 6.1a**) to 37°C (**Fig. 6.1b**), 25°C (**Fig. 6.1c**) and 4°C (**Fig. 6.1d**). The same result was found when this experiment was done by conventional dilution and plating.

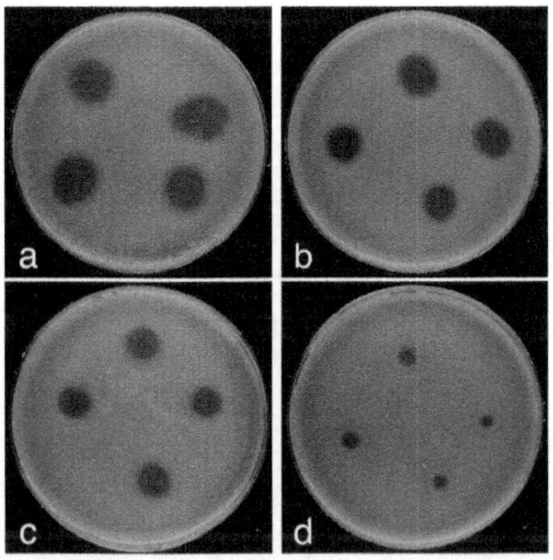

Fig. 6.1. Effect on plaque diameter of the temperature of agarose gelation before growth of bacteriophage 0305ϕ8-36. Host cell-containing molten agarose was added to a Petri plate previously equilibrated at either (**a**) 42°C, (**b**) 37°C, (**c**) 25°C, or (**d**) 4°C. The plates were then placed at room temperature (25 ±3°C) and inoculated after the agarose gelled. The plates were then incubated in thermal contact with each other for 8 h at room temperature.

3. In addition to providing a lawn for bacteriophage growth, the host bacteria in the molten agarose also suppress the growth of endogenous bacteria that are in the soil sample. If problems are encountered with endogenous bacteria overgrowing the lawn, the first solution to try is increasing the concentration of the host bacteria in the molten agarose.
4. As suggested in Note 3, new hosts can be isolated if the host bacteria in the molten agarose are omitted from the procedure for isolating bacteriophages. Omitting the host bacteria releases repression of the growth of endogenous bacteria of the soil sample. The pattern of endogenous bacterial growth varied, though it always included bacteria eventually growing to either confluence or near confluence. In one case, a swarming bacterium overgrew the Petri plate to form a lawn, except in restricted areas where colonies formed. Some colonies were surrounded with a zone of clearing (for example, arrow #1 in **Fig. 6.2**). The turbid continuous lawn (arrow #2) was formed by a sometimes swarming *B. pumilus*, as determined by single-colony propagation, followed by sequencing of the 16 s ribosomal RNA gene *(12)*. When plated on the *B. pumilus*, the zone indicated by arrow #1 in **Fig. 6.2** produced a *B. pumilus*-clearing *B. subtilis*, not a bacteriophage. The *B. subtilis* was also identified by single-colony propagation, followed by sequencing of the 16 s ribosomal RNA gene. The competition between the *B. pumilus* and the *B. subtilis* was replicated with single-colony purified strains (not shown). The colonies indicated by arrows #3 and #4 in **Fig. 6.2** were formed by *B. thuringiensis* and *B. megaterium*, respectively. The *B. pumilus* and *B. subtilis* strains from this plate have been used to isolate bacteriophages.

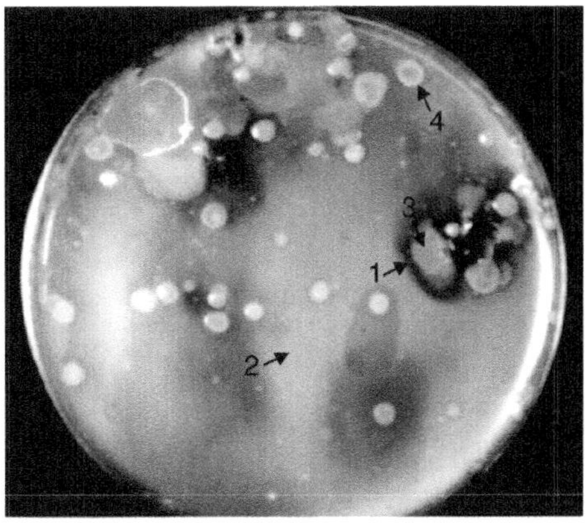

Fig. 6.2. Isolation of hosts. The procedure for bacteriophage isolation in **Section 2.1** was performed with the host cells omitted from the molten agarose. The Petri plate was incubated for 20 h at 25°C and then photographed.

5. For growth on a preparative scale, the procedure of steps 1–8 is repeated with the following changes: (a) The stabs used for bacteriophage inoculation are located throughout the bottom agar so that incubation and plaque formation results in confluent lysis. (b) After incubation, the entire agarose layer is removed with a sterile spatula and placed in a centrifuge tube with growth medium for extraction of bacteriophage particles. After further purification of bacteriophage particles, the optimal buffer used for short-term storage cannot be known in advance. A good place to start is the inclusion of 3% polyethylene glycol (PEG; molecular weight in the 1,500–8,000 range) to stabilize *(35)* the bacteriophage particles. Higher PEG concentrations may cause precipitation. In addition, magnesium cation should be present, as found for various bacteriophages over a period of over 50 years. Magnesium cation at a concentration of 0.05 M is a good place to start. An advantage of in-gel preparative growth is that aerosol production is dramatically reduced in comparison to preparative growth in liquid culture. Aerosols can create cross-contamination problems *(33)*.
6. To obtain an independent test of 0305ϕ8-36 aggregation after purification by buoyant density in a cesium chloride density gradient, analytical ultracentrifugation was used. Detection was performed by SYBR green staining and use of fluorescence detection optics previously described *(36)*. The resulting profile (**Fig. 6.3**) has sedimentation coefficients ranging from 350 to 1200, a confirmation of aggregation. For comparison, the same analysis is shown for bacteriophage T7 (**Fig. 6.3**), a

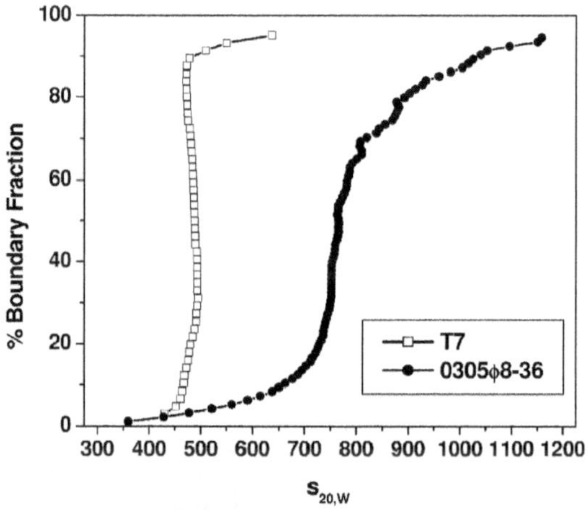

Fig. 6.3. Analysis of aggregation by analytical ultracentrifugation with fluorescence detection. The following were subjected to centrifugation at 3000 rpm, 20°C in a Beckman XLA analytical ultracentrifuge with fluorescence detection optics *(36)*: bacteriophage 0305ϕ8-36 (*filled circles*) and bacteriophage T7 (*empty squares*). Data processing was performed by the procedures in **ref.** *(37)*.

non aggregating bacteriophage that produces a mostly homogeneous sedimentation pattern with an average sedimentation coefficient of ~450 s.

Acknowledgments

We thank the following for support: The National Institutes of Health (GM24365 and GM069757), The Welch Foundation (AQ-764) and The Robert J. Kleberg Jr. and Helen C. Kleberg Foundation.

References

1. Wommack, K. E., K. E. Williamson, R. R. Helton, S. R. Bench and D. M. Winget. 2007., companion article.
2. Adams, M. H. 1959. Bacteriophages, N.Y. Interscience Publishers, New York.
3. Carlson, K. 2005. Appendix: Working with bacteriophages: Common techniques and methodological appproaches. In: Bacteriophages: Biology and Applications (Kutter, E. and Sulakvelidze, A., Eds.), CRC Press, Boca Raton, pp.437–494.
4. Guttman, B., R. Raya and E. Kutter. 2005. Basic phage biology. In: Bacteriophages: Biology and Applications (Kutter, E. and Sulakvelidze, A., eds.), CRC Press, Boca Raton, pp. 29–66.
5. Weinbauer, M. G. 2004. Ecology of prokaryotic viruses. FEMS Microbiol. Rev. *28*: 127–181.
6. Williamson, K. E., M. Radosevich and K. E. Wommack. 2005. Abundance and diversity of viruses in six Delaware soils. Appl. Environ. Microbiol. *71*: 3119–3125.
7. Ashelford, K. E., M. J. Day and J. C. Fry. 2003. Elevated abundance of bacteriophage infecting bacteria in soil. Appl. Environ. Microbiol. *69*: 285–289.
8. Dunigan, D. D., L. A. Fitzgerald and J. L. Van Etten. 2006. Phycodnaviruses: a peek at genetic diversity. Virus Res. *117*: 119–132.
9. Iyer, L. M., S. Balaji, E. V. Koonin and L. Aravind. 2006. Evolutionary genomics of nucleo-cytoplasmic large DNA viruses. Virus Res. *117*: 156–184.
10. Claverie, J, M., H. Ogata, S. Audic, C. Abergel, K. Suhre and P. E. Fournier. 2006. Mimivirus and the emerging concept of "giant" virus. Virus Res. *117*: 133–144.
11. Fangman, W. L. 1978. *Nucleic Acids Res.*, **5**, 653–665.
12. Serwer, P. S. J. Hayes, S. Zaman, K. Lieman, M. Rolando and S. C. Hardies. 2004. Improved isolation of under sampled bacteriophages: Finding of distant terminase genes. Virology *329*: 412–424.
13. Mesyanzhinov, V. V., J. Robben, B. Grymonprez, V. A. Kostyuchenko, M. V. Bourkaltseva, N. N. Sykilinda, V. N. Krylov and G. Volckaert. 2002. The genome of bacteriophage ϕKZ of *Pseudomonas aeruginosa*. J. Mol. Biol. *317*: 1–19.
14. Petrov, V. M., Nolan, J. M., Bertrand, C., Levy, D., Desplats, C., Krisch, H. M. & Karam, J. D. (2006) Plasticity of the gene functions for DNA replication in the T4-like phages. J. Mol. Biol., *361*: 46–68.
15. Donelli, G. 1968. Isolamento di un batteriofago di eccezionali dimensioni attivo su *B. megatherium*. Atti Accad. Naz. Lincei-Rend. Clas. Sci. Fis. Mat. Nat. *44*: 95–97.
16. Serwer, P., A. Estrada and R. A. Harris. 1995. Video light microscopy of 670 Kb DNA in a hanging drop: Shape of the envelope of DNA. Biophys. J. *69*: 2649–2660.
17. Serwer, P. 1983. Agarose gels: Properties and use for electrophoresis. Electrophoresis *4*: 375–382.
18. Griess, G. A., E. T. Moreno, R. A. Easom and P. Serwer, P. 1989. The sieving of spheres during agarose gel electrophoresis: Quantitation and modeling.Biopolymers *28*: 1475–1484.
19. Huang, S., S. J. Hayes, K. Lieman, G. A. Griess and P. Serwer. 2001. Strategies for analysis of the evolution of bacteriophages. Recent Res. Devel. Virol. *3*: 1–12.
20. Griess, G. A., D. M. Edwards, M. Dumais, R. A. Harris, D. W. Renn and P. Serwer. 1993. Heterogeneity of the pores of polysaccharide gels: Dependence on the molecular weight and

derivatization of the polysaccharide. J. Struct. Biol. *111*: 39–47.
21. Serwer, P. and G. A. Griess. 1999. Advances in the separation of bacteriophages and related particles. J. Chromatogr. B *722*: 179–190.
22. Matsuzaki, S., M. Rashel, J. Uchiyama, S. Sakurai, T. Ujihara, M. Kuroda, M. Ikeuchi, T. Tani, M. Fujieda, H. Wakiguchi and S. Imai. 2005. Bacteriophage therapy: a revitalized therapy against bacterial infectious diseases. Infection Chemotherapy *11*: 211–219.
23. Sulakvelidze, A. (2005) Phage therapy: an attractive option for dealing with antibiotic-resistant bacterial infections. Drug Discovery Today *10*: 807–809.
24. Clark, J. R. and J. B. March. 2006. Bacteriophages and biotechnology: vaccines, gene therapy and antibacterials. Trends Biotechnol. *24*: 212–218.
25. Liu, J., M. Dehbi, G. Moeck, F. Arhin, P. Bauda, D. Bergeron, M. Callejo, V. Ferretti, N. Ha, T. Kwan, J. McCarty, R. Srikumar, D. Williams, J. J. Wu, P. Gros, J. Pelletier and M. DuBow. 2004. Antimicrobial drug discovery through bacteriophage genomics. Nature Biotechnol. *22*: 185–191.
26. Miedzybrodzki, R., W. Fortuna, B. Weber-Dabrowska and A. Gorski. 2005. Bacterial viruses against viruses pathogenic for man? Virus Res. *110*: 1–8.
27. Brüssow, H. and E. Kutter. 2005. Genomics and evolution of tailed phages. In: Bacteriophages: Biology and Applications (Kutter, E. & Sulakvelidze, A., eds.), CRC Press, Boca Raton, pp. 91–128.
28. Casjens, S. R. 2005. Comparative genomics and evolution of the tailed-bacteriophages. Curr. Opin. Microbiol. *8*: 451–458
29. Hendrix, R. W. 2003. Bacteriophage genomics. Curr. Opin. Microbiol. *6*: 506–511.
30. Bennett, M. J., M. R. Sawaya and D. Eisenberg. 2006. Deposition diseases and 3D domain swapping. Structure *14*: 811–824.
31. Wang, H., I. Wu, Q. Yang, C. E. Catalano and P. Serwer. 2005. Single-particle visualization of assembly: I. Dimerization in a planar zone. J. Microscop. *217*: 83–92.
32. Serwer, P., S. J. Hayes, K. Lieman and G. A. Griess. 2007. *In situ* fluorescence microscopy of DNA bacteriophage aggregates. J. Microsc. *228*: 309–321.
33. Fortier, L.-C. and S. Moineau, (2006) Accompanying manuscript (**Chapter 16**).
34. Lillehaug, D. 1997. An improved plaque assay for poor plaque-producing temperate lactococcal bacteriophages.*J. Appl. Microbiol. 83*: 85–90.
35. Serwer, P., W. E. Masker and J. L. Allen. 1983. Stability and *in vitro* DNA packaging of bacteriophages: effects of dextrans, sugars, and polyols. J. Virol. *45*: 665–671.
36. MacGregor, I. K., A. L. Anderson T. M. Laue. 2004. Fluorescence detection for the XLI analytical ultracentrifuge. Biophys. Chem. *108*: 165–185.
37. Demeler, B. and K. E. van Holde. 2004. Sedimentation velocity analysis of highly heterogeneous systems. Anal. Biochem. *335*: 279–288.

Section 2

Bacteriophage Characterization

Chapter 7

Enumeration of Bacteriophages by Double Agar Overlay Plaque Assay

Andrew M. Kropinski, Amanda Mazzocco, Thomas E. Waddell, Erika Lingohr and Roger P. Johnson

Abstract

The determination of the concentration of infectious phage particles is fundamental to many protocols in phage biology, genetics, and molecular biology. In this chapter the classical overlay protocol is described.

Key words: Plaque, overlay technique, plaque-forming unit, pfu, plaque morphology, direct plating plaque assay.

1 Introduction

The determination of the concentration of infectious phage particles is a fundamental protocol for those working with bacterial viruses. The most common techniques are plaque assays, in which dilutions of the phage preparation are mixed with a permissive host bacterium and dispersed evenly onto solid medium. On incubation, the host bacterium forms a lawn on the solid medium, except where infectious phage particles lyse or inhibit the growth of the cells, resulting in a localized clear or translucent zone, termed a plaque. The infectious phage unit is thus termed a plaque-forming unit (pfu). The visualization of individual plaques permits far more than mere enumeration. It is the basis for the isolation of phages, their characterization by plaque morphology (clear versus turbid lysis, size of plaque, presence/absence of a halo), and the isolation of phage mutants.

Plaque assays rely on the ability of the permissive host bacterium to form a complete lawn on solid media, and on localized

Martha R. J. Clokie, Andrew M. Kropinski (eds.), *Bacteriophages: Methods and Protocols, Volume 1: Isolation, Characterization, and Interactions, vol. 501*, © 2009 Humana Press, a part of Springer Science+Business Media
DOI 10.1007/978-1-60327-164-6_7 Springerprotocols.com

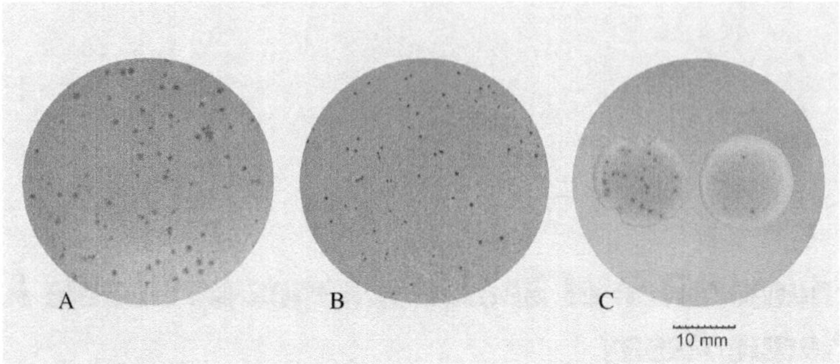

Fig. 7.1. The appearance of plaques formed by phage rV5 with the double overlay method (**A**), the direct plating method (**B**) and the drop method (**C**). The 10 mm bar applies only to the Drop Method (C).

expansion of the plaques until they become visible to the naked eye. Plaque expansion requires a sufficient yield of progeny phages from each infected bacterial cell to allow localized infection and lysis or altered growth of adjacent bacteria (for details on the theory and measurement of plaque development *see* **Chapters 14** and **15**). The techniques described below and in the subsequent chapters include the Double Agar Overlay method (**Fig. 7.1A**), the Direct Plating method (**Fig. 7.1B**), and a small scale Drop method (**Fig. 7.1C**). They have been employed effectively to study phages of;

This technique has been used by us with *Pseudomonas*, *Escherichia*, *Salmonella* and *Bacillus*, and with minor variations, should be applicable to most bacterial species that can grow as a confluent lawn. Phages requiring host bacteria that do not form lawns on solid agar, or yield insufficient progeny to form visible plaques, can be enumerated by methods other than plaque assays *(1)*, such as electron microscopy *(2, 3)*, fluorescent microscopy *(4)* or quantitative PCR *(5)*.

The double agar overlay plaque assay is also known as the "soft agar overlay," "double agar layer" or "double layer" method of plaque assay. The Double Agar Overlay method was originally outlined by André Gratia *(6, 7)* and formalized by Mark Adams *(3)*. Dilutions of phage suspension are mixed with host bacteria in a dilute, molten agar or agarose matrix (the "top agar" or "overlay"), which is distributed evenly to solidify on a standard agar plate (the "bottom agar" or "underlay"). After incubation, usually overnight, plaques are visualized as zones of clearing (or diminished growth) in the bacterial lawn, which grows in the overlay.

2 Materials

1. Supplies and equipment for preparation of broth and solid culture media, including 90 mm Petri dishes, 13 × 100 mm glass tubes with metal caps (Morton Culture Tube Closures, Fisher

Scientific) or plastic caps (Kaputs, Bellco Biotechnology, Inc.; Vineland, NJ; **Note 1**).

2. Variable or fixed-volume micropipettors for volumes of 1 ml, 100 μl, and 10 μl, with matching sterile pipette tips (**Note 2**).
3. Sterile dilution tubes (1.5 ml capped microcentrifuge tubes, capped 13 × 100 mm tubes, or similar).
4. Sterile repeating syringe (e.g., Wheaton Self-Filling Repetitive Syringe; Fischer Scientific or VWR), or similar device dispensing warm medium in volumes of 3 ml.
5. Heating block or waterbath for 13 × 100 mm tubes maintaining 48°C.

2.1 Media

Since both the overlay (top agar) and underlay (bottom agar) can usually be prepared from one broth medium with only different concentrations of agar or agarose, it is convenient to prepare the media from broth formulations rather than from agar formulations, adding the different amounts of agar during preparation.

For members of the bacterial genera listed above, Luria Bertani (LB) broth, Lennox, or tryptic soy broth (Difco Laboratories, Division of Becton Dickinson & Co.), work well. A modified nutrient agar (MNA, see Direct Plating method below), also works well for these phages.

For plaque assays of phages of bacteria not mentioned above, use the medium recommended for propagating the host bacterium [see for example, the American Type Culture Collection (ATCC) at http://www.atcc.org/Home.cfm, or Deutsche Sammlung von Mikroorganismen und Zellkulturen (DSMZ) at http://www.dsmz.de/].

Standard agar for bacteriological media is suitable for many phage – host systems. For certain organisms, such as *Caulobacter*, impurities in agar can inhibit bacterial growth, so the agar is best washed extensively with distilled water before use. Alternatively, molecular biology-grade agarose can be employed *(8)*.

Many phages require 1–10 mM divalent ions such as Ca^{2+} or Mg^{2+} for attachment or intracellular growth *(9–11)*. This is conveniently prepared as a sterile 1M $CaCl_2$ solution. When working with phage mixtures or uncharacterized phages, it is a good idea to include 1–2 mM Ca^{2+} in both the underlay and overlay media. To determine if the phage requires divalent ions for plaque development, test the phage in plaque assays with media without Ca^{2+} and supplemented with 50 mM citrate, as well as in media with 1–10 mM Ca^{2+}. If Ca^{2+} is required, a significant decrease in titer should result in the medium without Ca^{2+} *(12)*.

2.1.1 Underlay or Bottom Agar

1. Prepare the medium according to the manufacturer's instructions, adding agar (or agarose) to a final concentration of 15 g/l, and heat sterilize.

2. When the medium has cooled to 55–60°C, add sterile 1M $CaCl_2$ to the desired final concentration (1–10 mM, or 1–10 ml/l). Hold media at 48–50°C.
3. Dispense 18–25 ml of medium per plate, and when cool, bag and store the plates at 4°C. Bagged plates can be stored for up to 2 months.

2.1.2 Overlay or Top Agar

1. Prepare the broth medium according to the manufacturer's instructions, with the addition of agar (or agarose) to a concentration of 4–6 g/l. (For phages making small plaques, they are more visible at the lower agar concentration).
2. Heat to melt the agar or agarose, and then add 1M $CaCl_2$ to the molten broth/agar mixture to the desired concentration (1–10 mM, or 1–10 ml/l).
3. Promptly dispense 3-ml volumes of the medium into sterile 13 × 100 mm glass tubes, cap the tubes with metal or plastic caps, heat sterilize, and store at 4°C for up to 1 month (**Note 3**).
4. For use, heat the tubes containing the overlay medium to melt the agar and hold at 46–48°C in a waterbath or heating block. [N.B. It really is important that the agar be brought to a full boil, or there are likely to be crystalline areas remaining that make the plates very hard to interpret.] (**Note 4**)

2.1.3 Diluents

1. Phages can be diluted in the broth medium used to grow the host bacterium (without agar or agarose), or in the following:
   A. Saline – magnesium (SM) diluent plus gelatin (SMG) *(8)*: 50 mM Tris–HCl (pH 7.5) containing 100 mM NaCl, 8.1 mM $MgSO_4$ and 0.01% (w/v) gelatin.

Per liter:	NaCl	5.8 g
	$MgSO_4.7H_2O$	2.0 g
	1M Tris–HCl (pH 7.5)	50 ml
	2% (wt/vol) gelatin	5 ml
	Distilled water to:	1000 ml.

   B. Lambda (λ) diluent *(13)*: 10 mM Tris–HCl (pH 7.5), 8.1 mM $MgSO_4$.

Per liter:	$MgSO_4.7H_2O$	2.0 g
	1M Tris–HCl (pH 7.5)	10 ml
	Distilled water to:	1000 ml.

2. Dispense the diluent in 10 ml volumes (sufficient for one dilution series of phages) into screw-capped vials or tubes, apply caps, heat sterilize, and store at 4°C for up to 3 months (**Note 5**).

3 Methods

3.1 Double Agar Overlay Plaque Assay

The following description applies to the Double Agar Overlay plaque assay for a typical phage lysate containing approximately 10^6 to 10^{11} pfu/ml, with one plating per dilution. The range of dilutions and number of replicate plates for each dilution will alter the numbers of plates required.

1. Remove the required number of underlay agar plates from 4°C storage (typically 7 plates: 6 for dilutions of the phage and 1 for a bacterial control without phage), and dry them, inverted and tilted, in a 37°C incubator for 1–2 h or partially uncovered in a laminar flow hood for 10–15 min. (**Note 6**).
2. When the plates are dry, reassemble and number them sequentially with the dilutions to be plated (e.g., "–4" to "–9"), and a "control."
3. Remove the required number of tubes containing 3 ml of overlay medium from 4°C storage, melt the agar, and place the tubes in a waterbath or heating block set at 48°C (**Note 7**).
4. Set up a row of 9 sterile capped tubes or microcentrifuge tubes. Number them with the appropriate sequential 10-fold dilutions (e.g., "–1" to "–9") and aseptically add 900 μl diluent to each tube (**Note 8**).
5. Add 100 μl of phage lysate to the first tube, mix, change the pipette tip and transfer 100 μl to the second tube in the series (**Note 9**).
6. Using a fresh pipette tip for each transfer, continue making 10-fold dilutions. In the last tube, the phage preparation will have been diluted $1/10^9$ or 10^{-9} (**Note 10**).
7. Working quickly with one tube at a time, transfer 100 μl of the selected dilution of phages to a tube of warm overlay medium, immediately add 2 drops ($\sim$ 100 μl) of an overnight culture of the host bacterium, mix as above and pour the contents over the surface of a dried and labeled underlay plate (**Notes 11**, **12**, **13** and **14**).
8. Using a fresh pipette tip for each dilution tube, repeat Step 7 to prepare and pour the overlays for the remaining phage dilutions.
9. Allow the overlays to harden for 30 min, and then incubate the plates inverted in a single layer, at the desired temperature (**Note 15**).
10. Continue incubation for 18–24 h and then count the plaques on plates with 30–300 plaques (**Note 16**).
11. Determine the titer of the original phage preparation by using the following calculation: Number of plaques $\times 10 \times$ reciprocal of counted dilution = pfu/ml

4 Notes

1. The Kaputs are described as disposable but can be autoclaved and re-used many times.
2. Pipettors should be sterilized at least weekly and recalibrated periodically.
3. Alternatively, dispense the medium in 50–100 ml aliquots into screw-capped glass bottles, loosely apply the screw caps, heat sterilize, tighten the caps and store at 4°C for up to 3 months.
4. If the medium is stored in larger volumes in sealed glass bottles, boil the medium and dispense 3-ml volumes to the required number of sterile glass 13 × 100 ml tubes with caps.
5. Alternatively, dispense diluent in larger volumes in screw-capped bottles, apply caps, heat sterilize, and store at 4°C for up to 3 months. For use, transfer the required volume of this stock to a sterile tube, and work from this tube to avoid contamination of the stock diluent.
6. The state of the "underlay" medium is very important; it should not be too dry or too moist, since both of these factors will negatively affect plaque development.
7. The medium must be completely melted and without lumps.
8. Since most phage lysates contain between 10^6 and 10^{11} pfu/ml, dilutions of 10^{-5} to 10^{-9} will probably have phage concentrations that will lead to countable numbers of plaques. Experience will allow estimation of the probable concentration of phages in lysates. Note that during purification, the concentrations may exceed 10^{13} pfu/ml and dilutions will have to be continued beyond 10^{-9}.
9. Sufficient mixing is achieved with a gentle vortex created by flicking the tube with a finger, a low setting on a vortex mixer, or inversion in the case of sealed microcentrifuge tubes. Vigorous mixing of phage suspensions is not recommended.
10. These dilutions can be stored later at 4°C until the assay is completed and an accurate count has been obtained.
11. The volume of the overnight culture of the host bacterium required to obtain a confluent lawn with clearly visible plaques will become apparent with experience, and typically ranges from 50–200μl.
12. Mixing is designed to achieve an even distribution of the phages and host cells in the overlay medium without damaging the phage particles, or introducing bubbles, which later can be mistaken for plaques.
13. It is important to complete the additions of the diluted phages and the host bacterium, and the mixing and pouring of the molten overlay before the agar starts to solidify. It is better to stop distributing the overlay on the underlay

surface early than for partial solidification to occur. The latter can be visualized as a rough surface with small granules.

14. With experience, it is possible to reduce the number of dilutions for titration, and to prepare the dilution and plating of one tube simultaneously (i.e., using the same micropipette, transfer 100 μl to the next dilution tube and to a tube of overlay medium).
15. With fast growing bacteria such as *E. coli*, plaques are often visible within 4 h. In the case of Bacillus cereus plaque overgrowth occurs and counts should be made as soon as visible (T.El Arabi, personal communication).
16. Small plaques can be counted accurately into the hundreds. It is more difficult to accurately count large plaques such as those produced by phage T7.

5 Additional Points

Certain phage plaques are difficult to distinguish because of small size, incomplete lysis or the temperate nature of the phage. Relatively few studies have concentrated on the factors that influence plaque morphology and titer, and these have been predominantly on the viruses of the lactic acid bacteria *(13–15)*. While no general approaches can be described, the results of a number of studies may provide some insight on how to achieve more visible plaques.

In certain cases one can enhance plaque visibility by the incorporation of oxidation/reduction dyes. When filter-sterilized tetrazolium dyes, such as 2,3,5-triphenyltetrazolium chloride, are incorporated into the overlay at 50–300 μg/ml, the bacterial growth results in the reduction of the soluble colorless dyes to an insoluble red formazan. This activity has been used to enhance the visibility of the plaques, which are unstained in a surrounding red background *(16–18)*. Methylene blue can also be used, but in this case the plaques are stained blue on the straw-colored background of the agar medium. The optimal concentration of these reagents should be determined by experimentation. In the case of methylene blue, the plates should be incubated in the dark to prevent possible photodynamic effects.

Plaque size is influenced by numerous factors *(1)*. There are a number of approaches which have been reported to enhance plaque size:

1. Addition of sodium azide at 0.006% (for coliphage T2) and 0.03% (for *Staphylococcus* phage 42D) accompanied by an extended incubation period resulted in significantly larger plaques *(19)*.
2. Reducing the agar concentration from 0.6 to 0.3% may result in larger plaques.
3. McConnell and Wright *(20)* noted that with many enterobacterial phages, incubation under anaerobic conditions for 24 h

followed by 16 h under aerobic conditions resulted in plaques which were 2–8-fold larger.

4. While plaque assays commonly use overnight cultures of the host bacteria, in certain cases, log phase cells may give the best plaques *(13)*.

References

1. Carlson, K. 2005. Working with bacteriophages: Common techniques and methodological approaches., *In* E. Kutter and A. Sulakvelidze (Eds.), Bacteriophages: Biology and Applications. CRC Press, Baco Raton, FL.
2. Ackermann, H.-W. 2005. Electron microscopy., *In* E. Kutter and A. Sulakvelidze (Eds.), Bacteriophages: Biology and Applications. CRC Press, Boca Ratan, FL.
3. Adams, M.D. 1959. Bacteriophages. Interscience Publishers, Inc., New York.
4. Carlson, K. 2005. Working with bacteriophages: Common techniques and methodological approaches., *In* E. Kutter and A. Sulakvelidze (Eds.), Bacteriophages: Biology and Applications. CRC Press, Boca Ratan, FL.
5. Edelman, D.C. and J. Barletta. 2007. Real-time PCR provides improved detection and titer determination of bacteriophage. BioTechniques *35*:368–375.
6. Gratia, A. 1936. Des relations numeriques entre bactéries lysogenes et particules de bactériophage. Annales de l'Institut Pasteur *57*:652–676.
7. Gratia, J.-P. 2000. André Gratia: A forerunner in microbial and viral genetics. Genetics *156*:471–476.
8. Sambrook, J. and D.W. Russell. 2001. Molecular Cloning: A Laboratory Manual. Cold Spring Harbor Press, Cold Spring Harbor, New York.
9. Haberer, K. and J. Maniloff. 1982. Adsorption of the tailed mycoplasma virus L3 to cell membranes. Journal of Virology *41*:501–507.
10. Landry, E.F. and R.M. Zsigray. 1980. Effects of calcium on the lytic cycle of *Bacillus subtilis* phage 41c. Journal of General Virology *51*:125–135.
11. Mahony, D.E., P.D. Bell, and K.B. Easterbrook. 1985. Two bacteriophages of *Clostridium difficile*. Journal of Clinical Microbiology *21*:251–254.
12. Sechter, I. and C.B. Gerichter. 1968. Phage typing scheme for *Salmonella braenderup*. Applied Microbiology *16*:1708–1712.
13. Hongo, M. and A. Murata. 1965. Bacteriophages of *Clostridium saccharoperbutylacetonicum*. II. Enumeration of phages by application of the plaque-count technique and some factors influencing the plaque formation. Agricultural and Biological Chemistry *29*: 1140–1145.
14. Lillehaug, D. 1997. An improved plaque assay for poor plaque-producing temperate lactococcal bacteriophages. Journal of Applied Microbiology *83*:85–90.
15. Mullan, M.W.A. 1979. Lactic streptococcal bacteriophage enumeration. A review of factors affecting plaque formation. Dairy Industries International *44*:11–14.
16. Pattee, P.A. 1966. Use of tetrazolium for improved resolution of bacteriophage plaques. Journal of Bacteriology *92*:787–788.
17. Fraser, D. and J. Crum. 1975. Enhancement of Mycoplasma virus plaque visibility by tetrazolium. Applied Microbiology *29*:305–306.
18. Hurst, C.J., J.C. Blannon, R.L. Hardaway, and W.C. Jackson. 1994. Differential effect of tetrazolium dyes upon bacteriophage plaque assay titres. Applied & Environmental Microbiology *60*:3462–3465.
19. Qanber, A.A. and J. Douglas. 1976. Enhancement of plaque size of a staphylococcal phage. Journal of Applied Bacteriology *40*:109–110.
20. McConnell, M. and A. Wright. 1975. An anaerobic technique for increasing bacteriophage plaque size. Virology *65*:588–590.

Chapter 8

Enumeration of Bacteriophages by the Direct Plating Plaque Assay

Amanda Mazzocco, Thomas E. Waddell, Erika Lingohr, and Roger P. Johnson

Abstract

A method is described for determination of the concentration of infectious phage particles by the direct plating plaque assay, which is simpler and faster than the double agar overlay plaque procedure outlined in the previous chapter.

Key words: Plaque, overlay technique, plaque-forming unit, pfu, direct plating plaque assay.

1 Introduction

While the Double Agar Overlay method (**Chapter 7 Fig. 7.1A**) of plaque assay is generally considered the most accurate, and promotes larger plaques with well-defined morphology, reliable enumeration can be obtained for many phages by the Direct Plating method (**Chapter 7 Fig. 7.1B**), described originally by d'Herelle in 1917 *(1)*, and cited by Adams *(2)*. Since direct plating plaque assays do not require the overlay or top agar, it is simpler and faster to set up. The following method has been used successfully for enumeration of virulent *Escherichia* and *Salmonella* phages.

2 Materials

2.1 Equipment

1. Supplies and equipment for preparation of broth and solid culture media.
2. Variable or fixed-volume micropipettors for volumes of 1 ml, 100 μl, and 10 μl (**Notes 1** and **2**).

Martha R. J. Clokie, Andrew M. Kropinski (eds.), *Bacteriophages: Methods and Protocols, Volume 1: Isolation, Characterization, and Interactions, vol. 501*, © 2009 Humana Press, a part of Springer Science+Business Media
DOI 10.1007/978-1-60327-164-6_8 Springerprotocols.com

3. Sterile pipette tips for the micropipettors.
4. Sterile dilution tubes (1.5 ml capped microcentrifuge tubes, or sterile Titertubes® (BioRad) in 8 × 12 tube racks.
5. Sterile plate spreaders.

2.2 Media

The underlays (bottom agars) mentioned above, or the MNA given below, are suitable. This MNA contains salts and glucose used initially by Arber and others *(3)*, cited in *(4)*, to modify a LB underlay agar. It is more translucent than many other media, and hence enhances plaque visualization.

MNA: Per liter:	Nutrient Broth	180.0 g
	NaCl	76.5 g
	Agar No.1 (Oxoid)	90.0 g
	$CaCl_2$	74.7 mg
	$FeCl_3$	9.9 mg
	$MgSO_4.7H_2O$	4.5 g
	Distilled water to:	1000 ml
	30% sterile glucose	10 ml, after autoclaving

Dispense 20–30 ml into Petri plates and allow them to solidify at room temperature. Store in plastic sleeves at 4°C for up to 4 months.

2.3 Diluents

1. Phages can be diluted in the broth medium used to grow the host bacterium (without agar or agarose), or in the following:
   A: Saline–magnesium (SM) diluent plus gelatin (SMG) *(4)*: 50 mM Tris–HCl (pH 7.5) containing 100 mM NaCl, 8.1 mM $MgSO_4$ and 0.01% (w/v) gelatin.

Per liter:	NaCl	5.8 g
	$MgSO_4.7H_2O$	2.0 g
	1M Tris–HCl (pH 7.5)	50 ml
	2% (wt/vol) gelatin	5 ml
	Distilled water to:	1000 ml.

   B: Lambda (λ) diluent (21): 10 mM Tris–HCl (pH 7.5), 8.1 mM $MgSO_4$.

Per liter:	$MgSO_4.7H_2O$	2.0 g
	1M Tris–HCl (pH 7.5)	10 ml
	Distilled water to:	1000 ml.

2. Dispense the diluent in 10 ml volumes (sufficient for one dilution series of phages) into screw-capped vials or tubes, apply caps, heat sterilize, and store at 4°C for up to 3 months (**Note 3**).

3 Method

The following description applies to the direct plating plaque assay of a typical phage preparation containing approximately 10^8 to 10^{10} pfu/ml, with one plating per dilution. The range of dilutions and number of replicate plates for each dilution will alter the numbers of plates required. Plating from as few as three dilutions (10^{-6} to 10^{-8}) can be used for phage stocks containing 10^8 to 10^{10} pfu/ml. However, it is often beneficial to test a wider range of serial dilutions to verify the accuracy of the initial dilutions. Test other dilutions based on the expected titer of the sample, or a wider range of dilutions where the expected titer is unknown.

1. Remove 3 (or more) MNA plates from 4°C storage and dry them, inverted and tilted, in a 37°C incubator for 1–2 h or partially uncovered in a laminar flow hood at room temperature for 10–15 min. Number the plates according to the dilutions to be tested; in this example, 10^{-6}, 10^{-7}, and 10^{-8} (**Note 4**).
2. For each phage preparation, set up 8 dilution tubes numbered "–1" to "–8." Add 450 µl of sterile diluent to each tube.
3. Add 50 µl of the undiluted phage to the first tube, and mix with the diluent by moving the liquid up and down in the pipette tip at least thrice. Repeat this procedure for all dilutions, transferring 50 µl volumes and changing pipette tips between tubes.
4. Add 100 µl of an overnight culture of the host bacterium (typically containing 10^8 to 10^9 cfu/ml) to the dilutions to be plated (e.g., 10^{-6} to 10^{-8} dilutions). Use a clean pipette tip for each tube and mix by repeated aspiration, as in Step 3 (**Note 5**).
5. Incubate the tubes at 37°C for 15–20 min to allow the phages to attach to the bacteria (**Note 6**).
6. Promptly drop and spread 200 µl volumes of each selected, inoculated, dilution onto the correspondingly numbered agar plates (e.g., –6, –7, and –8; **Note 7**).
7. Allow the plates to dry for 30 min in a laminar flow hood, then cover, invert and incubate the plates at 37°C for 20–24 h.
8. Count plaques on plates with 30–300 plaques, and calculate the titer of the undiluted phage preparation as follows:

   Average number of plaques ×5× reciprocal of counted dilution = pfu/ml.

4 Notes

1. Multichannel micropipettors with up to 200 µl capacity, together with small tubes set up in 8 × 12 format (Titertubes®, BioRad Laboratories; Hercules, CA;

http://www.bio-rad.com/), are very efficient for diluting and inoculating large numbers of samples.

2. Pipettors should be sterilized at least weekly and recalibrated periodically.
3. Alternatively, dispense diluent in larger volumes in screw-capped bottles, apply caps, heat sterilize, and store at 4°C for up to 3 months. For use, transfer the required volume of this stock to a sterile tube, and work from this tube to avoid contamination of the stock diluent.
4. The state of the "underlay" medium is very important; it should not be too dry or too moist since both of these factors will negatively affect plaque development.
5. The volume of the overnight culture of the host bacterium required to obtain a confluent lawn with clearly visible plaques will become apparent with experience, and typically ranges from 50 – 100 μl.
6. The incubation time required for attachment will vary for different phage–host systems.
7. To ensure that an even bacterial lawn develops, use a bar or L-shaped spreader that facilitates rapid and even spreading of the inoculum over the agar without damaging the surface.

References

1. d'Herelle, F. 1917. Sur un microbe invisible antagoniste des bacilles dysentériques. Comptes rendus Académie Sciences *165*: 373–375.
2. Adams, M.D. 1959. Bacteriophages. Interscience Publishers, Inc., New York.
3. Arber, W.L., L. Enquist, B. Hohn, N.E. Murray, and K. Murray. 1983. Experimental Methods for use with Lambda., p. 433. *In* Lambda II. Cold Spring Harbor Laboratory Press, Cold Spring Harbor Laboratory, NY.
4. Sambrook, J. and D.W. Russell. 2001. Molecular Cloning: A Laboratory Manual. Cold Spring Harbor Press, Cold Spring Harbor, New York.

Chapter 9

Enumeration of Bacteriophages Using the Small Drop Plaque Assay System

Amanda Mazzocco, Thomas E. Waddell, Erika Lingohr and Roger P. Johnson

Abstract

The determination of the concentration of infectious phage particles is fundamental to many protocols in phage biology, genetics, and molecular biology. Described here is a drop plaque assay, which, being simpler, faster and more efficient than either the classical overlay or direct plating methods, enhances efficiency in processing large numbers of samples.

Key words: Plaque, overlay technique, plaque-forming unit, pfu, plaque morphology, direct plating plaque assay, drop plaque assay, spot plaque assay.

1 Introduction

We have previously described Double Agar Overlay method (**Chapter 7, Fig. 7.1A**), and the Direct Plating method (**Chapter 7, Fig. 7.1B**), here we describe a small scale Drop method (**Fig. 7.1C**).

Where numerous samples are to be titrated concurrently, greater efficiency and economy can often be achieved with very little loss of precision by using a small volume drop (or spot) method such as the one described below. The preparation and inoculation of dilutions of the phage stock with host bacteria are very similar to the methods for the direct plating plaque assay. After inoculation, mixing, and a brief incubation, small volumes (e.g., 20 μl) of the inoculated dilutions are dropped onto a suitable agar plate. Following overnight incubation, plaques become visible in the 10–12 mm circular bacterial lawns that develop from

Martha R. J. Clokie, Andrew M. Kropinski (eds.), *Bacteriophages: Methods and Protocols, Volume 1: Isolation, Characterization, and Interactions, vol. 501*, © 2009 Humana Press, a part of Springer Science+Business Media
DOI 10.1007/978-1-60327-164-6_9 Springerprotocols.com

the drops on the agar. The method relies on formation of readily visualized plaques, and in our hands, provides excellent results with virulent *E. coli* and *Salmonella* phages. As in the Direct Plating assay, using multichannel micropipettors and tubes arranged in 8 × 12 racks can enhance efficiency in preparing large numbers samples.

2 Materials

2.1 Equipment

1. Supplies and equipment for preparation of broth and solid culture media.
2. Variable or fixed-volume micropipettors for volumes of 1 ml, 100 μl, and 10 μl (**Notes 1 and 2**)
3. Sterile pipette tips for the micropipettors.
4. Sterile dilution tubes (1.5 ml capped microcentrifuge tubes, or sterile Titertubes® (BioRad) in 8 × 12 tube racks.
5. Sterile plate spreaders.

2.2 Media

The underlays (bottom agars) mentioned in **Chapters 7 and 8**, or the Modified Nutrient Agar (MNA) given below, are suitable. This MNA contains salts and glucose used initially by Arber and others *(1)*, cited in *(2)*, to modify a LB underlay agar. It is more translucent than many other media, and hence enhances plaque visualization.

MNA: Per liter:	Nutrient Broth	180.0 g
	NaCl	76.5 g
	Agar No.1 (Oxoid)	90.0 g
	$CaCl_2$	74.7 mg
	$FeCl_3$	9.9 mg
	$MgSO_4.7H_2O$	4.5 g
	Distilled water to:	1000 ml
	30% sterile glucose	10 ml, after autoclaving

Dispense 20–30 ml into Petri plates and allow them to solidify at room temperature. Store in plastic sleeves at 4°C for up to 4 months.

2.3 Diluents

1. Phages can be diluted in the broth medium used to grow the host bacterium (without agar or agarose), or in the following:
   A: Saline–magnesium (SM) diluent plus gelatin (SMG) *(2)*: 50 M Tris–HCl (pH 7.5) containing 100 mM NaCl, 8.1 mM $MgSO_4$ and 0.01% (w/v) gelatin.

Per liter:	NaCl	5.8 g
	$MgSO_4.7H_2O$	2.0 g
	1M Tris–HCl (pH 7.5)	50 ml
	2% (wt/vol) gelatin	5 ml
	Distilled water to:	1000 ml.

B: Lambda (λ) diluent (21): 10 mM Tris–HCl (pH 7.5), 8.1 mM $MgSO_4$.

Per liter:	$MgSO_4.7H_2O$	2.0 g
	1M Tris–HCl (pH 7.5)	10 ml
	Distilled water to:	1000 ml.

2. Dispense the diluent in 10 ml volumes (sufficient for one dilution series of phages) into screw-capped vials or tubes, apply caps, heat sterilize, and store at 4°C for up to 3 months (**Note 3**).

3 Method

The following description applies to the Drop Plaque Assay of a typical phage preparation containing approximately 10^6 or 10^{10} pfu/ml, with duplicate drops of each selected dilution being applied to the agar. The range of dilutions and number of replicate drops for each dilution may alter the number of plates required. About 8 to 10 drops of 20 µl can be spaced evenly on a standard 90 mm agar plate without coalescence. Typically, dilutions up to 10^{-8} are tested for phage stocks containing 10^6 to 10^{10} pfu/ml. Use other dilutions based on the expected titer of the sample, or a wider range of dilutions where the expected titer is unknown.

1. Remove 2 MNA plates from 4°C storage and dry them, inverted and tilted, in a 37°C incubator for 1–2 h or in a laminar flow hood at room temperature for 10–15 min. Label the plates with the identity of the phage.
2. Set up 8 dilution tubes, numbered "–1" to "–8," plus a "control" tube. Add 180 µl of sterile diluent to each tube.
3. Add 20 µl of the undiluted phage to the first tube, and mix with the diluent by moving the liquid up and down in the pipette tip at least 3 times. Repeat this procedure for all remaining dilutions, transferring 20 µl volumes and changing pipette tips between tubes. Do not add phages to the control tube.

4. Add 20 μl of an overnight culture of the host bacterium (typically containing 10^8 to 10^9 cfu/ml) to the control and the dilution tubes. Use a clean pipette tip for each tube and mix by repeated aspiration, as described in Step 3. Incubate the tubes at 37°C for 15–20 min to allow the phages to attach to the bacteria (**Notes 4 and 5**).
5. Take 20 μl of the dilution mixture in tube "–8" (i.e., the highest dilution to be tested) and drop it onto the labeled agar plate. Repeat for the same tube, dropping a duplicate 20 μl on the second plate (**Notes 6, 7, and 8**).
6. Repeat Step 5 for the other dilution tubes and for the control tube creating a set of drops from 8 dilutions and the bacterial control on each of the 2 plates (**Notes 9 and 10**).
7. Allow the plates to dry for 20–30 min in a laminar flow hood, then cover, invert and incubate at 37°C for 20–24 h.
8. Count the plaques in duplicate drops of one or more dilutions with 3–30 countable plaques, and average the counts from each dilution. Calculate the titer of the undiluted phage preparation as the average of the following calculation for each counted dilution:
   Average number of plaques × 50 × reciprocal of dilution = pfu/ml

4 Notes

1. Multichannel micropipettors with up to 200 μl capacity, together with small tubes set up in 8 × 12 format (Titertubes®, BioRad Laboratories; Hercules, CA; http://www.bio-rad.com/), are very efficient for diluting and inoculating large numbers of samples.
2. Pipettors should be sterilized at least weekly and recalibrated periodically.
3. Alternatively, dispense diluent in larger volumes in screw-capped bottles, apply caps, heat sterilize, and store at 4°C for up to 3 months. For use, transfer the required volume of this stock to a sterile tube, and work from this tube to avoid contamination of the stock diluent.
4. The volume of the overnight culture of the host bacterium required to obtain a confluent lawn in the drops and clearly visible plaques will become apparent with experience with a given phage–host system, and typically ranges from 10–30 μl.
5. The incubation time required for attachment may vary for different phage–host systems.
6. To apply the drop to the plate, express it gently so that it is hanging from the tip of the pipette. Gently release the drop by touching it to the surface of the agar. Do not express the

residual liquid in the pipette tip, which can result in spraying the phage/bacteria mixture beyond the smooth border of the drop.

7. The drop size may need to be adjusted if the phage produces large plaques.
8. The dryness of the agar is very important; if too moist, the drops will run and coalesce, if too dry, the bacteria will grow poorly.
9. Although advisable, changing tips between drops is not necessary provided the order of dropping proceeds from highest to lowest dilution.
10. Drop testing all dilutions is helpful at first, but can be reduced to 4 or 5 dilutions with experience.

References

1. Arber, W. L., Enquist, L., Hohn, B., Murray, N. E., & Murray, K. (1983) in *Lambda II* (Cold Spring Harbor Laboratory Press, Cold Spring Harbor Laboratory,NY), p. 433.
2. Sambrook, J. & Russell, D. W. (2001) *Molecular Cloning: A Laboratory Manual* (Cold Spring Harbor Press, Cold Spring Harbor, New York).

Chapter 10

Determination of Virus Abundance by Epifluorescence Microscopy

Alice C. Ortmann and Curtis A. Suttle

Abstract

Determination of virus abundance using epifluorescence microscopy is a rapid and accurate method. The protocol requires the concentration of virus particles by collection on a filter. The nucleic acid in the virus particles is then stained with a fluorescent stain and the sample viewed with an epifluorescence microscope. The method was originally developed to determine the abundance of virus particles in water samples, however the protocol has been adapted for cultures and sediment samples. Although the method provides total counts of all virus-sized particles, regardless of infectivity, the method can be used for rapidly screening samples for further study.

Key words: Epifluorescence microscopy, phage abundance, nucleic acid stains, SYBR, Yo-Pro-1.

1 Introduction

Epifluorescence microscopy can be a rapid and accurate method to determine the abundance of bacteriophages from both cultures and environmental samples. The method has been shown to have higher accuracy and precision than transmission electron microscopy (TEM) and flow cytometry for counting virus-sized particles *(1)*. The development of epifluorescence microscopy for counting viruses originated with studies of marine viruses. Originally, abundances were determined using TEM either with samples concentrated by ultrafiltration and spotted onto grids or with samples pelleted onto grids using ultracentrifugation *(2, 3)*. Both these methods allowed positive identification of virus-like particles and provided morphological information. However, the methods are time consuming, require access to a TEM and cannot

Martha R. J. Clokie, Andrew M. Kropinski (eds.), *Bacteriophages: Methods and Protocols, Volume 1: Isolation, Characterization, and Interactions, vol. 501*, © 2009 Humana Press, a part of Springer Science+Business Media
DOI 10.1007/978-1-60327-164-6_10 Springerprotocols.com

be carried out in the field. Because of these limitations, epifluorescence microscopy methods were developed.

Counting virus particles with epifluorescence microscopy requires that the viruses be collected onto a small pore-sized filter and stained with a fluorescent dye. The dye is excited with a specific wavelength of light, and emits light at a longer wavelength when bound to nucleic acids. This light results in a glowing particle larger than the actual size of the virus particles, enabling the viruses to be counted at a much lower magnification than is possible with unstained viruses.

The original methods for counting viruses in marine samples used DAPI (4′6-diamidino-2-phenylindole) *(4)*, which had previously been used for counting bacteria *(5)*. This dye is excited with UV light and emits a blue light. The light yield from this dye is relatively low, requiring the use of microscopes with very high quality optics for enumeration. Development of new dyes for nucleic acids resulted in the application of other stains for counting viruses including Yo-Pro-1 *(6)*, SYBR Green I *(7)*, and SYBR Gold *(8)*. All of these stains are excited by blue light and emit green light. Yo-Pro-1 is a very stable and bright fluorescent dye, but requires 48 h for complete stain penetration. SYBR Green I requires only 15 min for staining, but fades very quickly making counting difficult if the virus density is high *(9)*. SYBR Gold staining can be carried out using the same protocol as SYBR Green, but the dye is much more stable, making it a good choice for most samples. SYBR Gold emits a yellowish light compared to the other stains, and may not be the best choice when counting samples with humics or sediments. All of these stains have been shown to provide accurate estimates of virus abundance in aquatic environments and cultures, provided samples are immediately and carefully processed *(10)*.

2 Materials

2.1 Equipment

1. Epifluorescence Microscope, equipped with 100× objective and appropriate filter block (**Note 1**)
2. Filtration unit, to hold 25 mm diameter filters
3. Pipettes, from 2–1000 μl
4. Counter

2.2 Disposables

1. 25 mm dia filters, 0.45 μm pore-sized nitrocellulose
   0.02 μm pore-sized Al_2O_3 (Anodisc, Whatman)
2. Microfuge tubes, for dilution of stain and samples
3. Staining dish, plastic petri dish works well
4. Microscope slides
5. 25 mm × 25 mm coverslips
6. DF or FF grade immersion oil

2.3 Reagents

1. SYBR nucleic acid stain, 10,000× in DMSO (Invitrogen)
   Store at −20°C and protect from light. Use gloves and protect skin when using.
2. Glycerol
3. PBS (Phosphate buffered saline: 0.05M Na_2HPO_4, 0.85% NaCl, pH 7.5)
4. *p*-phenylenediamine

Store tightly capped and protect from air and light. Use gloves and protect skin.

3 Methods

This chapter provides the method for determining the abundance of viruses in either water samples or cultures using any of the SYBR stains. Information for counting viruses from soil or sediment samples and modifications necessary for staining with Yo-Pro-1 are included in the notes section. The basic steps include (1) preparation of the reagents, (2) preparation of the sample, (3) filtration and staining of the sample and (4) determining the abundance of viruses. These methods are modified from the original method developed for SYBR Green I staining *(7)*.

All steps involving the fluorescent stain should be done in very low light. If possible, set up an area in a dark room and use red or indirect light for preparation of the slides. Bright lights will cause bleaching of the stain and reduce the fluorescence of the virus particles.

3.1 Preparation of Reagents

For preparation of all reagents use 0.02 μm filtered deionized water. This is especially important in the dilution of the stain and the samples to prevent the introduction of virus particles.

3.1.1 Stain

Stock solution: Dilute the SYBR stain 1:10 with 0.02 μm filtered deionized water. Aliquot small volumes into plastic microfuge tubes and store at −20°C. SYBR stains tend to bind to plastics and glass, with the lowest binding to polypropylene.

3.1.2 Antifade

1. Prepare a solution of 50% glycerol and 50% PBS. Shake or vortex to ensure the two solutions mix together.
2. Prepare 10% stock solutions of *p*-phenylenediamine and store at −20°C in small volumes to minimize freeze/thaw cycles. When *p*-phenylenediamine oxidizes it turns brown and will no longer work well as an antifade. Check the color of the stock solution before making up the working solution; do not use it if it is tea colored or darker.
3. Prepare the working antifade solution immediately before making the slides. To determine the amount to make, estimate 50 μl per slide; this will be more than enough. To make the antifade solution add the 10% *p*-phenylenediamine solution

to the 50% glycerol/50% PBS so that the *p*-phenylenediamine final concentration is 0.1%.

3.2 Preparation of Sample

1. The sample, especially those from cultures, may need to be diluted to about 10^7 particles ml^{-1} prior to collecting the viruses on the filter. The abundance of viruses should be low enough so that individual particles can be identified, but not so low that the particles are <~ 10 per field (**Note 2**). Dilution of the sample should be done with $0.02\ \mu\mathrm{m}$ filtered solutions. It is important that the solution used for dilution does not contain viruses (**Note 3**).
2. Samples to be stained with SYBR should be fixed with 0.5% glutaraldehyde for up to 30 min at 4°C prior to preparing slides. For some samples this may improve the fluorescence of the particles and make counting easier. Storage of viruses fixed in aldehydes at 4°C is not recommended due to the loss of viruses during storage *(10–12)*. If fixation and storage of samples is necessary prior to the preparation of slides, the best method is to fix the samples in 0.5% EM grade glutaraldehyde and flash freeze in liquid N_2. The samples should then be stored at −86°C until the slides can be prepared *(10)* (**Note 4**).
3. The abundance of phages in soil or sediment samples can also be determined using epifluorescence microscopy, but further steps are necessary to prepare the sample and remove particulates that may interfere with counting (**Note 5**).

3.3 Filtration and Staining of Sample

1. Prepare the working solution of the stain in a plastic container. For staining, plastic petri dishes work well. Up to four filters can be stained in one dish and the dishes can be reused. To prepare the working solution add 2 μl of stock solution of SYBR stain to 78 μl of $0.02\ \mu\mathrm{m}$ filtered deionized water. One drop should be prepared for each filter (*see* **Note 6** for Yo-Pro-1 modifications).
2. Prepare the filtration unit, connecting it to a vacuum source. The vacuum should be no higher than 7 mm Hg.
3. Set up filtration unit using $0.45\ \mu\mathrm{m}$, nitrocellulose filters as a backing filter. This filter can be reused several times, as long as it has no holes and remains flat.
4. Apply a $0.02\ \mu\mathrm{m}$ Anodisc filter over the backing filter. Add a thin layer of deionized water to the backing filter before placing the Anodisc filter on top of it. Turn on the vacuum to pull this water through. It is important that there is no air trapped between the filters or the sample will not be pulled through the Anodisc. When handling the Anodisc filter, hold it by the plastic ring around the membrane. Because of the characteristics of the Anodisc membranes, the filters will not

bend, but crack. Check to make sure the membrane is not cracked before using the filter.

5. With the vacuum off, add the sample to the Anodisc. Without a filtration tower, water tension will allow 0.8–1.0 ml of sample to be placed on the filter. Make sure the entire volume is within the plastic ring, otherwise the sample will be pulled under the edge of the filter. Turn on the vacuum and filter the sample through the filter. If more than 1 ml needs to be filtered there are two options.
   a. If the total volume is 2 ml or less, add 1.0 ml of the sample and allow some to be filtered, adding more until all of the sample has been filtered. It is important that the filter does not dry out between additions of the sample.
   b. A filtration tower can be used for larger volumes. The tower must fit over the center of the Anodisc and not cross over the plastic ring. Measure the interior diameter of the tower so the diameter of the filtration surface is known. Ensure that the fields counted are within the area of filtration.
6. Once the sample has been entirely filtered, carefully remove the filter while the vacuum is still on. When lifting the Anodisc, touch only the plastic ring and be careful not to bend and crack the membrane.
7. Place the Anodisc, sample side up, on a drop of stain in a plastic petri dish. Allow the filter to stain for 15 min in the dark.
8. After 15 min, remove the Anodisc from the stain and place it on top of the 0.45 μm backing filter (add a small amount of deionized water between the filters to ensure no air is trapped). Turn on the vacuum and pull any fluid back through the filter.
9. Remove the Anodisc from the filtration unit while the vacuum is still on. Place the Anodisc, sample side up, on a Kim wipe and allow the filter to dry. When the filter is dry it will appear opaque.
10. When dry, mount the Anodisc onto a slide using the antifade solution. Place 12–15 μl of antifade on the slide and put the Anodisc on top. Add ~ 20 μl of antifade on top of the Anodisc and top with a coverslip. Press down on the coverslip to ensure that no bubbles are trapped between the filter and the coverslip (**Note 7**).
11. The slides can either be counted immediately or stored frozen at −20°C for as long as 4 months with no decrease in abundance of viruses or fluorescence *(6)*. Multiple freezing and thawing of the slides should be avoided.

4 Determining Abundance

1. Using either DF or FF grade immersion oil count the phages using the 100× objective. The particles will appear green (SYBR Green I or II) or yellowish (SYBR Gold) when excited with blue light.
2. Before starting to count the particles on the slide, check to ensure that the slide is "good." A good slide should have even staining with the sample evenly distributed across the entire filter. The sample should also be on a single plane with the fluorescent particles attached to the filter and not the coverslip or floating between the two.
3. There are two methods for counting the particles.
   a. The first method uses an ocular reticle with a grid divided into squares of known area. An individual then counts the number of particles within each field. The size of the field should be selected so that each field contains approximately 10 particles. Particles should be counted that are within each area. For particles touching the edges of the grid select 2 sides (i.e., left side and top) where particles are counted. Particles touching the other 2 sides (i.e., right side and bottom) will not be counted.
   b. The second method uses a CCD camera to obtain an image of a field, which is then either counted by an individual or analysed with computer software *(8)*. Using this method, the field is defined as the area of the image so the sample needs to be at the correct concentration to allow ease of counting.
4. The number of fields counted, the size of the field, the total number of particles, and the total volume of sample filtered needs to be recorded for each sample. The total area through which the sample was filtered must also be recorded. If no filter tower was used, this area is the total area of the filter (radius $= 9,500\ \mu m$), otherwise the radius of the filter tower needs to be determined.
5. The following equation can be used to calculate the abundance of phages in the sample *(13)*:

$$N_v = P_t \div F_t \times A_t \div A_f \div V_t$$

N_v = phage ml^{-1} P_t = total number of phage counted
F_t = total number of fields A_t = total area filtered (μm^2)
A_f = area of each field (μm^2) V_t = volume of sample filtered (ml)

6. The total number of particles counted will determine the size of the 95% confidence intervals that can be calculated for each sample. The 95% confidence intervals can be calculated using the following equations *(13)*:

$$\text{Upper } 95\% = P_t + 1.96 \times \sqrt{(P_t + 1.5)} + 2.42$$
$$\text{Lower } 95\% = P_t - 1.96 \times \sqrt{(P_t + 0.5)} + 1.42$$

5 Notes

1. The filter cube should provide excitation and emission spectra at the appropriate wavelengths for the stain chosen. For SYBR stains the following conditions apply:

   SYBR Green I and II excitation at 497 nm, emission at 520 nm
   SYBR Gold excitation at 495 nm, emission at 537 nm
   Check with the microscope manufacturer for the correct filter cube.
2. When sampling from an unknown sample, it may be useful to prepare several different dilutions of the sample to determine the best volume of sample to filter. It is best to dilute the sample to 800–1000 μl for filtration, as smaller volumes may not result in even filtering of the sample. An assumption of the method is that the viruses are evenly distributed on the filter.
3. When diluting the sample, tests may be required to determine the appropriate solution to use. If sampling from a culture, filtered media may be the best solution to dilute the sample with. Marine samples should be diluted with 0.02 μm filtered seawater to maintain the salinity of the sample and prevent bursting of cells and viruses, while lake samples can be diluted with filtered lake water or distilled water.
4. Fixation of samples with aldehydes following collection has been commonly used to preserve samples for preparation of slides at a later time. However, studies have shown that virus abundances decrease rapidly when fixed in either glutaraldehyde or formaldehyde and stored at 4°C *(10–12)*. If it is impossible to flash freeze fixed samples as described above and preservation is necessary, it may be possible to estimate the abundance of viruses by determining a decay curve and calculating the original abundance based on the length of time the sample was stored and the abundance determined. There is some evidence that the rate of decay is sample dependant, so errors in the estimate should be expected *(10)*.
5. Because viruses attached to soil or sediment particles can be impossible to see using epifluorescence microscopy, preparation of the sample requires that the viruses be dislodged from the particles. The following is a method tested on sand and mud from marine systems *(11)*.

   0.5 g of sample is mixed with 4.0 ml of 0.02 μm filtered distilled water and 1.0 ml pyrophosphate (10 mM final concentration). The mixture is sonicated for 3 min and then centrifuged at 800 × *g* for 1 min. The supernatant can then be diluted and the slides prepared as described above. For different types of sediments and soils, the amount of pyrophosphate and length of sonication may need to be tested and optimized.

6. Yo-Pro slides have the benefit of having stable fluorescence without the need for an antifade solution. The down side is that the staining requires 48 hr. The following modifications are necessary for staining with Yo-Pro-1 *(6)*.

   Prepare the sample and filter as above, but do not fix the sample as aldehydes interfere with staining. Place the filter on a 80 µl drop of Yo-Pro-1 solution (50 mM Yo-Pro-1, 2 mM NaCN). Place a wet filter (9 mm) in the top of the petri dish to prevent drying and stain in the dark for 48 hr. Place the filter back on the filtration unit and rinse twice with 0.02 µm filtered distilled water. Mount the filter on the slide using 100% glycerol. The slide is then ready to be counted as above. Yo-Pro-1 excites at a wavelength of 491 nm and emits at 509 nm. The staining time can be reduced to a few min by using microwave irradiation *(12)*. This method requires optimization for each microwave to obtain the desired results.

7. The exact amount of antifade solution used to mount the slide should be the minimum needed to fill the area between the filter and the coverslip. If the slides are going to be frozen before counting, more antifade solution should be used to compensate for shrinking during freezing.

8. Because the confidence intervals are determined from the total number of particles counted, it is important to count sufficient particles to have the desired accuracy. As the number of particles counted increases, the rate at which the accuracy increases is reduced. The number of particles counted is therefore a trade-off between accuracy and effort. For instance if $P_t = 100$, the 95% CI is 82–122, which is equivalent to $\pm \sim 20\%$. If the number of particles counted is increased to $P_t = 200$, the 95% confidence interval is 174–230, or $\pm \sim 14\%$. A further doubling of the particles counted to $P_t = 400$ only decreases the error to $\pm \sim 10\%$. In general, counting at least 200 particles in 20 fields is a recommended balance between accuracy and effort *(13)*.

References

1. Ferris, Matthew M., Stoffel, Carrie L., Maurer, Thain T., and Rowlen, Kathy L. (2002). Quantitative intercomparison of transmission electron microscopy, flow cytometry, and epifluorescence microscopy for nanometric particle analysis. *Anal. Biochem.* **304**, 249–256.
2. Børsheim, K.Y., Bratbak, G., and Heldal, M. (1990). Enumeration and biomass estimation of planktonic bacteria and viruses by transmission electron microscopy. *Appl. Environ. Microb.* **56**, 352–356.

3. Proctor, L.M. and Fuhrman, J.A., (1990). Viral mortality of marine bacteria and cyanobacteria. *Nature.* **343**, 60–62.
4. Hara, S., Terauchi, K., and Koike, I., (1991). Abundance of viruses in marine waters – Assessment by epifluorescence and transmission electron microscopy. *Appl. Environ. Microb.* **57**, 2731–2734.
5. Porter, K.G. and Feig, Y.S., (1980). The use of DAPI for identifying and counting aquatic microflora. *Limnol. Oceanogr.* **25**, 943–948.
6. Hennes, K.P. and Suttle, C.A., (1995). Direct counts of viruses in natural waters and laboratory cultures by epifluorescence microscopy. *Limnol. Oceanogr.* **40**, 1050–1055.
7. Noble, R.T. and Fuhrman, J.A., (1998). Use of SYBR Green I for rapid epifluorescence counts of marine viruses and bacteria. *Aquat. Microb. Ecol.* **14**, 113–118.
8. Chen, F., Lu, J.R., Binder, B.J., Liu, Y.C., and Hodson, R.E., (2001). Application of digital image analysis and flow cytometry to enumerate marine viruses stained with SYBR Gold. *Appl. Environ. Microb.* **67**, 539–545.
9. Bettarel, Y., Sime-Ngando, T., Amblard, C., and Laveran, H., (2000). A comparison of methods for counting viruses in aquatic systems. *Appl. Environ. Microb.* **66**, 2283–2289.
10. Wen, K., Ortmann, A.C., and Suttle, C.A., (2004). Accurate estimation of viral abundance by epifluorescence microscopy. *Appl. Environ. Microb.* **70**, 3862–3867.
11. Danovaro, R., Dell'anno, A., Trucco, A., Seresi, M., and Vanucci, S., (2001). Determination of virus abundance in marine sediments. *Appl. Environ. Microb.* **67**, 1384–1387.
12. Xenopoulos, M.A. and Bird, D.F., (1997). Virus à la sauce Yo-Pro: Microwave-enhanced staining for counting viruses by epifluorescence microscopy. *Limnol. Oceanogr.* **42**, 1648–1650.
13. Suttle, C.A., (1993). Enumeration and isolation of viruses, in *Handbook of Methods in Aquatic Microbial Ecology* (Kemp, P.F, Sherr, B.F., Sherr E.B., and Cole, J.J., eds.) Lewis Publishers, Boca Raton, FL, pp. 121–134.

Chapter 11

Enumeration of Bacteriophages Using Flow Cytometry

Corina P. D. Brussaard

Abstract

Rapid identification and enumeration of the numerically important bacteriophages has been till recently a major limitation for studies of virus ecology. The development of sensitive nucleic acid stains, in combination with flow cytometric techniques, has changed this. The flow cytometric method allows the detection and discrimination of a wide variety of viruses of different morphology, genome type, and size. The present paper describes an optimized protocol for the enumeration of bacteriophages using a standard benchtop flow cytometer.

Key words: Bacteriophage, enumeration, detection, flow cytometry, green fluorescence, nucleic acid-specific staining, SYBR Green.

1 Introduction

The study of viruses that are relevant pathogens for humans, animals or plants has received much attention for a long time already. Only recently, it became clear that viruses are very abundant in aquatic environments ($10^5 - 10^8$ ml^{-1}) and highly dynamic in both total numerical abundance and diversity *(1–5)*. Viruses appear important regulating components in population dynamics, diversity, succession, gene transfer, and geochemical cycling of elements *(2–4, 6–8)*. An assay allowing rapid enumeration of virus particles is logically most beneficial for studies of viral ecology. The more traditional methods for virus quantitation include transmission electron microscopy *(9–11)*, the use of antibodies, plaque counts, and most-probable-number assays. These techniques are, however, typically very time-consuming. And although the latter three assays have the advantage of detecting infectious virus particles, they are host-specific

Martha R. J. Clokie, Andrew M. Kropinski (eds.), *Bacteriophages: Methods and Protocols, Volume 1: Isolation, Characterization, and Interactions, vol. 501*, © 2009 Humana Press, a part of Springer Science+Business Media
DOI 10.1007/978-1-60327-164-6_11 Springerprotocols.com

and/or culture-based. The introduction of high fluorescence-yield nucleic acid-specific stains has allowed a more rapid identification and enumeration of the total virus community using epifluorescence microscopy *(11, 12)*. Recently, the development of an assay combining the use of these sensitive nucleic acid stains with flow cytometry *(13–16)* has resulted in rapid analysis with high precision and reproducibility. Direct comparison with epifluorescence microscopy and electron microscopy showed that total virus counts were highly comparable *(17)*.

The flow cytometric assay allows the discrimination of various virus groups in natural samples based on their green fluorescence *(16)*. But the method is also widely applicable as clearly shown by Brussaard et al. *(14)*. A large variety of viruses of different morphologies, genome type, and size could be detected by flow cytometry.

The protocol presented here is based on earlier studies *(14, 16)* executed on different viruses (including many bacteriophages) in order to provide an optimized and consistent method. That study clearly showed that optimal detection of virus particles depended on more than one factor, including type and concentration of fixative and dye, method of storage, type of solution used to dilute the sample, incubation temperature, and duration. Although recommended to specifically determine which conditions are optimal when analyzing specific phage species, one set of variables provided the best results for mixed bacteriophage samples. In summary, samples should be fixed with glutaraldehyde (0.5% final concentration, 15–30 min at 4°C), frozen in liquid nitrogen, stored below −80°C, at least 10-fold diluted in TE buffer (pH 8), stained with SYBR Green I at a final dilution of 5×10^{-5} of the commercial stock, incubated for 10 min in the dark at 80°C, and cooled for 5 min prior to analysis.

2 Materials

2.1 Preservation of Samples and Storage

1. Adjustable pipettes + tips: 100–1000 μl for sample, 2–20 μl for fixative.
2. Glutaraldehyde 25%, EM-grade (Merck). Aliquot (to prevent contamination) stored at 4°C in fixative fridge. *See* **Note 1** for safety measures.
3. Sterile cryovials, 1–2 ml (Greiner Bio-One International; Monroe, NC; http://gbo.com/)
4. Refrigerator (4°C), for fixed samples.
5. Liquid nitrogen.
6. Freezer, –80°C.

2.2 Working Stock Solutions

2.2.1 SYBR Green I

1. SYBR® Green I nucleic acid gel stain, 10,000× concentrate in DMSO, commercial stock (Invitrogen Molecular Probes; http://probes.invitrogen.com/). *See* **Note 2** for safety measures.
2. Sterile MilliQ (**Note 3**).
3. Microcentrifuge.
4. Sterile reaction tubes (Eppendorf vials), 1.5 ml.
5. Adjustable pipettes + sterile tips: 100–1000 μl, 2–20 μl.
6. Gloves.
7. Waste container for fluorescent dye solid waste.

2.2.2 Internal Reference

1. FluoSpheres carboxylate modified microspheres (beads), 1.0 μm, yellow–green fluorescent (Invitrogen Molecular Probes). Stored in the dark at 4°C.
2. Sonicator bath (or Vortex).
3. Sterile MilliQ.
4. Sterile tubes (5 and 15 ml).
5. Adjustable pipettes + tips: 100–1000 μl.

2.2.3 TE-Buffer

1. TRIS 1 M, pH=8.0 (Tris-base, m.w = 121.10).
2. EDTA 0.5 M, pH=8.0 (m.w = 372.24).
3. MilliQ.
4. Bottles with lid, 500 ml.
5. Autoclave.
6. Syringe, 50 ml.
7. Sterile membrane filter of 0.2 μm pore size (Sterile FP30/0.2 μm Schleicher & Schuell) (Whatman Inc.; Florham Park, NJ; http://www.whatman.com/).
8. Sterile bottle with lid, 50 ml.

2.3 Prestart

2.3.1 Getting ready of the Flow Cytometer (FCM)

1. Flow cytometer (FCM) with 488 nm Argon laser (benchtop FACSCalibur, Becton Dickinson, Inc.; Franklin Lakes, NJ; http://www.bd.com/). *See* **Note 4** for more information.
2. FCM tubes, 5 ml (Becton Dickinson, Inc.)
3. Tissues.
4. MilliQ.
5. Cleaning solutions for BD FCMs: BD™ FACSClean and BD™ FACSRinse

2.3.2 Calibration of flow rate

1. Becton Dickinson flow cytometer with 488 nm Argon laser.
2. FCM tubes, 5 ml (Becton Dickinson, Inc.).
3. Balance plus beaker glass (to put the sample tube in for weighing).
4. MilliQ.
5. Sterile and filtered TE-buffer 10:1, pH=8.0 (**Section 2.2.3**).
6. Chronometer.

2.3.3 Enumeration of viruses

1. Flow cytometer with 488 nm Argon laser.
2. FCM tubes, 5 ml (Becton Dickinson, Inc.).
3. Adjustable pipettes + tips: 100–1000 μl, 20–200μl, 5–50 μl.
4. Tube racks (should be able to stand 80°C).
5. Sterile and filtered TE-buffer 10:1, pH=8.0 (**Section 2.2.3**).
6. MilliQ.
7. Tissues.
8. Water baths, set at 35°C and 80°C.
9. Waste containers for fixative- and fluorescent dye-contaminated solid waste (e.g., tips) and liquid waste (samples).
10. Notation forms (**Note 5**).
11. BD™ FACSClean, BD™ FACSRinse and BD™ FACSFlow.

3 Methods

Viruses are too small in particle size to be discriminated solely on the basis of their light scatter properties using the standard commercially available benchtop flow cytometers. The new generation of fluorescent nucleic acid gel stains are high-sensitivity reagents, emitting intense fluorescence when intercalated with DNA and RNA. This qualifies them for many applications where the amount of nucleic acids is limiting, such as the detection of viruses. Their low background fluorescence, furthermore, adds to their usefulness. Indeed, the visible excitation maximum of SYBR Green I dye–stained nucleic acids near 497 nm is very close to the principal emission lines of many laser-scanning instruments, thus equipment such as the Argon-ion laser (488 nm) benchtop flow cytometers that are used to detect and enumerate viruses (**Note 6**).

Noteworthy to mention in this respect is the finding that different phage species revealed variable green fluorescence; sometimes comparable despite different genome size, and sometimes different despite comparable genome size *(16)*. The green fluorescence of stained phages is thus not linearly related to genome size. Although customary for stained bacteria, reference to high-DNA viruses and low-DNA viruses should be avoided at all times (better is to refer to high and low (green) fluorescent virus populations).

Because flow cytometers are basically not designed for the analysis of these small and abundant particles, attention to detail must be given in order to obtain data of high quality. An ultrasensitive stain and low background fluorescence (i.e., a high signal to noise ratio) are then of primary importance, but the aspect of optimal working conditions and thorough cleaning of the cytometer's flow cell should not be overlooked. Working close to the lower detection limit, it is, furthermore, essential to run with settings that do not generate electronic noise. It is, therefore, crucial to determine the level of background noise (**Fig. 11.1**). This can be accomplished through enumeration of sterile, $0.2\ \mu m$ pore-size filtered seawater (of comparable composition) or any other liquid equal to the actual sample. Ideally, one would use $0.02\ \mu m$ pore-size filtered natural sample for the blank, but often the filtration procedure is difficult and generates substantial background noise.

The addition of a mild surfactant at low concentration occasionally improves the coefficient of variance of the green fluorescent signal and thus may improve discrimination of different virus populations. Because the drawback with the addition of detergents is the generation of background noise, care should be taken using detergents when counting phage samples with relatively low green fluorescence (**Note 7**).

Because reproducibility and accuracy are important when analyzing relatively low-fluorescent phage particles, it is essential to work fast in order to keep the time between staining and cooling

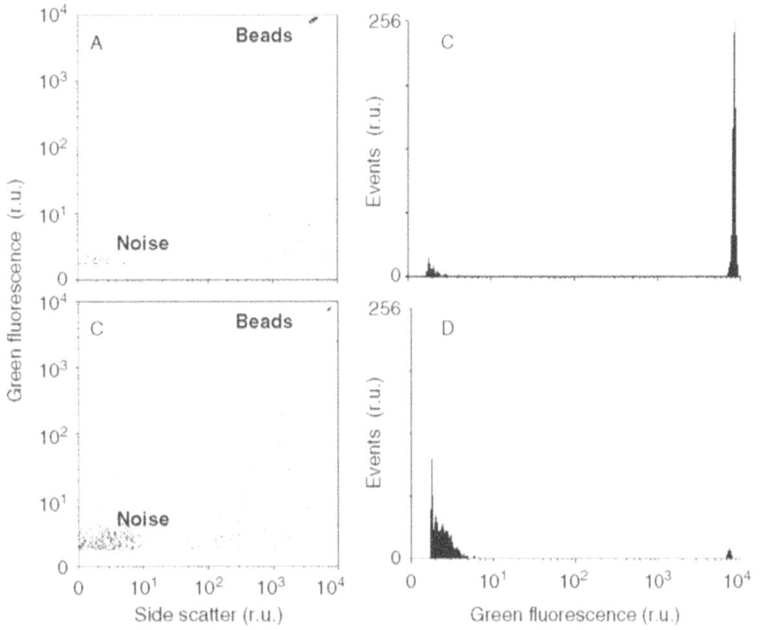

Fig. 11.1. Biparametric plots of side angle scatter versus green fluorescence. (**A**) FluoSpheres carboxylate modified $1.0\ \mu m$ microspheres (Molecular Probes, Inc.) form a tight cluster and are used as internal standard. (**B**) Blank for natural seawater samples: $0.2\ \mu m$ TE-buffer (pH 8) with the appropriate volume of sterile 0.2 μm seawater (100-fold diluted), beads as internal standard ($10\ \mu l$ per 1 ml total volume), and SYBR Green I as dye (final concentration 5×10^{-5} the commercial stock).

of the sample, and actual analysis as short as possible. Reproducibility should also be determined (**Note 8**). Standard deviations for total virus counts are generally < 5%, determined for a variety of virus samples, including natural samples with mixed phage communities (Brussaard, unpublished data).

Preservation of Samples and Storage

1. Carefully pipette 0.5–1 ml of sample into a cryovial. If possible prepare replicates to have the opportunity to reanalyze if necessary.
2. Add 25% EM-grade glutaraldehyde (**Note 9**) to a final concentration of 0.5% (20 μl to 1 ml sample).
3. Allow fixation to proceed for 15–30 min at 4°C.
4. Freeze sample in liquid nitrogen (N_2) and store at −80°C (**Note 10**).

Working Stock Solutions

SYBR Green I

1. Thaw the commercial stock solution of SYBR Green I in the dark at room temperature, mix by vortexing for 10 s, followed by a short spin in a microcentrifuge.
2. Prepare a working stock solution by adding 5 μl of the commercial stock to 995 μl sterile MilliQ in sterile Eppendorf vials. Work in dimmed light (SYBR Green I is light sensitive). Prepare several working stock vials at once and store at −20°C till use.

Internal Reference

1. Mix the commercial stock of beads rigorously as the beads tend to aggregate: preferably sonicate briefly or otherwise use a Vortex shaker.
2. Prepare a primary stock solution by adding 1–2 drops of the commercial stock into 10 ml sterile MilliQ in a sterile tube (e.g., 15 ml plastic Greiner or Falcon tube). This primary stock can be stored at 4°C for long periods of time, but do check the quality of the beads prior to use by running a small subsample of the working stock diluted in TE-buffer through the FCM at the appropriate settings (**Section 3.4**). The beads must give a tight population with a specific green fluorescence, as shown in **Fig. 11.1**.
3. Prepare for daily use a working stock by adding 10 μl primary stock to 2.5 ml sterile MilliQ or TE-buffer (10:1, pH=8.0) in a 5 ml sterile tube with lid. The working stock can be kept outside the fridge for the entire day. Do mix every time before use.

TE-Buffer 10:1 (10 mM Tris, 1 mM EDTA)

1. Prepare 500 ml TE-buffer 10:1 by adding 5 ml of 1 M Tris (pH=8.0) and 1 ml of 0.5 M EDTA (pH=8.0) to 494 ml of MilliQ. Mix well and check pH (should be 8.0).

2. Divide over two bottles of 500 ml and autoclave for 20 min with the lids closed. Although it is this way rarely necessary to adjust the pH afterwards, do check the pH before use (adjust if necessary to pH=8.0). Store at room temperature.
3. Filter a small volume of the TE-buffer (50 ml) through a 0.2 μm pore-size filter into a sterile bottle (or tube) prior to use. Check the quality of the solution before use by running a nonheated, as well as a heated subsample of the TE-buffer through the FCM at the appropriate settings (**Section 3.4**). Both samples should not give more than 30 events s^{-1} at a flow rate of approximately 35 $\mu l\,min^{-1}$.

Prestart

Getting Ready of the Flow Cytometer

1. Start with checking the sheath and waste containers: empty the waste container and fill the sheath container with freshly prepared MilliQ. Because the samples are fixed, MilliQ can be used as sheath fluid. It contains insignificant numbers of particles and is cheaper than TE (the solution the virus samples are diluted in).
2. Turn on the FCM, pressurize and wipe the outer sleeve of the sample inlet (sampler). Place a tube with MilliQ under the sampler and let the machine run for at least 10 min.
3. Take away the sample tube, wipe the sampler and replace with a new tube newly filled with MilliQ. Check whether the machine is clean enough to allow the enumeration of viruses by setting the trigger on side scatter (SSC) and the voltage below, but close to, the level where instrumental noise starts to become significant (**Note 11**). Typically, the SSC voltage is set around 300; the event rate at a flow rate of 35 $\mu l\,min^{-1}$ should then be < 75 events s^{-1}.
4. Only when the FCM is considered clean according to the above criteria is precise enumeration of phages secured. If the event rate is higher, the machine is still too dirty and the machine should be cleaned before use: run BD™ FACSClean for 10 min, followed by 10 min of BD™ FACSRinse, and 10 min of MilliQ. Test again with a new subsample of MilliQ. If the machine is still dirty, try a longer cleaning run, primer several times (after removing sample tube), and/or contact the responsible person for advice about a more rigorous cleaning procedure. Depending on what other type of samples are analyzed, the FCM can be rather dirty according to the standards of those interested in enumerating viruses. In this case, be prepared that cleaning may become an important activity prior to the actual use of the machine, which can take quite some time.

Calibration of Flow Rate

1. Make sure the waste container is empty and sheath fluid container is filled with MilliQ.

2. Select the appropriate flow rate. The typical flow rate is MED, between 25 and 35 μl min^{-1}, which is a good intermediate between speed, statistical number of phages counted, and precision (the lower the flow rate the better the precision).
3. Fill a FCM tube with 2–3 ml of TE and determine its weight (X_0).
4. Remove the outer sleeve of the sampler carefully and wait until a droplet falls. Before the next droplet forms, put the tube under the sampler and place the sample arm in the 'run' position. Simultaneously, start the chronometer.
5. Run the sample for at least 15 min. Remove the tube and stop the chronometer.
6. Weigh the tube and calculate the flow rate (μl min^{-1}) using the formula:

$$\frac{(X_i - X_f) \times 1000}{t} = flowrate$$

where X_0 = initial weight, X_1 = final weight, and t = time (min). **Note 12**.

Enumeration of Viruses

1. Turn on the 80°C and the 35°C water bath.
2. Allow time for the SYBR Green I working solution to thaw (in the dark at room temperature). Although the SYBR Green I working stock solution can theoretically withstand repeated freezing and thawing, it is recommended to reuse the same working stock vial only once in order to maintain optimal staining quality.
3. Take a set of samples out of the −80°C freezer and thaw them relatively quickly (1–2 min) in water of about 35°C (thawed samples should still be cool). A set of eight samples at a time works very well (**Note 13**).
4. Prepare a dilution series (0.5–1 ml per tube) for each sample in TE-buffer in order to optimize the staining (**Note 14**) and to prevent coincidence of the phages during analysis (**Note 15**). Make sure to minimize the error due to low-volume pipetting, and dilute at least 10-fold (and preferably >25-fold) when working with seawater samples (**Note 16**). However, very high dilutions not only require longer analysis time, but also result in loss of the emission signal of the nucleic acid–dye complex. Typically, the optimal event rate is between 200–600 events s^{-1}.
5. Always add a blank of TE-buffer with, according to the dilutions prepared, the appropriate volumes of sterile 0.2 μm filtered reagent equal to the actual sample but without the phages (e.g., seawater, PBS-buffer). Ideally, one would filter the actual sample through a 0.02 μm pore-size filter (the

majority of the natural phages will pass through larger pore sizes) before fixation and freezing, but often this filtration is difficult and generates substantial unwanted background noise.

6. Add beads as internal standard to the tubes (5 μl of working stock to 500 μl sample; **Note 17**). As these beads have specific characteristics when analyzed one can check for variation and errors introduced by the flow cytometer. When there is specific interest in the mean green fluorescence of a certain population of phages, normalize the signal to the internal beads.
7. Finally, mix the working stock of SYBR Green I well, spin briefly in a microcentrifuge, and add the dye to a final concentration of 0.5×10^{-4} of the commercial stock (5 μl to 500 μl sample; **Note 18**). Process in dimmed light as SYBR Green I is light sensitive.
8. Incubate the samples at 80°C (for optimal staining characteristics) in the dark for 10 min, after which the samples are allowed to cool in the dark for approximately 5 min before analysis (**Note 19**).
9. Acquisition: Using MilliQ as sheath fluid, run the sample at an event rate below 1000 events s^{-1} (preferably between 200 and 600 events s^{-1}), at a flow rate between 20 and 50 $\mu l\,min^{-1}$ for 1–2 min. Before starting data acquisition, however, make sure the discriminator is set on green fluorescence and the voltage level is such that no significant electronic noise is generated (**Note 20**). Furthermore, wait for the sample flow rate to stabilize before allowing acquisition of the data; this typically takes about 15 s, but waiting a little longer will allow a better flush of the flow cell with the sample of interest.
10. Wipe sample needle between each analysis with moist tissue in order to reduce contamination. Change tissue regularly.
11. Analysis: In order to be able to optimally analyze the majority of the particles present one should collect the parameters on logarithmic scales (four-decade dynamic range). Data are collected as list-mode files, which can easily be analyzed by a wide range of software (**Note 21**). Viruses are discriminated on the basis of the scatter and fluorescence obtained after staining; green fluorescence *vs.* side scatter (**Figs. 11.2** and **11.3**). Correct the raw data for the blank (background noise) before calculating the total virus abundance per ml taking into account the dilution factor.
12. When ready with the analysis of the virus samples, make sure the machine is clean by rinsing with BD™ FACSClean and BD™ FACSRinse (generally 10 min each is sufficient), and followed by rinsing with MilliQ (or BD™ FACSFlow).

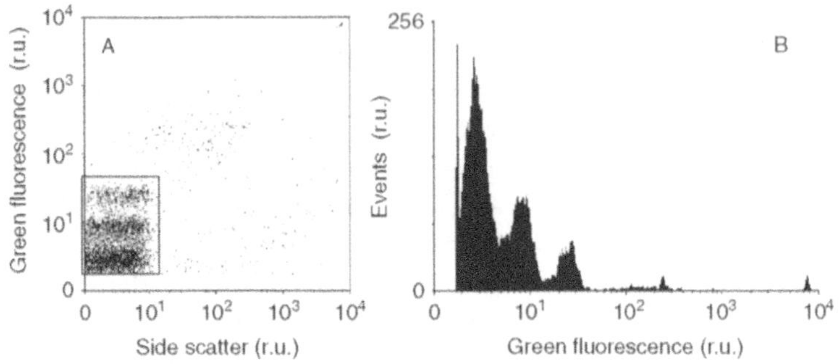

Fig. 11.2. Typical flow cytometric distribution of a natural seawater sample from the Atlantic Ocean ($37°07'$ N, $22°0'$ W), taken at 70 m depth. (**A**) Dot plot of side scatter versus green fluorescence. Window represents the total community of natural viruses. (**B**) Histogram of green fluorescence. The two subpopulations with the lowest fluorescence represent mostly the numerically dominant bacteriophages. The subpopulation with relatively high fluorescence may include algal viruses.

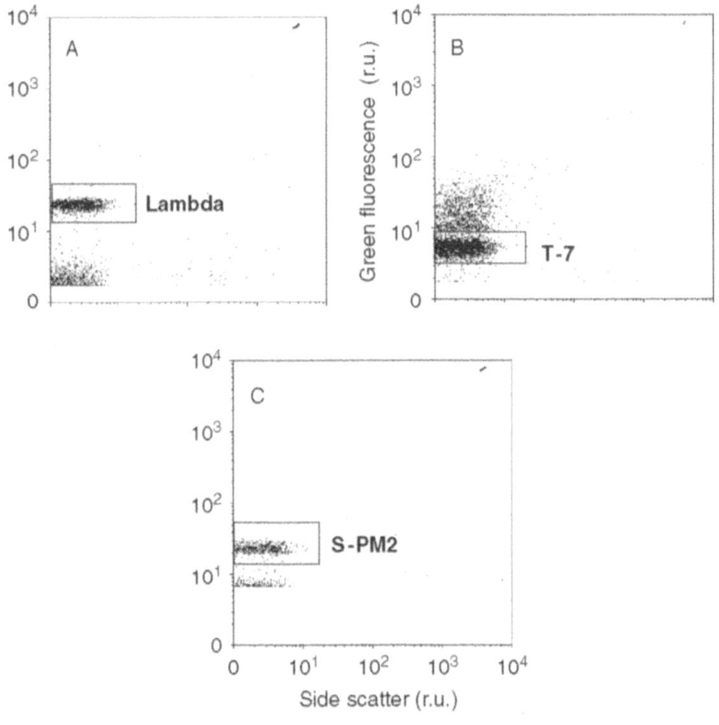

Fig. 11.3. Histograms of green viral nucleic acid fluorescence of three phage species, (**A**) Coliphage Lambda, (**B**) Bacteriophage T-7, and (**C**) photosynthetic marine cyanophage Synechococcus sp. virus (S-PM2). The Bacteriophage T-7 sample showed particles with higher green fluorescence, which represent phage aggregates as could be confirmed using epifluorescence microscopy. The green fluorescence threshold has been raised for the cyanophage S-PM2 sample in order to reduce the number of irrelevant events (other bacteriophages from contaminating bacteria in the culture) and, thus, improve the acquisition of this phage.

4 Notes

1. Glutaraldehyde should be treated with care, as it is highly toxic. Wear personal protection (gloves, lab coat) and in the hood (good ventilation).
2. Always work in the hood and wear gloves when working with the commercial stock of SYBR Green I as it is dissolved in DMSO, which is also carcinogenic. Trash waste in special container.
3. Unless stated otherwise, all primary and working stock solutions should be prepared in water that has a resistivity of $18.2\ \mathrm{M\Omega}$-cm and total organic content of less than five parts per billion. This standard is referred to as "MilliQ" in this text. The water should be freshly prepared.
4. For detection and discrimination of phages, a sensitive flow cytometer is needed. In the present paper reference is made to Becton Dickinson benchtop flow cytometers (e.g., FACSCalibur), which was used for the development of the phage enumeration method. This bench top machine is not too expensive and can be easily transported, which allows enumeration of viruses during field campaigns and/or on board of research vessels (in the case of aquatic virus enumeration).
5. Standard notation forms are very handy when analyzing high numbers of samples and can easily be made in MS-Excel: 14 columns headed 'operator, folder, file name, description, flow rate, time analysis, event rate, gated events total, FSC, SSC, green fluorescence, orange fluorescence, red fluorescence, and trigger'.
6. Staining phage samples with SYBR Green I in combination with flow cytometry provided similar or higher fluorescence intensities and total count as compared to other cyanine dyes (SYBR Green II, SYBR Gold, OliGreen, PicoGreen), making it the most suitable dye for phage enumeration *(16)*.
7. The addition of mild detergents has the potential of facilitating the permeabilization of the phage particles and subsequently enhancing the green fluorescence. Brussaard *(16, 17)* tested several ionic and anionic surfactants, such as Triton X-100, Tween 80, NP-40, and SDS, at low final concentrations (0.1% v/v final concentration). Although some improved the virus signal at times, all surfactants generated unsolicited background noise. Detergents should, therefore, be used with care and in addition sufficient blanks should be taken along.
8. It is important to know the accuracy of your virus measurements. Determine the variation in sampling and handling by

manifold sampling of the original virus batch and one sample tube, respectively.

9. Fixation has a positive effect on the staining properties of phages as compared to unfixed samples *(16)*. Testing different fixatives at various final concentrations revealed that glutaraldehyde clearly gave the best results. At final concentrations >1% a reduction of total phage abundance was observed. The use of good quality, high purity glutaraldehyde, such as EM-grade, is strongly recommended.
10. The temperature the phage samples are stored at is found to be crucial for reliable enumeration of the phages *(12, 16)*. Storage of fixed, unfrozen samples results in a rapid loss of phage, in contrast to fixed samples frozen in liquid N_2 and stored at −80°C *(16)*.
11. To check at which voltage instrumental noise is generated, one triggers the appropriate parameter (Green Fluorescence) and while running a pure solution such as MilliQ, the voltage is slowly increased from zero while keeping an eye on the event rate. The voltage at which the event rate suddenly increases rapidly is the level the PMT starts generating electronic noise. Make sure to set the voltage for checking whether the machine is clean enough somewhat below this critical value.
12. Calibration of the sample flow rate is essential for reliable counts and should be determined regularly during analysis of samples (at least prior and after a day's session of running samples, but preferably more often). Fluorescent microspheres with a known concentration may be used to estimate the actual flow rate but one should have checked whether this is a valid approach for the sample fluid used as these beads are electrostatic. For example, seawater makes these microspheres sticky, changing the expected concentration. Good results are obtained for the BD FCMs by weighing the sample before and after analysis, which can be extended to other brands of FCMs. In case of FCMs with relatively large dead-volumes Beckman, a calibration curve should be established in order to determine the actual flow rate. Do realize that seagoing acquisition demands different methods because weighing is not an option at sea; either one does use beads as rough estimate, or one prepares tubes with the appropriate liquid, weighs and seals them, and when on board these tubes are run for a certain time, sealed again to have their weight determined when back on land. A combination of these methods is recommended.
13. Because the abundance of phage declines in the fixed thawed samples it is recommended to analyze only several samples at a time (sets of eight samples each work well). Upon thawing and preparation of the dilution series, store the

samples at 4°C in a fixative fridge, allowing reanalysis within a few hours when needed. Never re-freeze the thawed samples, but trash them after several hours instead. Preferably work in a climate-controlled room (18°C) to avoid warming of the samples by the heat coming from the FCM, and subsequent enhanced loss of phage from the fixed thawed samples.

14. TE-buffer gave the most optimal staining and total count as compared to other choices of dilution solution (water, Tris, PBS, seawater) to dilute the phage samples in *(16)*. Diluted staining reagent SYBR Green I is more stable in buffer than in water, and TE is commonly used in combination with SYBR Green I at a pH of 7–8.5 (Molecular Probes Inc.). Since SYBR Green I is pH sensitive, make sure to set the pH of the TE-buffer at a constant value (pH 8.0 gives optimal sensitivity for SYBR Green I).
15. At low dilutions coincidence, i.e., two or more virus particles are simultaneously within the sensing zone of the FCM, becomes a real problem. On average, coincidence occurs for viruses above 800–1000 events s^{-1}. Working with specific phage species, it is important to determine the optimal event rate because there are certain phages that form doublets or aggregate relatively easily.
16. Previous tests showed that the ratio of TE-buffer to seawater sample could affect the green fluorescence signal and/or the total phage count. Salts can, indeed, profoundly influence the differential absorption values and fluorescence of the complex of SYBR Green and dsDNA. The samples should, therefore, be diluted at least 10-fold and preferably more. In case the original sample contains only a very low total abundance of phage, use a higher flow rate. When working with another type of solutions the phages are suspended in, one should test if and to what extent the ratio sample to buffer influences the phage detection and enumeration.
17. Not having to change the pipette tip every sample, add the beads to the empty tubes before adding the TE-buffer and the sample.
18. Not having to change the pipette tip for every sample, add the drop of dye high on the side of the tube and tick the drops down into the TE-buffer carefully when all samples have received the dye. Make sure no phage-containing sample has touched that specific spot of the tube's wall (in doubt, change tips!). Addition of SYBR Green I after heating the sample results in reduced staining and lower total counts and is, therefore, not recommended.
19. Using lids on the sample tubes is recommended. Use a water bath instead of a heating block to prevent evaporation. Furthermore, when using a lid on the water bath make sure

no condense droplets coming from the inside of the lid fall into the tubes.

20. Although you like to have your FCM set as sensitive as possible, you have to make sure the voltage settings for the green fluorescence is ALWAYS set below the level where instrumental noise starts to interfere (*see* for procedure **Note 10**). The typical setting for the green fluorescence on the BD FACSCalibur FCM in my experience is between 475 and 600. This value is machine- and PMT-dependent and should, therefore, be checked when any manipulation with this PMT has occurred. Typical settings on a FACSCalibur FCM are FSC (forward scatter) = E02, SSC (side scatter) = 550, FL1 (green fluorescence) = 520, FL2 (orange fluorescence) = 500, FL3 (red fluorescence) = 500.
21. In order to allow rapid acquisition of the collected data files using any available software, it is best to provide short coding names to the files. A simple, but very efficient filing system is to code the first few characters for the type of parameter analyzed, followed by characters for the year month and date. Effective freeware is the program named "CytoWin" (http://www.sb-roscoff.fr/ Phyto/index.php), which is especially easy to learn (an advantage when having many students go through the lab).

References

1. Bergh, Ø., Børsheim, K.Y., Bratbak, G. and Heldal, M. (1989) High abundance of viruses found in aquatic environments. Nature **340**, 467–468.
2. Proctor, L.M. (1997) Advances in the study of marine viruses. Microsc. Res. Tech. **37**, 136–161.
3. Wommack, K.E. and Colwell, R.R. (2000) Virioplankton: viruses in aquatic ecosystems. Microbiol. Mol. Biol. Rev. **64**, 69–114.
4. Weinbauer, M.G. (2004) Ecology of prokaryotic viruses. FEMS Microbiol. Rev. **28**, 127–181.
5. Børsheim, K.Y., Bratbak, G. and Heldal, M. (1990) Enumeration and biomass estimation of planktonic bacteria and viruses by transmission electron microscopy. Appl. Environ. Microbiol. **56**, 352–356.
6. Fuhrman, J.A. and Suttle, C.A. (1993) Viruses in marine planktonic systems. Oceanography **6**, 51–63.
7. Wilhelm, S.W. and Suttle, C.A. (1999) Viruses and nutrient cycles in the sea. BioScience **49**, 781–788.
8. Jiang, S.C. and Paul, J.H. (1998) Gene transfer by transduction in the marine environment. Appl Environ Microbiol **64**, 2780–2787.
9. Hara, S., Terauchi, K. and Koike, I. (1991) Abundance of viruses in marine waters: Assessment by epifluorescence and transmission electron microscopy. Appl. Environ. Microbiol. **57**, 2731–2734.
10. Hennes, K.P. and Suttle, C.A. (1995) Direct counts of viruses in natural waters and laboratory cultures by epifluorescence microscopy. Limnol. Oceanogr. **40**, 1050–1055.
11. Noble, R.T. and Fuhrman, J.A. (1998) Use of SYBR Green I for rapid epifluorescence counts of marine viruses and bacteria. Aquat. Microb. Ecol. **14**, 113–118.
12. Wen, K., Ortmann, A.C. and Suttle, C.A. (2004) Accurate estimation of viral abundance by epifluorescence microscopy. Appl. Environ. Microbiol. **70**, 3862–3867.
13. Marie, D., Partensky, F., Vaulot, D. and Brussaard, C.P.D. (1999) Enumeration of phytoplankton, bacteria, and viruses in marine samples. In: Current Protocols in Cytometry (Robinson, J.P.e.a., Ed., Vol. Supplement 10,

pp. 11.11.11–11.11.15. John Wiley & Sons, Inc., New York.
14. Brussaard, C.P.D., Marie, D. and Bratbak, G. (2000) Flow cytometric detection of viruses. J. Virol. Methods **85**, 175–182.
15. Chen, F., Lu, J.-R., Binder, B.J., Liu, Y.-C. and Hodson, R.E. (2001) Application of digital image analysis and flow cytometry to enumerate marine viruses stained with SYBR Gold. Appl. Environ. Microbiol. **67**, 539–545.
16. Brussaard, C.P.D. (2004) Optimization of procedures for counting viruses by flow cytometry. Appl. Environ. Microb. **70**, 1506–1513.
17. Marie, D., Brussaard, C.P.D., Thyrhaug, R., Bratbak, G. and Vaulot, D. (1999) Enumeration of marine viruses in culture and natural samples by flow cytometry. Appl. Environ. Microbiol. **65**, 45–52.
18. Zipper, H., Brunner, H., Bernhagen, J. and Vitzthum, F. (2004) Investigations on DNA intercalatioi and surface binding by SYBR Green I, its structure determination and methodological implications. Nucl. Acids Res. **32**, e103.

Chapter 12

Basic Phage Electron Microscopy

Hans-W. Ackermann

Abstract

Negative staining of purified viruses is the most important electron microscopical technique in virology. The principal stains are phosphotungstate and uranyl acetate, both of which have problems and advantages. Particular problems are encountered in photography, calibration of magnification, measurements, and interpretation of artifacts.

Key words: Artifacts, contrast, magnification, negative staining, positive staining, phosphotungstate, uranyl acetate.

1 Introduction

The first electron micrographs of bacteriophages were published as early as 1940 *(1, 2)*. The technique of negative staining was introduced in 1959 *(3)* and revolutionized the study of viruses. Electron microscopy (EM) became a basis of comparative virology and classification and numerous electron microscopic techniques are currently available. In a general way, the electron microscope can be considered as a huge camera and electron micrographs are often the only permanent record of an investigation. The quality of micrographs depends less on stains and techniques than skill and dedication of the individual electron microscopist.

Transmission EM (TEM) includes negative and positive staining of isolated viral particles, thin sectioning, shadowing, immuno-EM, cryo-EM with three-dimensional image reconstruction, enzymatic particle digestion on the grid, and the so-called Kleinschmidt technique of DNA spreading on protein films *(4)*. Most of these techniques are practiced in a few specialized

Martha R. J. Clokie, Andrew M. Kropinski (eds.), *Bacteriophages: Methods and Protocols, Volume 1: Isolation, Characterization, and Interactions, vol. 501*, © 2009 Humana Press, a part of Springer Science+Business Media
DOI 10.1007/978-1-60327-164-6_12 Springerprotocols.com

laboratories or are outdated; for example, shadowing is rarely done anymore. Similarly, scanning electron microscopy (SEM), which allows observation of bacteriophages adsorbed to bacteria, has little importance in phage research. Negative staining however, a technique of extreme simplicity, is of universal importance in virology. It allows instant comparison, classification, and identification of viruses, sometimes down to the species level.

Phage electron microscopy has been discussed elsewhere in detail *(5)*. It suffered a sharp decline in quality in the last 20 years *(6)*; indeed, the high technical development of electron microscopy contrasts with the often-dismal phage descriptions in the recent literature. Negative staining has been the object of a book *(7)* and numerous papers and book chapters *(5, 7–11)*. Its importance is illustrated by the fact that over 5,000 bacteriophages have been examined in the electron microscope *(12)*.

The principle of negative staining is to mix particles and an electron-dense solution of a metal salt of high molecular weight and small molecular size, into which particles are embedded like bacteria in Chinese (Indian) ink and appear white in a dark background. The stains used in phage research include tungstates (potassium and sodium phosphotungstate, lithium tungstate, sodium silicotungstate), uranyl salts (acetate, formate, magnesium acetate, nitrate, oxalate), ammonium or vanadium molybdate, and molybdic acid *(5)*. Many technical variants of the basic staining method exist: "direct extraction" of phages from an agar surface, distribution of stains and viruses with a nebulizer, successive staining, staining from below, and the pseudoreplica technique *(10)*. These techniques have few applications or are of historical interest only. The advent of digital electron microscopy and printing has created a completely new situation (**Section 4**).

This chapter is concerned exclusively with the two most common stains, phosphotungstate (PT) and uranyl acetate (UA), and the most common technique of staining. On the other hand, phage EM is a multi-step procedure, which, besides simple staining, involves a succession of techniques that must be discussed:

1. Purification of phage particles.
2. Staining.
3. Observation.
4. Photography.
5. Magnification control.
6. Measurements.
7. Interpretation of errors and artifacts.

2 Materials

2.1 Purification

1. A medium size centrifuge able to reach $25,000 \times g$, equipped with a fixed-angle rotor.
2. Centrifuge tubes (1–2 ml) with conical bottoms and caps to prevent evaporation.
3. A solution of 0.1 M ammonium acetate, pH 7 *(13)*.
4. Pasteur pipettes made pointed by drawing out their ends in the flame.

2.2 Staining

1. Phosphotungstate (potassium or sodium), 2%, pH 7.2.
2. Uranyl acetate, 2%, pH 4–4.5.
3. PT and UA supplemented with 2–3 drops/ml of bacitracin solution (50 μg in 100 ml of water).
4. Pointed Pasteur pipettes.
5. Pointed tweezers.
6. Electron microscopical grids, 200–400 square mesh (Athene type), copper or steel, carrying a Formvar or collodion film stabilized with a 2–10 nm thick carbon layer.
7. Strips of filter paper.
8. Petri dishes laid out with filter paper or boxes with slots to store and transport grids.

Phosphotungstate solutions are prepared by dissolving phosphotungstic acid in distilled water and neutralizing the strongly acidic solutions with KOH or NaOH. They can be kept in stoppered bottles at 4°C for at least 2 years. Over time, solutions may become acidic by absorption of CO_2 *(7)*. Tailed phages stained with aged PT have unsharp outlines and tails without transverse striations. Ammonium molybdate, used with some frequency in phages *(5)*, has more or less the same staining properties as phosphotungstate.

Uranyl acetate is dissolved in distilled water. This chemical is toxic and radioactive and should be handled with care. UA dissolves slowly. The solution is adjusted to the desired pH with 1 M KOH or NaOH and keeps well for 2 years in a stoppered bottle at 4°C.

Hydrophilic grids show poor adsorption of phages and unequal distribution of phages and stain. Grids can be made hydrophilic by high-voltage glow discharge (4×10^{-1} mbar, 50 kV for 60 s). Glow discharge ionizes molecules and atoms in the vacuum. This induces negative charges on the grid and makes the latter hydrophilic *(15, 16)*. Alternatively, grids may be rinsed with a wetting agent (Alcian blue, bacitracin, poly-*L*-lysine, serum albumin, sucrose) or treated with UV light *(16, 17)*.

I prefer staining with PT or UA solutions supplemented with 2–3 drops/ml of bacitracin *(17)*. I also have found that a grid that appears hydrophobic in the electron microscope may simply be re-charged with phages and stain. The irradiation to which it is subjected by the beam, usually makes it hydrophilic again.

2.3 Darkroom or "Wet" Photography

All procedures for darkroom photography are well known *(8)* and countless photographical products are (or were) available. Their detailed description is both redundant and impossible. The subject is in constant evolution as, due to the introduction of digital recording and printing, certain suppliers recently changed or eliminated part of their product line. For example, Eastman Kodak (Rochester, NY) formerly produced excellent papers and chemicals for automatic development that are no longer available. For manual development, use:

1. Basic darkroom equipment: safety lights, trays for films and paper, pincers.
2. Fine grain positive film of 35 mm or 70 mm, best nonperforated. Photographic plates are now rarely used in electron microscopy. They are bulky and fragile, and have no advantages over films.
3. Chemicals for films: high-speed and high-contrast developer, stop bath (3% acetic acid), fixer, hyposulfite clearing agent.
4. Soap or detergent solution.
5. High-quality enlarger.
6. Set of graded gelatine filters.
7. Paper: of the glossy, polycontrast type, with or without resin coating.
8. Chemicals for paper: developer, stop bath (3% acetic acid), fixer, hyposulfite clearing agent.
9. For paper without resin coating, a washing drum and a drying machine.
10. Miscellany: turpentine oil for eliminating scratches, a Freon duster, carbon tetrachloride for wiping off fingerprints and dirt specks from films, a cardboard or a spoon for selective exposure ("dodging"), test fluid for checking fixer exhaustion, retouching fluids for paper.

2.4 Magnification Control (Calibration)

The lens current of an electron microscope and the resulting magnification may fluctuate within the same day. The specifications of manufacturers on magnification are unreliable and require frequent controls and adjustments. Unfortunately, many test specimens are unsuitable for high magnification (notably diffraction grating replicas) or even shrink in the beam (latex spheres). The only acceptable, though imperfect, standards for high-magnification control are beef liver catalase crystals, which have parallel lines of subunits with a periodicity of 8.8 nm *(18)*, and extended T4 phage tails. T4 tails, as determined on catalase-controlled preparations, measure 114 nm in length including the base plate.

1. A grid with PT-stained beef liver catalase crystals. Crystal suspensions and ready made grids are commercially available. Grids with PT-stained catalase crystals can be stored for at least 2 months.
2. Alternatively, a grid with T4 tails. T4 phages are prepared in the laboratory as described below (**Section 3.1** and **3.2**). Both PT- and UA-stained grids are reusable. PT grids may be used for a month and UA grids for a year, respectively. T4 tails are much easier to use than catalase crystals.

3 Methods

3.1 Purification

This is one of the most important steps in phage EM because lysates contain proteins and other impurities. Both interfere with stains and particle observation. Purification is also required for detection of abnormal and contaminant phages. Crude phages from lysed areas on agar may occasionally be examined for identity checks. In all other circumstances, purification is mandatory; in a general way, the examination of crude lysates is useless and to be proscribed.

Purification is best achieved by centrifugation and washing in buffer. This can be done in the ultracentrifuge using a swinging-bucket rotor ($70 - 80,000 \times g$ for 4–6 h). However, a fixed-angle rotor reduces the distance that phage particles have to travel during sedimentation, permitting a considerable reduction of g forces and centrifugation times and hence the use of relatively inexpensive medium size centrifuges. The following is a fail-safe procedure:

1. A sterile high-titer lysate (10^8 viable phages/ml or better) is prepared.
2. Phages are sedimented for 60 min at $25,000 \times g$.
3. The supernatant is discarded and replaced by ammonium acetate solution.
4. Phages are sedimented again at for 60 min at $25,000 \times g$.
5. The procedure is repeated once or twice. The final sediment is ready for staining.
6. Sediments may be stored for further examination: for 2 weeks at 4°C and for months, even years, as long as some fluid remains in the tube, at –20°C. They will tolerate one thawing only without damage for phage particles.

Purified phage preparations usually contain phage and cellular debris (cell wall, pili, flagella) and even complete bacteria. This is not a problem as the electron microscopist is normally able to seek out areas without these elements. Depending on the initial state of purity of the phage lysate, variable amounts of protein may persist in the preparation. This is sometimes beneficial because proteins are wetting agents and enhance spreading of phages and stains. In some cases, it is necessary to wash phages three times, but it should be kept in mind that any manipulation increases the danger of damaging and losing phages.

Purification by density gradient (CsCl, Cs_2SO_4, sucrose, metrizamide) centrifugation is a standard biochemical technique and is not described here. As an example, tailed phages band in CsCl at 1.4–1.45 g/ml and filamentous phages at 1.3 g/ml. The technique is much more complicated and time consuming than simple centrifugation and washing. In experienced hands, it yields a pure, beautiful preparation, but care must be taken to dialyze phage preparations adequately.

3.2 Staining

In the classical version, a drop of phage suspension is deposited on a grid. Phages are allowed to adsorb for 1 min and a drop of stain is added. After 1 min more, the liquid is drained off with filter paper. The grid dries instantly and is ready for examination. The limit of detection is 10^5 particles per milliliter. If phage suspensions are too concentrated for examination, the procedure is inverted (stain first, phages later).

PT and UA form glassy films around particles and penetrate empty capsids. Neither stain causes DNA ejection or tail contraction. The stains are complementary and not equivalent. Both have desirable and undesirable properties (**Table 12.1**) and one should always use both (on different grids please). Stained grids are stored in Petri dishes or special small boxes. PT-stained preparations generally deteriorate after a month and cannot be stored for long times. When stored in special grid boxes, UA-stained grids keep for as many as 28 years. Their longevity is essentially limited by the stability of the supporting Formvar film. However, PT and UA grids stored in a dish may be destroyed overnight by warm, humid weather through recrystallization of the stains.

3.2.1 Phosphotungstate (5,7)

PT is a neutral stain, does not act as a fixative, causes negative staining only, gives consistent results, and does not crystallize on the grid. The achieved contrast is variable. Virus capsids may be flattened and appear enlarged. Tails are thin and well defined. PT often accumulates around phage heads, leaving tails poorly stained. PT is excellent for detection of aberrant forms and contaminating phages.

3.2.2 Uranyl Acetate (5,7,19)

UA is acidic, acts as a fixative, inactivates phages, and is unpredictable because it produces, without apparent reason, both negative and positive staining on the same grid (**Fig. 12.1(1)**). Contrast is generally excellent. Negatively stained heads are often

Table 12.1
Comparison of PT and UA

Parameters			PT	UA
General	pH		Neutral	Acidic
	Crystallizes on grid		–	+/–
	Fixation		–	+
	Contrast		+/–	+
	Adhesion to grids		+/–	+
	Negative staining		+	+/–
	Positive staining		–	+/–
	Halo around capsid		–	#
	Precipitate around phages		–	+
	Longevity of grids		Months	Years
Preservation	Capsid	Angular	+/–	+
		Rounded	+/–	–
		Flattened	+/–	–
		Shrunken	–	#/–
		Swollen	–	+/–
	Tail	Straight	+/–	+
		Swollen	–	+
		Striations	+/–	+
		Fibers	+/–	+/–
Detection	Pentagonal capsids		+/–	+
	Polyheads		+	–
	Polytails		+	+/–
	Contaminant phages		+	+/–

PT, phosphotungstate; UA, uranyl acetate. +, yes; –, no; #, positively stained particles only. Modified from ref. 5.

more regular than PT-stained capsids, but tails and other proteinic structures (e.g., empty capsids) appear thickened (**Fig. 12.2(5)**). Pentagonal capsids are better visible than in PT. Positive staining is due to the strong affinity of UA for double-stranded DNA (dsDNA) *(20)*. Positively stained phage heads are deep black, have no edges, and are up to 30% smaller than negatively stained capsids. They should not be measured. However, UA positive staining is valuable in environmental research when the only objective is phage counts, because deep-black heads are easily

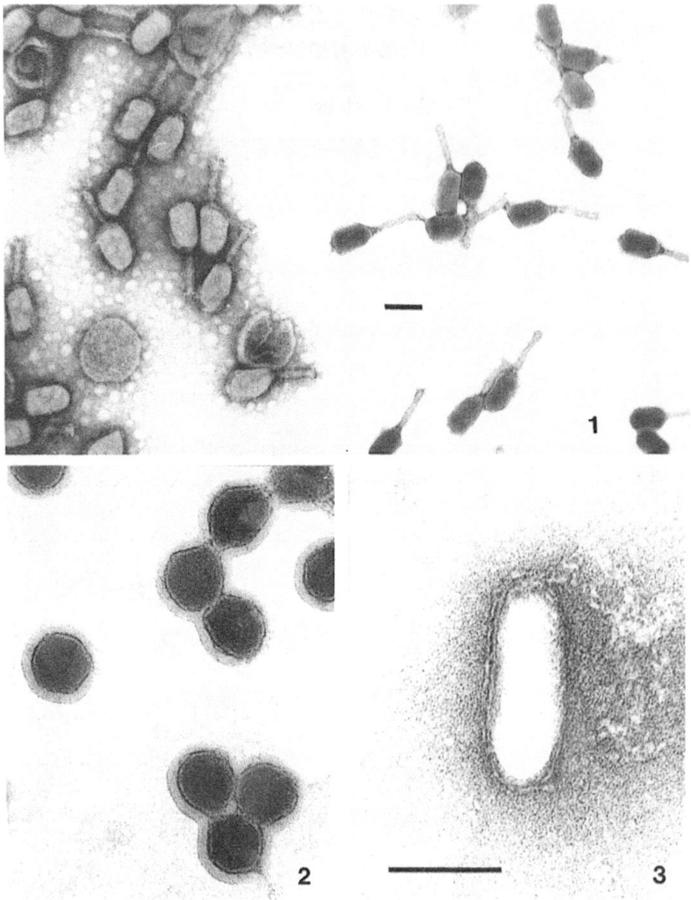

Fig. 12.1. Staining artifacts with uranyl acetate. Final magnification of Fig.1, × 92,400; Figs. 2 and 3, × 3,297,000. Bars indicate 100 nm. **(1)** Negative and positive staining within the same area. Positively stained phage heads are black. *Aeromonas salmonicida* phage 65. **(2)** Positive staining with false capsule (halo), *Vibrio cholerae* typing phage I. **(3)** Stain precipitate suggesting a capsule around *Salmonella newport* phage 7–11.

detected and counted at low magnification. Two serious disadvantages of UA are its tendency to crystallize on the grid and to produce poor negative staining with a mealy background.

3.3 Observation

Grids are examined at approximately 60 kV. To limit contamination by hydrocarbon molecules, which may obscure fine details of phages, the object stage of the electron microscope may be cooled with liquid nitrogen. However, this is not necessary for a skilled operator.

3.4 Photography

3.4.1 Darkroom Photography

The principles of film and paper development are age-old and will not be described here. The exact concentrations and working conditions of reagents depend on the products used. In films, a developing bath is followed by a stop bath, a fixation bath, and a bath that neutralizes the fixer. The film is then rinsed for

5 min, dipped into a soap solution to avoid the formation of water droplets, wiped dry with a squeezer, and left to harden for 20 min. In paper development ("printing"), the paper is exposed for 2–6 s and transferred successively into a developer, a stop bath, and a fixer. In modern resin-coated films, no neutralization of the fixer is required. The paper is washed and left to dry. Papers without resin coating must be washed for 30 min and dried and glazed in a special machine. The contrast of photographs can be dramatically improved by using filters. Details such as tail striations and tail fibers are sometimes better seen in low rather than high contrast.

3.4.2 Digital Photography (see Section 4)

3.5 Magnification Control

In conventional TEM, two pictures of catalase crystals or T4 phages are taken for every series of 20–50 exposures and at the highest magnification used for measurements. The head of the enlarger is then adjusted. For example, at a magnification of $300,000\times$, 20 parallel lines of a catalase crystal correspond to 5.2 cm. Suitable catalase crystals may be hard to find because the crystals tend to bend or to break and dissolve into individual subunits. At the same magnification, T4 tails, which are 114 nm long, measure 3.4 cm. Their advantage is that they are easily prepared in any phage laboratory and easy to measure. However, they are relatively small (thus inviting errors of measurements) and sometimes stretched after UA staining (not in PT).

3.6 Measurements

As a rule, phage dimensions should be so complete that the reader can build a model. In conventional EM, at least 10 particles per phage should be measured on prints (not films or plates). Digital EM allows measurements on the screen, but this does not lend itself to measure small phage components such as tail fibers. Particles should be intact and well preserved. Capsids should be angular and have parallel sides. Isometric capsids are measured between opposite apices (which is easier than between opposite sides). This gives three diameters per phage capsid, of which the average is calculated. Positively stained capsids should not be measured in any case.

3.7 Errors and Artifacts (See Section 3.2)

In addition to positive staining and shrinking of phage capsids, UA may produce false envelopes in the form of haloes and precipitates (**Fig. 12.1**) and cause swelling of phage tails (**Fig. 12.2(4)**). Furthermore, the observation of inovirus-like structures may be suggested by the presence of bacterial pili or filaments of slime (**Fig. 12.2(5)**). Tailless phages with cubic symmetry, e.g., members of the *Tectiviridae* family, are simulated by (a) tailed phages which have lost their tails, (b) phage heads which are tailless by nature (defective lysogeny), (c) phages whose tails are very short and difficult to see, and (d) tailless heads with DNA con-

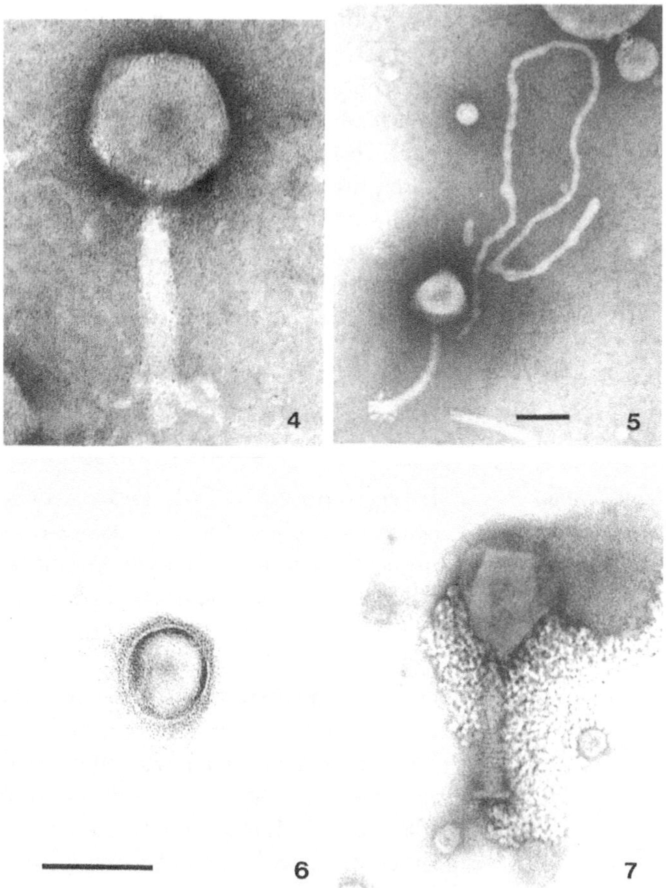

Fig. 12.2. Various artifacts. Fig. 6, × 148,500, all others × 297,000. UA, uranyl acetate; PT, phosphotungstate. Bars indicate 100 nm. **(4)** Abnormally swollen particle, UA, *Pseudomonas aeruginosa* phage ϕKZ. Tail striations and fibers are barely discernible. **(5)** False filamentous phage and *Bacillus licheniformis* myovirus BL1, PT. The mucous nature of the filament is evident in its variable diameter. **(6)** False tectivirus, tailless head of *Synechococcus* myovirus S-PM2. Its content is indistinguishable from a tectivirus internal vesicle; PT. **(7)** NaCl precipitate around coliphage T4, PT.

densates (**Fig. 12.2(6)**). Washing in saline generates salt precipitations around phages that also interfere with observation (**Fig. 12.2(7)**).

4 Digital Electron Microscopy: Buyers Beware

Digital TEMs are slowly replacing conventional TEMs. This is partly due to pressure from electron microscopists, especially unskilled newcomers, who are now able to correct focus and astigmatism with a touch of a button and to bypass darkroom photography entirely. Simultaneously, major manufacturers of photographic material are reducing or phasing out the production of films and papers for electron microscopy. Digital TEMs may have excellent resolution (as good as conventional TEMs), but the quality of the pictures produced is often shameful. Most laboratories with digital EMs that I have contacted continue to use "wet" photography for prints. Even a company representative whom I consulted, considered darkroom photography as the *Nec plus ultra* of imagery.

4.1 Advantages

Adjustment of focus, astigmatism, and alignment are automated. Image recording, printing, and storage are extremely simple. Images can be immediately viewed, shared with other laboratories, provided with scale markers, and combined with other micro-

Table 12.2
Sources of Errors in Phage Electron Microscopy

Error	Cause
Shrunken capsids	Uranyl acetate positive staining
Swollen tails	Acidity of uranyl acetate
False envelopes	- Halo around UA-positively stained phage heads
	- UA precipitates around phages
False inoviruses	- Pili
	- Bacterial slime made filamentous by centrifugation
False cubic phages	Damaged tailed phages, tailless defective phages, podoviruses with very short tails (some cholera phages)
False tectiviruses	DNA condensates in tailless phage heads

Table 12.3
Do and Do not

Do	Do not
Purify phages	Examine crude lysates
Use two stains (PT and UA)	Wash phages in saline
Use filters for contrast	Fix phages
Calibrate with catalase crystals or T4 tails	Calibrate with diffraction grating replicas
Measure 10 particles or more	Measure positively stained phages
Give complete dimensions	Use grids without a carbon layer
Add a scale marker	

graphs. No darkroom work is required and teaching in the EM room is simplified.

4.2 Problems

1. Digital electron microscopes are expensive. Their introduction resulted in a quantum leap in prices. To the basic coast of the EM, one must now add that of a high-quality camera (4 × 4 K= 16 megapixels, about $300,000; 2 × 2 K = 4 MP; about $100,000), a high-quality printer ($5–6,000) and equally expensive special paper. Less expensive cameras and ordinary printers will produce poorer, if not very poor pictures. For comparison, the camera of an old conventional Philips EM300 costs $2,000.
2. The electron microscopist has little control over the EM and depends totally on company technicians, service contracts, and camera software. This is particularly worrisome for magnification calibration and presages high repair costs for electron microscopes not covered by service contracts.
3. I have seen very few satisfactory high-magnification digital images of viruses. It is as if EM companies targeted inexperienced users and are selling mediocre instruments at top prices. What shall the virologist do? The situation is bound to evolve. Until cameras become better and less expensive, it seems best:
    1. To keep conventional electron microscopes as long as possible.
    2. If a digital electron microscope is purchased, to buy only the very best equipment available, insist on the presence of conventional plate or film cameras, and use "wet" photography for important micrographs.

5 Notes

5.1 Purification

1. Do not wash in saline.
2. Glutaraldehyde fixation is sometimes used in widely varying conditions *(5)*. It complicates manipulations and is unnecessary since UA preserves phages. Photographs of glutaraldehyde-fixed phages are of generally poor quality. However, fixation with 2.5% glutaraldehyde may be indicated for marine samples which cannot be examined immediately and have to travel weeks before reaching a laboratory *(15)*.
3. For particle counts, phages may be centrifuged directly onto grids. The Beckman Coulter Airfuge® (Fullerton, CA; http://www.beckmancoulter.com/) uses very small (180–240 μl) volumes and short times (3–5 min). Phages T4 and T7 survive sedimentation at $115,000 \times g$, but are generally damaged at $185,000 \times g$ *(5)*. For larger volumes, especially in environmental studies, conical centrifuge tubes are pro-

vided with a flat epoxy bottom onto which a grid is deposited. Phages are then sedimented at $80,000 \times g$ for 90 min without further purification and stained with UA *(21)*.

5.2 Staining

1. PT and UA are used in concentrations of 0.5–4%.
2. Grids with support films can be prepared by the electron microscopist *(8, 11, 14)*, but this is a cumbersome procedure. It is more expedient to purchase ready made grids, commercially.
3. A short cut, only to be used for quick identity checks, is to deposit a drop of PT on a lysed area on an agar plate, to agitate it gently to let phages float off, and to collect phages with a grid. This technique works best for large phages and cannot replace a regular purification procedure.

5.3 Darkroom Photography

1. Store films at 4°C and dry them in a desiccator before use.
2. In films, the development temperature is critical. Old developers must be discarded. Applying turpentine oil to the film can eliminate scratches.
3. In paper development, the age of the developer is not critical, but stop baths must be renewed after 20–25 full size paper sheets. Fixers must be checked and discarded if exhausted.
4. Selective exposure of structures, e.g., of phage tails, is achieved by covering phage heads with the help of a cardboard or a spoon.
5. White specks, scratches, and other small imperfections can be corrected on paper with a retouching fluid.

5.4 Capsid Shape

Proof for the icosahedral shape of isometric virus capsids is obtained by the simultaneous observation of particles with hexagonal and pentagonal outlines. The simple observation of hexagons is insufficient since octahedra, dodecahedra, and icosahedra may all present hexagonal outlines.

References

1. Pfankuch, E. and Kausche, G.A. 1940. Isolierung und übermikroskopische Abbildung eines Bakteriophagen. *Naturwissenschaften 28*,46.
2. Ruska H. 1940. Über die Sichtbarmachung der bakteriophagen Lyse im Übermikroskop. *Naturwissenschaften 28*, 45–46.
3. Brenner, S. and Horne, R.W. (1959) A negative staining method for high resolution electron microscopy of viruses. *Biochim. Biophys. Acta 34*, 103–110.
4. Kleinschmidt, A.K. (1968) Molecular weight and conformation of DNA, in *Nucleic Acids, Meth. Enzymol. 12B*, (Grossman L, Moldave, K, eds.), Academic Press, New York, NY, pp. 361–372.
5. Ackermann, H.-W. and DuBow, M.S. (1987) *Viruses of Prokaryotes*,Vol. 1. *General Properties of Bacteriophages, CRC* Press, Boca Raton, FL, pp. 103–130.
6. Ackermann, H.-W. (2004) Declining electron microscopy. *Lab. News*2004 (12), 25. **See also:** BEG News 21, 2–3, http://www.phage.org/
7. Hayat, M.A. and Miller, S.E. (1990) *Negative Staining*,McGraw-Hill, New York, NY, pp. 1–50.

8. Dykstra, M.J. (1992) *Biological Electron Microscopy. Theory, Techniques, and Troubleshooting.*Plenum Press, New York, NY, pp. 103–105, 183–208, 218–221.
9. Horne, RW. (1965) Negative staining methods, in *Techniques for Electron Microscopy,* 2nd ed. (Kay, D.H., ed.), Blackwell Scientific Publications, Oxford, UK, pp. 328–355.
10. Nermut, M.V. (1973) Methods of negative staining, in *Methodensammlung der Elektronenmikroskopie* (Schimmel, G. and Vogell, W, eds.), Wissenschaftliche Verlagsgesellschaft Stuttgart, Germany, Section 3.1.2.3.
11. Tikhonenko, A.S. (1970). *Infrastructure of Bacterial Viruses,* Plenum Press, New York, pp. 1–22.
12. Ackermann, H.-W. (2007) 5500 Phages examined in the eletron microscope. *Arch. Virol.* ***152,*** 277–243.
13. Bradley, D.E. (1967) Ultrastructure of bacteriophages and bacteriocins. *J. Bacteriol.* ***31,*** 230–314.
14. Bradley, D.E. (1965) The preparation of specimen support films, in *Techniques for Electron Microscopy,* 2nd ed. (Kay, D.H., ed.), Blackwell Scientific Publications, Oxford, UK, pp. 58–74.
15. Cochlan, W.P., Wikner, J., Steward, G.F., Smith, D.C., and Azam, F. 1993. Spatial distribution of viruses, bacteria and chlorophyll *a*in neritic, oceanic and estuarine environments. *Mar. Ecol. Prog. Ser.* ***92,*** 77–87.
16. Gentile, M. and Gelderblom, H.R. (2005) Rapid viral diagnosis: role of electron microscopy. *New Microbiol.* ***28,*** 1–12.
17. Gregory, D.W. and Pirie, B.J.S. (1973) Wetting agents for biological electron microscopy. I. General considerations and negative staining. *J. Microsc.* ***99,*** 251–205.
18. Luftig, R.B. (1967) An accurate measurement of the catalase crystal period and its use as an internal marker for electron microscopy. *J. Ultrastruct. Res.* ***20,*** 91–102.
19. Ackermann, H.-W., Jolicoeur, P., and Berthiaume, L. 1974. Avantages et inconvénients de l'acétate d'uranyle en virologie comparée: étude de quatre bacteriophages caudés. *Can. J. Microbiol.* ***20,*** 1093–1099.
20. Huxley, H.E. and Zubay, G. (1961) Preferential staining of nucleic acid-containing structures for electron microscopy. *J. Biophys. Biochem. Cytol.* ***11,*** 273–296.
21. Børsheim, K.Y., Bratbak, G., and Heldal, M. (1990) Enumeration and biomass estimation of planktonic bacteria and viruses by transmission electron microscopy. *Appl. Environ. Microbiol.* ***56,*** 352–356.

Chapter 13

Phage Classification and Characterization

Hans-W. Ackermann

Abstract

Prokaryote viruses include 14 officially accepted families and at least five other potential families awaiting classification. Approximately 5,500 prokaryote viruses have been examined in the electron microscope. Classification has a predictive value and is invaluable to control experimental techniques and results. In describing viruses, the choice of methods depends on structure and taxonomical position of viruses. The study of isometric, filamentous, and pleomorphic viruses requires more detailed investigations than that of tailed species.

Key words: *Caudovirales, Myoviridae, Podoviridae, Siphoviridae, Corticoviridae, Tectiviridae, SH1,* from "Serpentine-Lake-Hispanica", STIV, "Sulfolobus-Icosahedral-Turreted-Virus", *Leviviridae, Cystoviridae, Inoviridae, Lipothrixviridae, Rudiviridae, Plasmaviridae, Fuselloviridae, Salterprovirus, Guttaviridae, Ampullaviridae, Bicaudaviridae, Globuloviridae.*

1 Introduction

1.1 General

"Phages" or bacteriophages are usually defined as viruses of bacteria. Most of them have heads and tails. The term "phage" has now to be broadened since several viruses of halophilic and methanogenic Archaea are tailed. The term "Prokaryote viruses" appears more appropriate since it also encompasses a number of viruses, mostly of hyperthermophiles, which do not resemble any conventional bacterial virus. In this chapter the term "phage" will be used for the sake of convenience.

The purpose of this chapter is to give an introductory overview of "phage" properties and taxa. No universal method for virus classification exists. Numerical taxonomy *(1, 2)*, which goes back to the eighteenth century French botanist Adanson who advocated the use of at least 60 unweighted characters, never

Martha R. J. Clokie, Andrew M. Kropinski (eds.), *Bacteriophages: Methods and Protocols, Volume 1: Isolation, Characterization, and Interactions, vol. 501*, © 2009 Humana Press, a part of Springer Science+Business Media
DOI 10.1007/978-1-60327-164-6_13 Springerprotocols.com

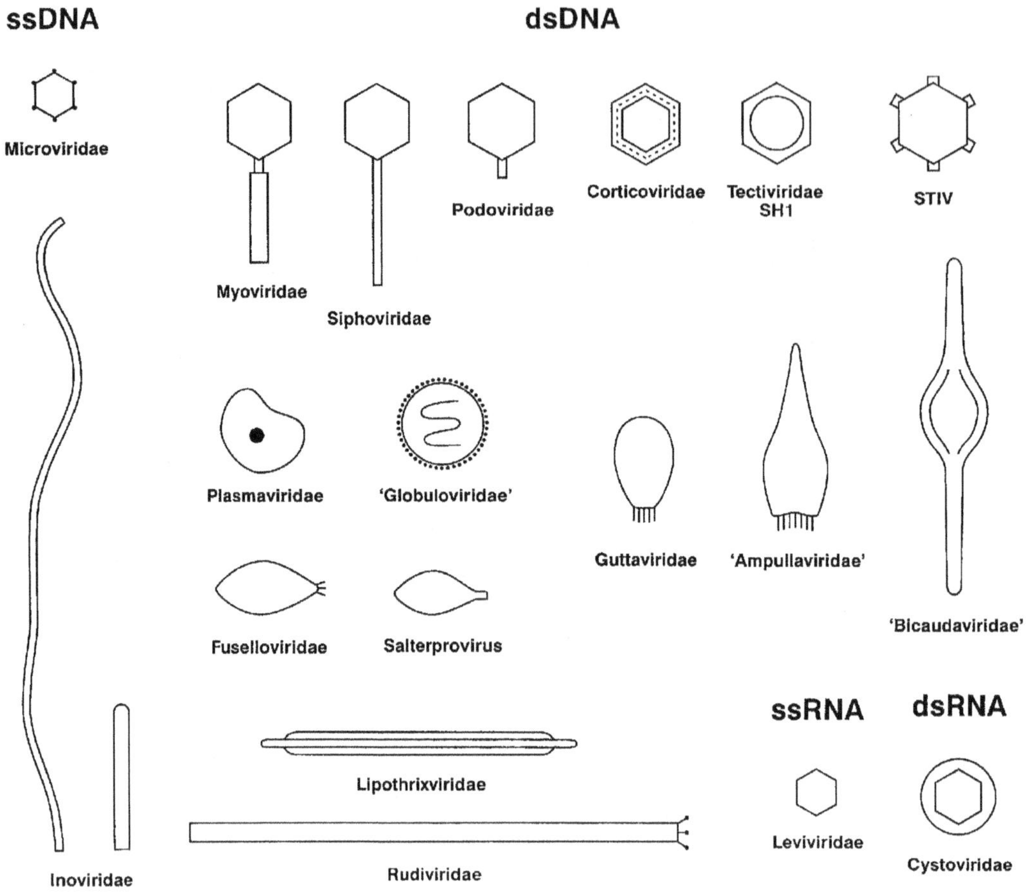

Fig. 13.1. Prokaryote Virus Morphotypes.

took hold in virology. The advent of computers did not change things. There is no accepted general computer program for classifying all bacteriophages, just for small groups such as T7-like phages or individual proteins. Thus, phage classification is still as much an art as a science.

1.2 History

Félix d'Hérelle, co-discoverer of bacteriophages, maintained that there was only one bacteriophage with many races, the *Bacteriophagum intestinale (3)*. After the advent of electron microscopy, the first phage classification was proposed in 1943 by Ernst Ruska as part of a general scheme for viruses *(4)*. He distinguished three morphological types of phages. In 1948, Holmes *(5)* classified phages as a suborder within the order Virales. They included one genus and 46 species. This system, essentially based on host range, never won acceptance. In 1962, Lwoff, Horne, and Tournier *(6)* proposed that viruses be classified by their nucleic acid type (DNA or RNA), capsid shape, presence or absence of an envelope, and number of capsomers. Tailed phages were given order rank and named *Urovirales*. A virus classification committee was founded in 1966 *(7)* and adopted the principle of classifying viruses by the properties of the virion and its nucleic acid instead of host range and pathogenicity. It developed into the ICTV or International Committee of Taxonomy of Viruses. The ICTV has so far issued eight reports *(8)*. More complete accounts of the development and problems of phage classification may be found elsewhere *(8–10)*.

1.3 The Present State

The present phage classification is derived from a scheme proposed by Bradley in 1967 *(11)*. It included six basic morphological types, exemplified by phages T4, λ, T7, ϕX174, MS2, and fd. The edifice of phage classification grew slowly over the years by accretion of new families and genera. It now includes one order, 14 families, and 37 genera *(8)*. The ICTV has adopted the polythetic species concept, meaning that a species is defined by a set of properties that may or may not all be present in a given member *(12)*. Phage classification is open-ended since new phages are discovered daily and the ICTV is far behind in its classification schedule.

2 Prokaryote Viruses

The number of phages of known morphology is over 5,500 *(13)*. Eubacterial phages occur in over 150 host genera, including anaerobes, endospore-formers, actinomycetes, cyanobacteria, mycoplasmas, and spirochetes. Archaeal viruses are found in ten genera of halophiles, methanogens, or extreme thermophiles (**Chapter 5**). Phages have colonized every conceivable habitat. Phages are virulent (lytic) or temperate (lysogenic); carrier states exist in some groups. The genomes of temperate phages persist within their hosts as integrated prophages or as plasmids. The ability to lysogenize is found in several different phage families.

3 Phage Taxa

Phages include viruses with double-stranded DNA (dsDNA; the vast majority), single-stranded DNA (ssDNA), single-stranded RNA (ssRNA), and double-stranded RNA (dsRNA; very rare). Most virions (96%) are tailed; other types (herein called CFP) are "cubic," filamentous, or pleomorphic (~200 representatives, less than 4%). The term "cubic" denotes cubic symmetry and icosahedral shape. Some types contain lipids in envelopes or internal constituents. The latter are invariably ether- and chloroform-sensitive.

Table 13.1
Overview of Prokaryote Viruses

Shape	Nucleic acid	Family	Genera	Particulars	Example	Members
Tailed	dsDNA (L)	*Myoviridae*	6	Tail contractile	T4	1,320
		Siphoviridae	7	Tail long, noncontractile	λ	3,229
		Podoviridae	4	Tail short	T7	771
Polyhedral	ssDNA (C)	*Microviridae*	4	Conspicuous capsomers	ϕX174	40
	dsDNA (C, S)	*Corticoviridae*	1	Complex capsid, lipids	PM2	3?
	dsDNA (L)	*Tectiviridae*	1	Double capsid, lipids, pseudo-tail	PRD1	19
	dsDNA (L)	SH1*		Double capsid, lipids	SH1	1
	dsDNA (C)	STIV*		Turret-shaped protrusions	STIV	1
	ssRNA (L)	*Leviviridae*	2	Poliovirus-like	MS2	39
	dsRNA (L, M)	*Cystoviridae*	1	Envelope, lipids	ϕ6	3
Filamentous	ssDNA (C)	*Inoviridae*	2	Long filaments, short rods	M13	67
	dsDNA (L)	*Lipothrixviridae*	4	Envelope, lipids	TTV1	7
	dsDNA (L)	*Rudiviridae*	1	Stiff rods, TMV-like	SIRV-1	3
Pleomorphic	dsDNA (C, S)	*Plasmaviridae*	1	Envelope, no capsid, lipids	L2	5
	dsDNA (C, S)	*Fuselloviridae*	1	Lemon-shaped, envelope, lipids?	SSV1	11
	dsDNA (L, S)	—	1**	Lemon-shaped, envelope	His1	1
	dsDNA (C, S)	*Guttaviridae*	1	Droplet-shaped	SNDV	1
	dsDNA (L)	Ampullaviridae*		Bottle-shaped, helical NC	ABV	1

(continued)

Table 13.1 (continued)

Shape	Nucleic acid	Family	Genera	Particulars	Example	Members
	dsDNA (C)	Bicaudaviridae*		Two-tailed, development cycle, helical NC	ATV	1
	dsDNA (L)	Globuloviridae*		Envelope, spherical, lipids, helical NC	PSV	1

C, circular; L, linear; M, multipartite; NC, nucleocapsid; S, supercoiled; —, no name; *, nonclassified; **, genus *Salterprovirus*. Members indicate numbers of phages examined by electron microscopy, excluding phage-like bacteriocins and known defective phages (based on computations from January 2006; from reference **13**).

CFP phage groups are mostly small and often have only a single representative.

Several interesting groups of archaeal viruses were discovered in recent times and are waiting for classification. The novel groups comprise viruses of Euryarchaeota and Crenarchaeota and are listed here for completeness (STIV type, *Ampullaviridae, Bicaudaviridae*, *Globuloviridae)*. The hosts of crenarachaeote viruses are hyperthermophiles of the genera *Acidianus, Pyrobaculum, Sulfolobus*, or *Thermoproteus*. These viruses occur in hot, acidic environments such as boiling hot springs, solfataras, and geysers. Host–virus relationships are generally unclear since these viruses are generally excreted and plaque assays are not available.

3.1 Tailed Phages, Order Caudovirales (dsDNA)

Tailed phages infect Eubacteria and Archaea and are probably extremely ancient, predating the separation of these superkingdoms. Virions contain dsDNA and have icosahedral or elongated heads. Tails are helical and generally provided with fixation structures (baseplates, spikes, fibers). There is no envelope. Particles adsorb to their hosts and infect them from the outside. The progeny phages are assembled via complex pathways, with phage DNA entering preformed capsids. Tailed phages are the most numerous and ubiquitous of all viruses and extremely varied in size and structure, DNA content and composition, genome structure, proteins, antigenic and biological properties. Virions are virulent or temperate. Tailed phages are divided into three families:

3.1.1 Myoviridae

Tails consist of a neck, a contractile sheath, and a central tube. Myoviruses tend to be larger than other groups and include some

Table 13.2
Particle Dimensions, Lipid Contents, and Nucleic Acid Molecular Weight

Family or genus	Capsid size, nm	Tail length, nm	Lipids, %	Nucleic Acid	
				%	MW, kb
Tailed phages	67	153	–	46	79
Range	30–160	10–800		30–62	17–498
Microviridae	27		–	26	4.4–6.1
Corticoviridae	60		13	14.3	9.0
Tectiviridae	63		15	14	15–16
SH1	70		+		30.9
STIV	74		–		17.6
Leviviridae	23		–	30	3.5–4.3
Cystoviridae	75–80		20	10	13.4
Inoviridae (*Inovirus*)	760–1950 × 7		–	6–21	5.8–7.3
(*Plectrovirus*)	85–250 × 7		–		4.5–8.3
Lipothrixviridae	410–2200 × 24–38		22	3	16–42
Rudiviridae	610–900 × 23		–		33–36
Plasmaviridae	80		11		12
Fuselloviridae	85 × 55		10		15
Salterprovirus	74 × 44				15
Guttaviridae	110–185 × 70–95		–		20
Ampullaviridae	230 × 75		–		24
Bicaudaviridae	130–150 × 56–70	260–760 × 24			62.7
Globuloviridae	100		+		28.3

kb, kilobases; MW, molecular weight; nm, nanometers; +, present, –, absent.

of the largest and most highly evolved tailed phages. (~1300 observations, 25% of tailed phages).

3.1.2 Siphoviridae

Tails are simple, noncontractile, flexible or rigid tubes. Siphoviruses are the most numerous of tailed phages (over 3,200 observations, 61%).

3.1.3 Podoviridae

Tails are short and noncontractile. Podoviruses may be more related to siphoviruses than to myoviruses. (~750 observations, 14.5%)

3.2 "Cubic" Phages

3.2.1 Microviridae (ssDNA)

Virions have no envelope and are very small and provided with 12 knob-like capsomers. Their DNA, like that of the filamentous *Inoviridae*, replicates via the "rolling circle" model and becomes temporarily double-stranded in the process. Phages of this type have been isolated against a variety of different hosts including enterobacteria, *Bdellovibrio, Chlamydia*, and *Spiroplasma*.

3.2.2 Corticoviridae (dsDNA)

Virions have a multilayered, lipid-containing capsid and circular DNA. The single certain representative known, *Alteromonas* phage PM2, was isolated from seawater.

3.2.3 Tectiviridae (dsDNA)

Virions consist of a proteinic outer shell and an inner lipoprotein vesicle. During infection or shaking with chloroform, the vesicle produces a "pseudo-tail" tube of ~60 nm in length, which acts as a DNA injection device. Tectiviruses infect enterics, bacilli, and *Thermus* species. Some tectiviruses are temperate and have a plasmid prophage state.

3.2.4 SH1; from "Serpentine-Lake-Hispanica" (dsDNA)

Particles have the same structure as tectiviruses and contain lipid. The only known isolate (SH1), is a still unclassified representative of this group that infects halobacteria. Tail-like tubes have not been shown *(14)*.

3.2.5 STIV, "Sulfolobus-Icosahedral-Turreted-Virus" (dsDNA)

A single representative is known (STIV). The virus infects the hyperthermophilic Archaeon *Sulfolobus* and is characterized by apical protrusions. STIV has structural relationships to tectiviruses, adenoviruses, and phycodnaviruses *(15)*.

3.2.6 Leviviridae (ssRNA)

Plus-stranded RNA and a resemblance to polioviruses characterize this group. Representatives of this family of phages infect enterobacteria, pseudomonads, acinetobacters, and caulobacters. Many leviviruses adsorb to plasmid-depended bacterial pili.

3.2.7 Cystoviridae (dsRNA)

Viruses of this type have an envelope, are the only phages with a segmented genome, and infect *Pseudomonas syringae* only. They resemble reoviruses in capsid structure, their segmented genome, and the presence of RNA polymerase.

3.3 Filamentous Phages

3.3.1 Inoviridae (ssDNA)

Viruses replicate via the "rolling circle" model and generate double-stranded intermediate DNA. Particles are excreted from infected cells without killing the host. Within the genus *Inovirus* (long filaments, ~ 40 viruses) are found viruses infecting enterics, pseudomonads, vibrios, xanthomonads, and *Thermus* species. The DNA of some of its members is able to integrate, in its double-stranded replication form, into the genome of some host bacteria. These temperate inoviruses are major virulence factors in *Vibrio cholerae*. Members of the genus *Plectrovirus* (short rods, ~ 15 members) are specific for mycoplasmas.

3.3.2 Lipothrixviridae (dsDNA)

Particles are long rods with a lipoprotein envelope. Virions inhabit hot springs and infect hyperthermophilic Archaea *(Acidianus, Sulfolobus, Thermoproteus)*.

3.3.3 Rudiviridae (dsDNA)

Virions are straight rods without envelopes and resemble the tobacco mosaic virus. They inhabit hot springs and infect the hyperthermophile, *Sulfolobus*.

3.4 Pleomorphic Phages

3.4.1 Plasmaviridae (dsDNA)

This family has only one member (*Acholeplasma* phage L2) and is found in mycoplasmas only. Viruses infect mycoplasmas by membrane fusion, are excreted by budding and are almost unique in the viral world as they consist of an envelope and a nucleoprotein granule (no capsid).

3.4.2 Fuselloviridae (dsDNA)

Virions are spindle-shaped and have no capsid. The genomes of at least some fusselloviruses persist as plasmids and integrated prophages. These viruses inhabit hot springs and infect *Acidianus, Pyrococcus, Sulfolobus.*

3.4.3 Salterprovirus (dsDNA)

Only two members of this spindle-shaped genus are known, His1 and His2. They were isolated from hypersaline lagoons, infect halobacteria of the genus *Haloarcula* and are morphologically related to fuselloviruses. It differs from them by its linear DNA with terminal proteins, lytic nature, and the presence of DNA polymerase. The virus constitutes a "floating genus" without family affiliation *(10a)*.

3.4.4 Guttaviridae (dsDNA)

Virions are droplet-shaped and have a unique beehive-like structure with a "beard" of fibers. They were found in a solfatara in New Zealand and are active on *Sulfolobus.*

3.4.5 Ampullaviridae (dsDNA)

Particles have a very characteristic and unique structure. They consist of a bottle-shaped mantle, a cone-shaped inner body, and a helical nucleocapsid. They have been observed in hot springs, and are active against *Acidianus (16)*.

3.4.6 Bicaudaviridae (dsDNA)

The salient feature of these viruses is an extracellular development cycle. Particles start as oval or arrow-shaped entities that contain a helical nucleocapsid and grow tail-like appendages at both ends. They are the largest of all archaeal viruses, occur in hot springs, and infect *Acidianus (17)*

3.4.7 Globuloviridae (dsDNA)

These viruses consist of a spherical, lipid-containing envelope and a helical nucleocapsid. Except that they contain DNA, they resemble paramyxoviruses. They occur in hot springs, and hosts are *Pyrobaculum* and *Thermoproteus* *(18)*.

4 Alternative Systems

The sudden availability of numerous genomic sequences has showed that enormous amounts of horizontal gene transfer have occurred during phage evolution. Tailed phages in particular appeared as a web of entities, a reticulate complex with a criss-cross pattern of relationships *(19, 20)*. The "phenetic" classification used by the ICTV was seen as unsatisfactory and should be replaced by a classification based on the phage genomics. Tailed phages were to be classified into "modes" composed of modules (genes or groups of genes) *(20)*. What was overlooked was that this would result in an astronomical number of combinations and that genomic relationships are an expression of the genome and in fact phenetic *(2)*. In a variant proposal, it was suggested to classify tailed phages by their head or tail modules *(21)*. Neither proposal was further developed and no phage classification of any extent was attempted. It now appears that even frequent horizontal gene transfer does not preclude tree-like hierarchical phylogenies of microorganisms *(22)*.

A different approach was used by Forest Rohwer and Rob Edwards (San Diego State University) by comparing phage genomes in a tree- and distance generating computer programs *(23)*. This resulted in the "Phage Proteomic Tree." A close examination of this tree shows a few potential inconsistencies: e.g., the tectivirus PRD1 grouped with φ29 – a member of the *Podoviridae*; and, phages λ (*Siphoviridae*) and P22 (*Podoviridae*) formed a single clade. Interestingly, many groups of the Proteomic Tree are identical to phage families or genera described by the ICTV. The Proteomic Tree is thus an independent confirmation of these phage groups. It is worth noting that the "Proteomic Tree" is not a classification scheme.

5 Criteria for Phage Characterization

5.1 General

The practical applications of phage classification are the identification of novel phages and prediction and control of experimental results. The ICTV uses any property of a virus for classification. Table 3 lists some 70 properties used for phages. Nature

Table 13.3
Criteria for Phage Classification

Parameter		Level of discrimination		Notes
		Tailed	CFP	
Nucleic acid				
Nature	DNA, RNA, ss, ds		Family	
Anatomy	Linear, circular, supercoiled		Family	
	Number of segments		Family	RD
	Terminal repeats	Genus, species	Family	
	Cohesive ends	Genus, species?		
	Single-stranded gaps	Genus?		RD
	Terminal proteins	Genus		RD
Physical	Molecular weight, length	Genus, species	Family, genus	
Chemical	Percent of particle	Genus	Family	RD
	G+C	Species	Genus, species?	
	Bases: ATGC, AUGC	Species	Genus, species?	RD
	Unusual bases			RD
	Sequence of bases			
	DNA-associated proteins			O
	DNA-associated sugars			O
Genome	Hybridization	Species		O
	Order of genes	Genus, species?	Family	
	Number of genes	Genus, species	Family, genus	
	Genome map	Genus, species	Family, genus	
	RE map	Species, strain		
	RE pattern	Species, strain		
	Number of nucleotides	Species, strain?	Family to species	
	Toxin-encoding genes	Species	Species	
Virion				
Morphology	Shape	Family	Family, genus	
	Capsid symmetry	Genus	Family	
	Number of capsomers	Genus	Family	RD
	Presence of envelope		Family	
	Fine structure	Family to species	Family, genus	

(continued)

Table 13.3 (continued)

Parameter		Level of discrimination		Notes
		Tailed	CFP	
	Dimensions	Genus, species	Family, genus	
	Internal proteins	Genus		RD
Physical	Weight	Species	Family	
	Buoyant density	Species, strain	Family	
Chemical	Protein %, lipid %		Family	
	Number or proteins	Species?	Family, genus	
	Protein molecular weight	Species?	Family, genus	
	Amino acid composition		Genus?	O
	Amino acid sequence	Species?		
	Lipid composition		Family?	RD
Serology	Neutralization	Species	Species	RD
Inactivation tests				
Physical	Heat, UV light	Species, strain	Family to species	
Chemical	Chloroform, ether	Species, strain	Family	
Replication and assembly	Assembly site		Family	RD
	Assembly pathway	Family, genus	Family	RD
	Concatemers	Genus		RD
	DNA circularization	Genus		RD
	DNA translocation	Genus		RD
	Integration	Genus		
	Nucleic acid packaging	Genus		RD
	Protein-primed	Genus		RD
	Rolling circle	Genus	Family	RD
	Recombination-repair	Genus		RD
	Transposition	Genus, species?		RD
Physiology	Adsorption site	Species, strain	Family	
	Adsorption velocity	Species, strain	Species, strain	
	Burst size	Species, strain	Species, strain	
	Complementation	Species		RD
	Conversion	Genus, species	Species	

(continued)

Table 13.3 (continued)

Parameter	Level of discrimination		Notes
	Tailed	CFP	
Helper function	Genus?		RD
Host range	Species, strain	Family to strain	
Latent period	Species, strain	Species, strain	
Virulent or temperate	Genus, species	Family	
Mutual exclusion	Species		RD
Plasmid stage	Genus, species	Family?	RD
Release		Family	
Thin section			RD
Transduction	Genus? Species		

CFP, cubic-filamentous-pleomorphic; O, obsolete; RD, rarely detemined; SDS, sodium dodecylsulfate; empty spaces in columns 3 and 4, not applicable or uncertain.

of nucleic acid and gross morphology are "more equal than others" and carry a heavy weight. They determine generally the order and family appurtenance of a virus. At the other end of the scale are inactivation tests, restriction endonuclease digestion patterns, and host range, which are identifiers for species and strains. Depending on the virus, a criterion can be high-level and low-level.

There is a huge gap between tailed and CFP phages. For example, the presence of dsDNA, common to all tailed phages, is no criterion for discrimination within this group, but is very useful for CFPs. Most available criteria are of low-level (species–strain) for tailed phages and of high-level (family) for CFPs. Some properties are rarely determined or even obsolete; e.g., the presence of DNA-associated sugars and amino acid composition of phage proteins. This does not mean that they are useless; it merely signifies that nobody investigates them anymore. It is regrettable that particle weights and the complete base composition of phage nucleic acids are now in this category (*See* **Volume 2 Chapter 2**).

It is out of question to (a) base phage identification on a single property and (b) determine all possible properties of a given phage. Further, identification must be cost-effective, especially in an industrial context. A property that is determined with great cost of labor and time in a few specialized laboratories will not easily become an important criterion. The selection of criteria has been discussed elsewhere *(9)*. The easiest, fastest, and least expensive way of phage identification is electron microscopy. It is comparable to Gram staining in bacteriology. Generally, it allows one in a few minutes to:

1. attribute phages to families.
2. suspect the presence of lipids.
3. identify novel phages.
4. attribute novel phages to known species, for example in the well-investigated tailed phages of enterobacteria, pseudomonads, or bacilli.
5. decide on the course of investigation which has to follow.
6. estimate DNA size from capsid diameters.

5.2 A Step-Wise Procedure

Any unknown phage should first be examined in the electron microscope (**Chapter 10**). There are three situations:

1. Phage is cubic, filamentous or pleomorphic: Investigate, as a minimum, nucleic acid type and the presence of lipids (sensitivity to ether and chloroform) if this suspected from phage morphology. Because of the large number of high-level criteria, CFPs must be more thoroughly investigated than tailed phages.
2. Phage is tailed: Investigate any property according to your fancy or the possibilities of your lab, in particular DNA molecular weight. There are no rules.
3. Phage is tailed and of industrial importance. Investigate biological properties that can be used for phage control (host range, adsorption velocity, latent period, burst size, inactivation by disinfectants).

Genome sequencing has brought enormous possibilities and novel problems. At the time of writing, to the author's reckoning, the genomes of approximately 160 tailed phages and 40 CFPs have been completely sequenced. In addition, there are complete sequences of prophages and a very few of phages of unknown morphology, plus a sea of genome fragments. The main problems, which hopefully will be solved, are poor identification of phages in genome databases and difficulties related to gene identification, genome mosaicism, genome comparison, and presentation of data:

1. Databases contain many misidentified, unpublished, or poorly annotated virus genomes *(24)*. This is particularly true of prophage genomes.
2. Gene function is often unknown. Gene identifications are frequently limited to morphogenetic and lysis genes.
3. The genomes of many (or possibly all) phages are mosaics, with parts of the puzzle occurring in the genomes of bacteria, eukaryote viruses, and eukaryotes.
4. Genomes and genes are painstakingly compared one-by-one.
5. There are gradual relationships and transition forms.
6. Genomic maps are presented in widely divergent ways, making comparisons difficult.

References

1. Sneath, P.H.A. and Sokal R.R. (1973) *Numerical taxonomy, The Principles and Practice of Numerical Classification*. W.H. Freeman, San Francisco, pp. 23 and 110.
2. Sneath, P.H.A. (1989) Analysis and interpretation of sequence data for bacterial systematics: the view of a numerical taxonomist. *System. Appl. Microbiol.* **12,** 15–31.
3. d'Hérelle, F., (1918) Technique de la recherche du microbe filtrant bactériophage *(Bacteriophagum intestinale). C.R. Soc. Biol.,* **81,** 1160–1162.
4. Ruska, H. (1943) Versuch zu einer Ordnung der Virusarten. *Arch. Ges. Virusforsch.* **2,** 480–498.
5. Holmes, F.O. (1948) Order Virales; the filterable viruses, in *Bergey's Manual of Determinative Biology,* 6th ed. (Breed, R.S., Murray, E.G.D., and Hitchens, A.P., eds.), Williams & Wilkins, Baltimore, pp. 1126–1144.
6. Lwoff, A., Horne, R.W., and Tournier P. (1962) A system of viruses. *Cold Spring Harbor Symp. Quant. Biol.* **27,** 51–62.
7. P.C.N.V. (1965) Proposals and recommendations of the Provisional Committee on Taxonomy of Viruses (P.C.N.V.). *Ann Inst Pasteur* **109,** 625–637.
8. Fauquet, C.M., Mayo, M.A., Maniloff, J., Desselberger, U., and Ball L.A., eds. (2005) Virus Taxonomy: VIIIth Report of the International Committee on Taxonomy of Viruses. Academic Press/Elsevier, London, pp. 35–116, 279–295, 443–446, 741–750.
9. Ackermann, H.-W. and DuBow, M.S. (1987) *Viruses of Prokaryotes,* Vol. 1, *General Properties of Bacteriophages,* CRC Press, Boca Raton, pp. 13–28 and 130–135.
10. Ackermann H.-W. (2005) Bacteriophage classification. In: Bacteriophages – Biology and Applications, eds. E. Kutter, A. Sulakvelidze. CRC Press, Boca Raton, FL, pp. 67–89.

10a. Bath, C., Cukalac, T., Porter, K., Dyall-Smith, M.L. (2006) His1 and His2 are distantly related, spindle-shaped haloviruses belonging to the novel virus group, Salterprovirus. Virology. **350**, 228–39.

11. Bradley, D.E. (1997) Ultrastructure of bacteriophages and bacteriocins. *Bacteriol. Rev.* **31,** 230–314.
12. Van Regenmortel, M.H.V. (1990) Virus species, a much neglected but essential concept in virus classification. *Intervirology* **31:**241–271.
13. Ackermann, H.-W. (2007) 5500 Phages examined in the electron microscope. *Arch. Virol.* **152:**277–243.
14. Porter, K., Kukkaro, P., Bamford, J.K.H., Bath, C., Kivelä, H.M., Dyall-Smith, M.L., and Bamford D.H. (2005) SH1: A novel, spherical halovirus isolated from an Australian hypersaline lake. *Virology* **335,** 22–33.
15. Rice, G., Tang, L., Stedman, K., Roberto, F., Spuhler, J., Gillitzer, E., Johnson, J.E., Douglas, T., and Young, M. (2004) The structure of a thermophilic archaeal virus shows a double-stranded DNA viral capsid type that spans all domains of life. *Proc. Natl. Acad. Sci. USA* **101,** 7716–7720.
16. Häring, M., Vestergaard, G., Rachel, R., Chen, L., Garrett, R.A., and Prangishvili, D. (2005) Independent virus development outside a host. *Nature* **436,** 1101–1102.
17. Häring, M., Rachel, R., Peng, X., Garrett, R.A., and Prangishvili, D. (2005) Viral diversity in hot springs of Pozzuoli, Italy, and characterization of a unique bottle-shaped archaeal virus, *Acidicanus* bottle-shaped virus, from a new family, the *Ampullaviridae. J. Virol.* **79,** 9904–9911.
18. Häring, M., Peng, X., Brügger, K., Rachel, R., Stetter, K.O., Garrett, R.A., and Prangishvili, D. (2004) Morphology and genome organization of the virus PSV of the hyperthermophilic archaeal genera *Pyrobaculum* and *Thermoproteus*: a novel virus family, the *Globuloviridae. Virology* **323,** 233–242.
19. Hendrix, R.W., Smith, M.C.M., Burns, R.N., Ford, M.E., and Hatfull, G.F. (1999) Evolutionary relationships among diverse bacteriophages and prophages: all the world's a phage. *Proc. Natl. Acad. Sci. USA* **96,** 2192–2197.
20. Lawrence, J.G., Hatfull, G.F., and Hendrix, R.W. (2002) Imbroglios of viral taxonomy: genetic exchange and failings of phenetic approaches. *J. Bacteriol.* **184,** 4891–4905.
21. Proux, C., van Sinderen, D., Suarez, J., Garcia, P., Ladero, V., Fitzgerald, G.F., Desiere, F., and Brüssow H. (2002) The dilemma of phage taxonomy illustrated by comparative genomics of Sfi21-like *Siphoviridae* in lactic acid bacteria. *J. Bacteriol.* **184,** 6026–6036.
22. Ge, F., Wang, L.-S., and Kim J. (2005) The cobweb of life revealed by genome-scale estimates of horizontal gene transfer. *PloS Biol* **3:**e316 (8 pp.).
23. Rohwer, F. and Edwards, R. (2002) The Phage Proteomic Tree: a genome-based taxonomy for phage. *J. Bacteriol.* **184,** 4529–4535.
24. Adams M.J. and Antoniv J.F. (2006) DPVweb: a comprehensive database of plant and fungal virus gene and genomes. *Nucleic Acids Res.* **34,** D382–385 (Database Issue).

Chapter 14

Phage Host Range and Efficiency of Plating

Elizabeth Kutter

Abstract

The host range of a bacteriophage is defined by what bacterial genera, species and strains it can lyse; it is one of the defining biological characteristics of a particular bacterial virus. Because of host factors such as masking by O antigens that affects injection and the presence of restriction endonucleases, the relative efficiency of plating (EOP), that is, the titer of the phage on a given bacterial cell line compared to the maximum titer observed, may vary considerably. This chapter describes rapid procedures for determining the host range and relative EOP on each host of any phage.

Key words: Bacteriocin, plaque, host range, broad host range, efficiency of plating, EOP, *Pseudomonas*, *Escherichia coli*, sewage, lysis from without, CEV1, spot test, ECOR, Felix d'Herelle Reference Center for Bacterial Viruses.

1 Introduction

This chapter expands on the concepts and techniques in the chapters on phage plating and phage typing to look at the broad question of a phage's relative ability to infect a wide range of different host strains.

The isolation of each new phage involves the use of one specific host strain on which an environmental sample, such as sewage, water, soil or feces, is spread, forming plaques each of which should come from a single phage particle. While there may well be millions or billions of phage in that sample, the only ones that will be seen are those that can form plaques on that particular strain. Occasionally samples contain enough phage against a particular bacterium to form plaques when plated directly using that strain as host, or even to require some dilution before individual plaques can be seen. However, much more frequently the

Martha R. J. Clokie, Andrew M. Kropinski (eds.), *Bacteriophages: Methods and Protocols, Volume 1: Isolation, Characterization, and Interactions, vol. 501*, © 2009 Humana Press, a part of Springer Science+Business Media
DOI 10.1007/978-1-60327-164-6_14 Springerprotocols.com

concentrations are low enough that phage against a particular host can only be obtained by extensive enrichment techniques, first killing or filtering out most bacteria in a sample (or a liquid extraction of a solid sample (**Chapter 1**)) and then adding concentrated nutrients along with several potential target bacteria of interest, incubating at optimal growth temperature for those bacteria for one or more days, lysing the culture with chloroform and plating out samples using each of these potential target bacteria as hosts (**Chapter 2**).

For example, it is possible to isolate phages which infect virtually all *Escherichia coli* and*Pseudomonas* strains using half a liter of clarified sewage enriched with 1/10 volume of 10 × tryptic soy broth (TSB) medium. However, far more phages attacking *E. coli* O157 are obtained by enriching from 10 g of fecal material macerated in broth than are obtained from sewage. For example, in studies of the distribution of phage attacking O157 in samples taken from different pens in a cattle feedlot, none were seen without enrichment, but increasingly extensive enrichment techniques finally yielded phage against O157 from 39/60 samples, while 58/60 yielded phage against *E. coli* B *(9)*. A variety of approaches for isolating phages under different circumstances have been described by Carlson, who also details a range of other techniques for working with phages *(1)*.

Some phages obtained in this fashion may have a very narrow host range, whereas for others it may be broad. In general, when phages are obtained from the ATCC or from the Felix d'Hérelle Reference Center for Bacterial Viruses (**Chapter 16**; http://www.phage.ulaval.ca/index.php), they are supplied with a host that was deposited along with the phage. It is seldom made clear whether this is simply the host on which they happened to be isolated, an active host selected from some arbitrary set of bacteria under study, or a common lab strain on which they grow well, such as *E. coli* B or K12 for a large fraction of T4-like phages.

Certain phages display broad specificity, infecting or lysing, for example, all tested *Pseudomonas aeruginosa* strains (phage PB1, D. Bradley, personal communication), all tested fluorescent pseudomonads (φS1) *(2)* or salmonellae (Felix O1) *(3)*, while others have a far more restrictive host range *(4)*.

1.1 Developing the Relevant Bacterial Collections

The first step in determining the host range for a given phage is to have a large collection of bacteria to test. Ideally, this is a well-characterized collection, such as the *E. coli* collection of reference (ECOR) *(5, 6)*. Most frequently, however, it is simply a large collection of bacteria collected for some specific purpose, such as a collection of pathogenic strains from a particular set of patients or hospitals. In the latter case, one often has no idea of how varied the strains in question are, or how representative they are of bacteria of that species. Although it is useful to apply the prevalent typing scheme for that particular bacterium, e.g., Pulsed Field Gel Electrophoresis (PFGE), Restriction Fragments Length Polymorphism (RFLP), Multi Locus Sequence Typing (MLST) or sequence data from the 16S rRNA, this data will not give information regarding their susceptibility to bacteriophage infection. Therefore, testing them with a set of typing phages (**Chapter 38**) can be very helpful in determining their breadth and relatedness. It can also let one choose a subset of the strain collection that can be used for initial host-range testing of new phages. Here, one selects one or more representatives of each "phage type," as determined by a set of about 6–8 typing phages, and tests each new phage of interest against that set of hosts.

1.2 Spot Testing Exploration of Host Range

The second step is to spot test each of the phages on the selected set of hosts. It is easiest to test multiple phages at the same time; using square plates, up to 36 phages can be tested simultaneously. If one is using a set of hosts that have already been characterized using typing phages, then one can include the set of typing phages as controls on each plate and be sure that the strains are behaving as predicted.

The technique here is to grow up each of the bacterial strains to be tested either as a fresh overnight culture or one grown up to mid-log phase and kept on ice, depending on what works well for that species of bacteria. For *E. coli* and *Pseudomonas* we have used Difco TSB medium as well as TSB plates and top agar, prepared as described in **Chapter 7**.

1.3 Determining the Efficiency of Plating (EOP) of Phage on the Susceptible Strains

Many phages, such as T4, have an efficiency of plating of 100% in optimum conditions— every phage particle attaching to a host cell can enter and make a plaque on the appropriate strains under ideal conditions. However, a number of factors can affect plating efficiency, including the specific host strain, so it is very important to also check the relative efficiency of plating on various susceptible hosts.

2 Materials

2.1 Developing the Relevant Bacterial Collections

See discussion above and **Chapters 7 and** 16. You should have appropriate media, Petri dishes and growing conditions for the bacteria in question.

2.2 Spot Testing Exploration of Host Range

1. A 10-μl micropipette with appropriate tips
2. Two plates containing bottom agar per bacterial strain marked on the bottom with the strain name and date. (Square plates, as seen in **Fig. 14.1**, are useful if you are doing many phages; if you are doing just a few, round plates with a grid marked on the back suffice.)
3. Two glass plating tubes per host; these may already contain the top agar (4 ml for square plates, 3 ml for round plates), which is then melted thoroughly by placing them in a beaker of water and bringing it to a boil.
4. Heating block set at 46°C in which you place plating tubes after they have been boiled or tubes for freshly boiled top agar.
5. Plating bacteria for each strain to be tested. For large-scale host-range studies, we routinely grow 5-ml overnight cultures in large test tubes with gentle shaking.
6. A rack with an aliquot of each phage at a concentration of about 10^8 plaque-forming units (pfu) per ml in a microcentrifuge (Eppendorf) tube. These are set up such that each can be moved to an adjacent row after it has been used, avoiding potential problems of getting confused as to where you are.

2.3 Determining the Efficiency of Plating (EOP) of Phage on the Susceptible Strains

Materials are as above and in **Chapter 7** and **16**.

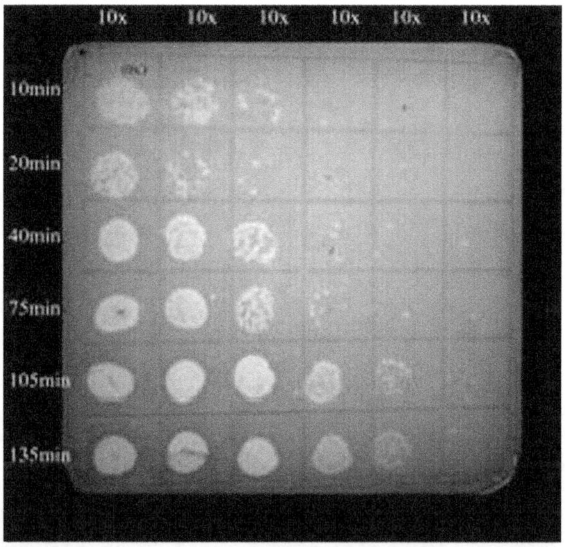

Fig. 14.1. Spotting of 6 sets of serial phage dilutions on a square plate. This method is used to systematically compare the plating efficiencies of various phages on particular host strains as well as, here, to immediately visualize phage titers over major segments of a phage infection experiment.

3 Methods

3.1 Developing the Relevant Bacterial Collections

See discussion above and **Chapters** 7 and 16.

3.2 Spot Testing Exploration of Host Range

1. Bring the top agar cautiously to a boil in the beaker of water or in a microwave oven, stopping periodically to mix it gently. (It is important that all microcrystals be dissolved, or the resulting uneven texture will make counting plaques difficult.)
2. Put the tubes in the heating block, add agar, and wait at least 10 min for the agar to cool to the 46° block temperature before proceeding to adding the bacteria and pouring the lawn to avoid killing the bacteria.
3. Add 0.1–0.3 ml of the first bacterium to be tested to each of the first two plating tubes, flick gently with the finger several times to mix, and immediately pour on the plates one at a time, swirling a bit for even distribution. (Check for your system what amount of bacteria works best to produce a good lawn.)
4. Allow top agar to set at least 15 min and then sequentially spot 10 μl of each of the phages on the plates. Use a fresh pipette tip for each spot, and consider using filtered tips; you don't want to take any chance of contaminating your phage samples with any of the bacteria that you are using, which could lead to confusion.
5. Spot testing should be done in duplicate, on separate plates, in case one of your plates has some problem.
6. Put the phage in exactly the same order on each plate to facilitate analysis and reduce errors.
7. The next day, examine and classify the spots. A common system for assessing the success of infection by the phage is:

   +4 complete clearing
   +3 clearing throughout but with faintly hazy background
   +2 substantial turbidity throughout the cleared zone
   +1 a few individual plaques
   0 no clearing – but you may see a spot where the pipette tip touched the agar

It is also a good idea to take digital photographs of the plates for future consultation.

In the case of the ECOR collection, a phylogenetic tree has been determined based on analysis of the electrophoric and/or functional patterns of 35 different enzymes (**Fig. 14.2**). In preliminary ECOR host-range tests of 59 phages from worldwide

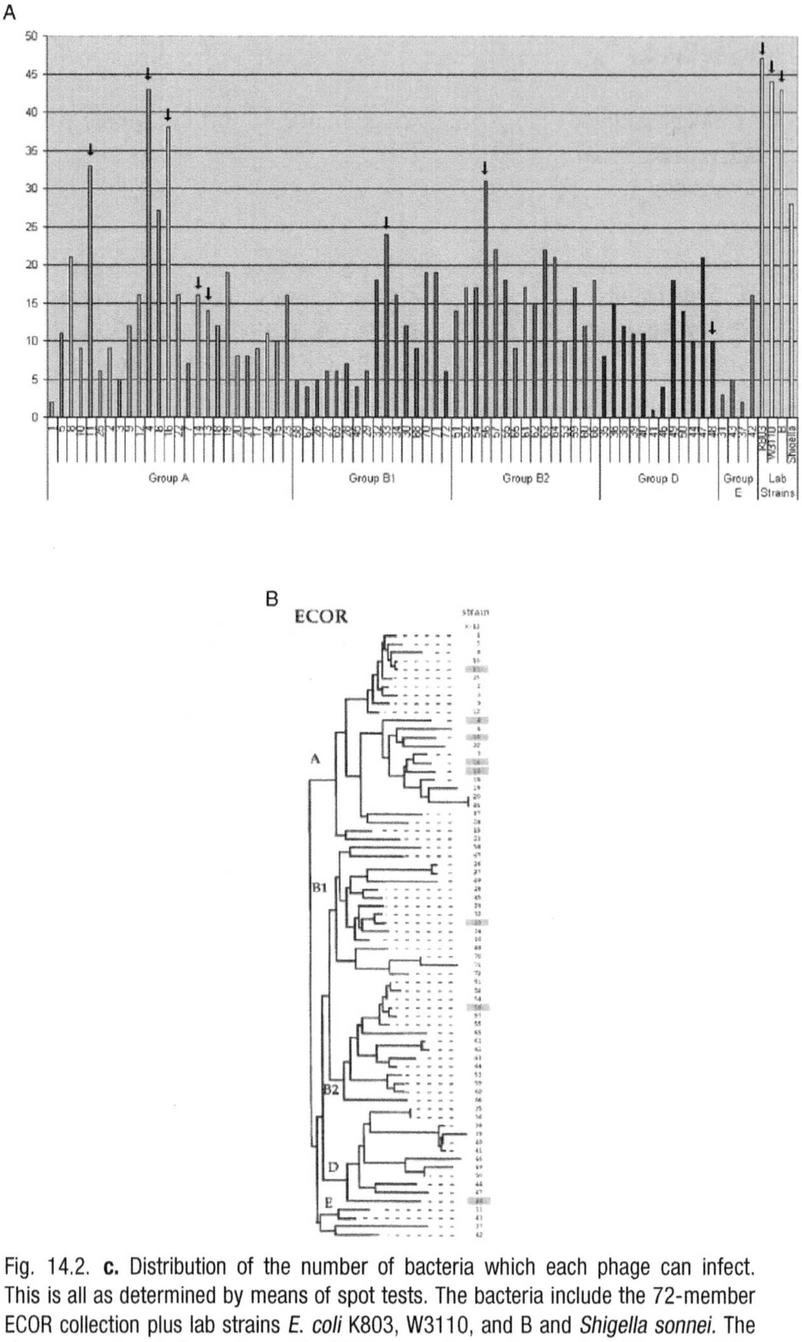

Fig. 14.2. **c.** Distribution of the number of bacteria which each phage can infect. This is all as determined by means of spot tests. The bacteria include the 72-member ECOR collection plus lab strains *E. coli* K803, W3110, and B and *Shigella sonnei.* The phages are almost all members of the T4 family and come from various labs around the world, including therapeutic phages used in the Republic of Georgia. **b.** Genetic relationships of the 72 ECOR strains, based on 35 enzyme loci. Different phylogenetic groups are designated by the letters A through E. Adapted from Selander et al. *(6)*. **a.** Number of phage from a given set that can infect each member of the ECOR collection, displayed in the same order as the phylogenetic alignment presented in **Fig. 14.2b**. The arrows (and the highlighting in **b**) indicate strains that can be infected by one particular phage, CEV1.

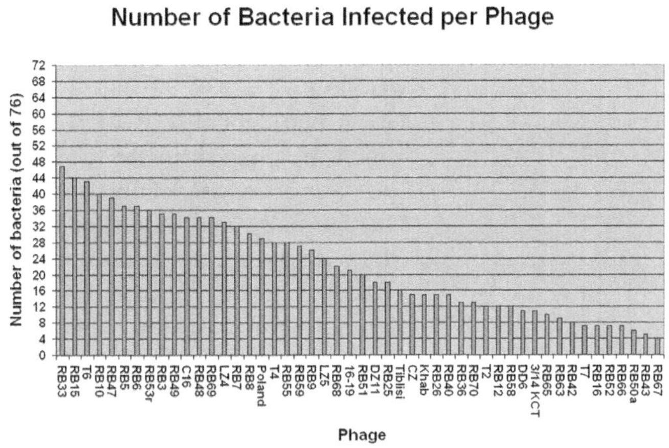

Fig. 14.2. (continued).

sources, we found that there was a great deal of variability in host sensitivity (as determined by a +3 or +4 level of clearing in the spot tests). There is very little correlation between the patterns of sensitivity and phylogenetic relationship, as seen in **Fig. 14.2a**, where the number of these phages to which each host is sensitive is plotted in the same pattern as the phylogenetic tree. Different phages hit very different numbers of bacteria in this collection, as seen in **Fig. 14.2c**. The arrows in **Fig. 14.2a** and highlighting in **Fig. 14.2b** mark strains that are susceptible to phage CEV1, a T4-like phage we isolated from the feces of a flock of sheep totally resistant to inoculation with *E. coli* O157:H7 *(7)*. As can be seen, only about 10% of the ECOR strains are sensitive to CEV1, but those are distributed among all of the main phylogenetic groups of *E. coli*. CEV1, like most of the phages in this collection, can also infect the common lab *E. coli* strains K12 and B, both of which are lacking O antigens and are very good for isolating phages with relatively broad host ranges *(8, 9)*, as well as a lab Shigella strain.

Interestingly, we found that it was easy to isolate phage using enrichment techniques from 500 ml of local sewage even against the bacteria hit by only one member of our classical collection; however, those phages generally had very narrow host ranges. It thus appears that the surfaces of some *E. coli* strains have changed very substantially, presumably due to environmental pressures, as compared to the rest of the cell.

3.3 Determining the Efficiency of Plating (EOP) of Phage on the Susceptible Strains

Many phages, such as T4, have an efficiency of plating of 100% in optimum conditions – that is, every phage particle attaching to a host cell can enter and make a plaque on the appropriate strains under ideal conditions. However, a number of factors can affect plating efficiency, including the specific host strain, so it is very important to also check the relative efficiency of plating on various susceptible hosts.

1. Check whether exponential or overnight cultures give better phage counts for the general system with which you are working.
2. Grow plating cultures of the various strains of bacteria for which the phage you are testing are positive in the host-range spot-test assay above, and also for the host strain(s) on which you normally grow and titer your phages.
3. Make a series of 10-fold dilutions in Eppendorf tubes of each of the phages to be tested, using a fresh pipette tip for each step and going down to a predicted titer of 100 phage per ml on the basis of the titer on the host on which you grew that phage.
4. Prepare your plates for spot testing as above; here, square plates are particularly useful. We normally check 6 phages per plate, using each row for spotting the last 6 dilutions of one phage, and make duplicate plates for each set. (Again, use the same order of phage on the plates for each of the bacteria to facilitate accurate comparisons.)
5. Count your phage. For phages that make large plaques, check the plates after a few hours and count as soon as the plaques are clearly visible. For phages that make small plaques, generally up to 20 plaques/spot can be counted the next day. This dilution series should give you reasonable data even if your phage plates 10 times as efficiently on a particular strain as on your original host; make and plate further dilutions if your last one is still not countable. A range of up to 10-fold difference in titer is not uncommonly seen over a group of hosts. Multiple orders of magnitude differences bring into question whether a given strain is in fact meaningfully susceptible to the particular phage.
6. For more precise plaque counts and comparison of the plaque sizes and morphologies on each strain, plate the right dilution you have now determined for each phage on individual round plates (**Chapter 7**).

The importance of such EOP testing rather than relying on the simple single-concentration spot test was again emphasized in recent work with a new *Pseudomonas* phage. This phage made clear, +4 spots on all of the hosts tested in an extensive initial panel. However, subsequent EOP studies showed that while clear lysis was obtained in the first two dilutions, no lysis at all was observed at further dilutions, and no individual plaques were ever seen.

The conclusion here is that the phage were probably able to bind to this strain and cause bacterial death, through an abortive infection and/or via lysis from without, but were not able to produce sufficient progeny phage in this strain to form plaques. For example, some prophages have specific systems that lead to the breakdown of cellular energetics when a variety of lytic phages infect the cell, thus committing suicide but protecting the remaining host population from infection. This is what leads to the totally abortive infection of *E. coli* CR63(λ) by T4 rII mutants, and to the absence of lysis inhibition and enlarged plaques when these rII mutants infect *E. coli* B, which carries a different prophage *(10)*. Alternatively, some unrelated bacteriocin could be causing the lysis. When the block is due to an incompatible restriction-modification system, plaques are often seen but at a level several orders of magnitude lower. The few infecting phage that manage to get modified rather than restricted now carry the new host's modification and thus can make a normal-looking plaque.

On page 460 Carlson *(1)* gives instructions for carrying out and interpreting a killer-titer assay, which can be used to determine the actual concentration of phage particles lethal to the host when the phage appears to enter the cell but produces too few particles to make a plaque.

References

1. Carlson, K (2005) Appendix: Working with Bacteriophages. In Kutter, E. and A. Sulakvelidze. Bacteriophages: Biology and Applications. CRC Press 2005.
2. Kelln, R. A. & Warren, R. A. (1971). Isolation and properties of a bacteriophage lytic for a wide variety of pseudomonads. *Can. J. Microbiol.* **17:** 677–682.
3. Kuhn, J., Suissa, M., Chiswell, D., Azriel, A., Berman, B., Shahar, D., Reznick, S., Sharf, R., Wyse, J., Bar-On, T. *et al.* (2002). A bacteriophage reagent for *Salmonella*: molecular studies on Felix 01. *International Journal of Food Microbiology* **74:** 217–227.
4. Bigby, D. & Kropinski, A. M. (1989). Isolation and characterization of a *Pseudomonas aeruginosa* bacteriophage with a very limited host range.*Can. J. Microbiol.* **35:** 630–635.
5. Ochman, H. and R. K. Selander, (1984). Evidence for clonal population structure in *Escherichia coli.* Proc. Natl. Acad. Sci. U.S.A. **157:** 690–693.
6. Selander, R. K., Caugant, D. A. and Whittam, T.S. (1987). Genetic Structure and Variation in Natural Populations of *Escherichia coli.* In (F.C. Neidhardt, Editor in Chief) *Escherichia coli* and*Salmonella Typhimurium* CRC Press.
7. Raya, R.R., Varey, P., Oot, R.A., Dyen, M.R., Callaway, T.R., Edrington, T., S., Kutter, E.M., and Brabban, A.D. (2006). Isolation and Characterization of a New T-Even Bacteriophage, CEV1, and Determination of Its Potential To Reduce Escherichia coli O157:H7 Levels in Sheep. Appl Environ Microbiol. **72:** 6405–64103.
8. Chibani-Chennoufi, S., Sidoti, J., Dillmann, M.-L., Bruttin, A., Kutter, E., Krisch, H., Sarker, S., and Brüssow, H. Isolation of Bacteriophages from the Stool of Pediatric Diarrhea Patients. J. Bacteriol. **186:** 8287–8294.
9. Oot, R.A., Raya, R.R., Callaway, T.R., Edrington, T.S., Kutter , E.M. and Brabban, A.D.. (2007) Prevalence of Escherichia coli O157 and O157:H7-infecting bacteriophages in feedlot cattle feces. Letters in Applied Microbiology. **45:** 445–453.
10. Paddison, P., Abedon, S.T., Dressman, H..K., Gailbreath, K., Tracy, J., Mosser, E., Neitzel, J., Guttman, B., and Kutter, E. (1998). The Roles of the Bacteriophage T4 *r* Genes in Lysis Inhibition and Fine-Structure Genetics: A New Perspective. Genetics **148:** 1539–1550.

Chapter 15

Measurement of the Rate of Attachment of Bacteriophage to Cells

Andrew M. Kropinski

Abstract

Practical methods are described for studying the adsorption rate of bacteriophages to cells and the interaction between these viruses and their surface receptors.

Key words: Adsorption, neutralization, inactivation, receptor, T4, M13, chloroform, Tectiviridae, lipopolysaccharide.

1 Introduction: Phage Adsorption to Bacterial Cells

The determination of the rate of attachment in bacterial cells is fundamental to studies on phage ecology for estimating the impact of predator (phage) on prey (host cell) populations. It is also important for assuring synchronicity of phage infection in the one-step growth experiment and in studies of transcription in phage-infected cells. The following techniques are also useful in defining the nature of the cellular receptor.

In 1931, Krueger [*(1)* and **Fig. 15.1**] demonstrated that the binding of phage particles to bacterial cells, living or dead, followed first-order kinetics (as illustrated in the **Fig. 15.1** by the line with the squares) that could be defined by the subsequent equation:

$$k = \frac{2.3}{\mathrm{B}t} \log \frac{\mathrm{Po}}{\mathrm{P}}$$

Martha R. J. Clokie, Andrew M. Kropinski (eds.), *Bacteriophages: Methods and Protocols, Volume 1: Isolation, Characterization, and Interactions, vol. 501*, © 2009 Humana Press, a part of Springer Science+Business Media
DOI 10.1007/978-1-60327-164-6_15 Springerprotocols.com

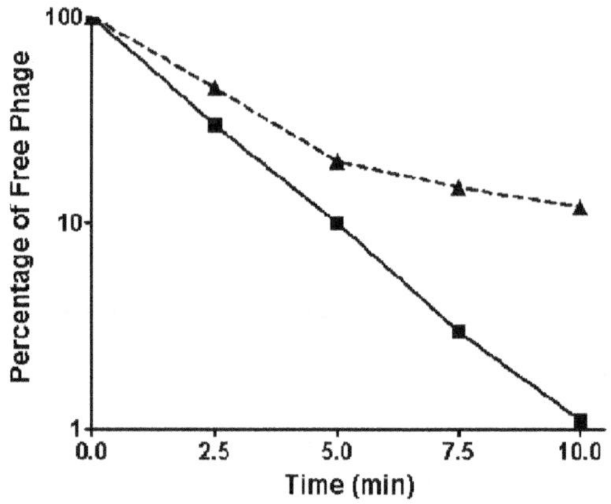

Fig. 15.1. Two examples of the adsorption kinetics of bacteriophages to cells: ■__■ in which all the phages follow first-order kinetics; and, ▲__▲ in which a slower adsorbing subfraction exists.

Where *k* is the adsorption rate constant, in ml/min; B is the concentration of bacterial cells; *t* is the time interval in which the titer falls from Po (original) to P (final).

This work was extended by Schlesinger *(2)* to show that *k* was independent of the concentration of bacterial cells or phage over wide ranges of concentrations. He also demonstrated that phage preparations were not homogenous with respect to their adsorptive properties. In all cases the adsorption assays reveal the presence of a slower or nonadsorbing subfraction (for example the line with triangles in the above figure). While this usually represents less than 5% of the phage (e.g., coliphages T4 and T7) it can also be a sizeable fraction of the population, as in the case of *Pseudomonas* phage ϕS1 *(3)*.

A large number of factors, including the growth phase of the cells, presence of salts (particularly divalent ions), organic compounds (phage T4 requires tryptophan for adsorption), agitation, temperature, cell size, and density of surface receptors all influence *k (4, 5)* (**Chapter 14 and 15**). It has been pointed out by Kasman et al., *(6)* that coliphage T4 which recognizes several hundred receptor sites per cell displays a *k* of 2.4×10^{-9} ml/min, while M13 which only binds to the tip of F pili, at most will find two or three receptors per cell, and it's adsorption rate constant is concomitantly low (3×10^{-11} ml/min).

In the following protocol the dilution into cold broth containing chloroform has a triple affect. The cold often abolished reversible binding, while the dilution factor slows adsorption rate by a factor of twenty. In addition, the chloroform treatment kills the bacterial cells thus effectively removing infective centers (i.e., phage-infected cells which otherwise would produce plaques). Needless to say, this technique will not work with chloroform-sensitive viruses such as members of the *Tectiviridae*. In these cases it is recommended that one dilute the phage–host mixtures into cold broth, remove the phage–cell complexes by centrifugation ($10,000 \times g/10$ min), and titer the supernatant. In addition, if one is studying the adsorption rate of a phage to multiple hosts it is useful to ascertain whether these cells contain prophages active on the titration host, since the latter may interfere with determination of accurate plaque counts for the phage being studied.

2 Materials

1. 1.0, 0.1, and 0.01 ml micropipetters such as Finnpipette or Eppendorf pipetters (Fisher Scientific). These should be periodically recalibrated.
2. Sterile tips.
3. Sterile capped 13×100 mm test tubes.
4. Two sterile 125 ml Erlenmeyer flasks.
5. Heating block or waterbath set at 48°C.
6. Water bath shaker set at the growth temperature of your bacterium.
7. Visible spectrophotometer set at 650 nm.
8. Overnight culture of your bacterium grown in medium of choice supplemented with 1–10 mM $CaCl_2$, subcultured and grown to mid-log phase.
9. Phage diluted in growth medium (plus Ca^{2+}) to a titer of $1 - 3 \times 10^5$. Prewarm to the assay temperature just prior to the experiment.
10. Bucket or styrofoam box containing crushed ice.
11. Agar plates and overlays (**Chapter 7**).

3 Method

1. Add 0.95 ml of medium to 12 capped 13×100mm test tubes, and number these A1–A10 and C1 and C2 (**Note 1**).
2. Using a Pasteur pipette add 3 drops of chloroform to each tube and then place them in numeric order on ice.
3. Chill for at least 10 min before proceeding.
4. Take a mid-log phase culture of your bacterium and dilute it to give at least 10 ml with an OD_{650} of 0.1–0.2 (**Note 2**).
5. Label the two 125 ml flasks "A" and "C", and add 9 ml of the cell suspension to flask "A", and 9 ml of medium to flask C (**Note 3**).
6. Place the flasks in a waterbath shaker at 60 rpm and allow a temperature equilibration period of 5 min before adding the phage (**Note 4**).

7. Place the remainder of the cell suspension on ice.
8. At t = 0 add 1 ml of warmed phage suspension with a titer of $1 - 3 \times 10^5$ pfu/ml to flask A and start the timer (**Note 5**).
9. Immediately afterwards add 1 ml of phage to flask C.
10. At 1 min interval remove 0.05 ml aliquots from flask A to the chilled tubes.
11. Mix the contents vigorously for 10 s using a vortex-type mixer and place the tubes back on ice.
12. Continue to remove samples from flask A until t = 10 (**Note 6**).
13. Remove 0.05 ml samples from flask C to tubes C1 and C2. Mix and place on ice.
14. Sequentially, and with fresh pipette tips, remove 0.1 ml from each chloroformed tube to tubes of molten overlay medium, add host cells, mix and pour over the surface of plates (for full details of this aspect of the experimental protocol *see* **Chapter** 7).
15. Dilute the bacterial cell suspension to 10^{-6} and plate 0.1 ml aliquots from the 10^{-4} to 10^{-6} dilutions on appropriate solidified medium so as to obtain isolated colonies.
16. Incubate all plates at the appropriate incubation temperature and after a suitable incubation period count the plaques (step 14) and the colonies (step 15). Record your data including dilutions, volumes plated, and counts.
17. From the results of steps 15 and 16 you will be able to calculate the number of bacteria per $OD_{650\,nm}$ (cfu/OD). This will be useful in subsequent experiments.
18. Using a sheet of 2 or 3 cycle semi-log paper, label the X axis with the time coordinates 0–10 min., and the Y axis with the number of plaques (**Note** 7).
19. For t = 0 average the number of plaques on the plates corresponds to C1 and C2.
20. Fill in the data, and using a transparent ruler draw the most logical line (or lines) through the points (**Note 8**). Determine the time required for a decrease in 50% of the unadsorbed phage.
21. The adsorption rate constant (k) can be calculated from the bacterial concentration determined in steps 15–17 (N.B. the bacterial density in flask "A" is actually 90% of this value), and the time required to achieve 50% adsorption according to the formula given above.

4 Notes

1. The medium should be that which you use to grow your bacterium and propagate your phage. If you are starting a new project I recommend that you use the medium

recommended for the propagating the host bacterium (see for example American Type Culture Collection (ATCC at http://www.atcc.org/) or Deutsche Sammlung von Mikroorganismen und Zellkulturen (DSMZ at http://www.dsmz.de/). N.B. many phages require 1–10 mM divalent ions (such as Ca^{2+} or Mg^{2+}) for optimal adsorption and the medium should be supplemented accordingly *(4, 7)*.

2. After the first trial run you will know what changes to make to the cell density or sampling time points.
3. Extremely accurate pipetting and uniformity of approach is required with this procedure in order to obtain interpretable results.
4. It is recommended that you do not place the cell suspension on ice or dilute with cold medium since this may affect the ability of the bacteria to support phage adsorption.
5. This titer is based upon phage which produce medium to small plaques. i.e., where you can easily and accurately count 100–300 plaques on a plate. For phage like T7 which produce very large plaques you will require a lower initial titer or some means of minimizing plaque size.
6. I would not recommend taking samples after 10 min because you may run into problems with phages with short latent periods.
7. It is unnecessary to calculate the titer.
8. It is probably unnecessary to run statistical analyses on the data points.

References

1. Krueger,A.P. 1931. The sorption of bacteriophage by living and dead susceptible bacteria. Journal of General Physiology *14*:493–503.
2. Schlesinger, M. 1965. Adsorption of phages to homologous bacteria. II. Quantitative investigations of adsorption velocity and saturation. Estimation of the particle size of the bacteriophage. Z. Hyg. Immunitaetsforsch. 114:149–160 (1932). Translated from German., p. 26–36. *In* G.S. Stent (Ed.), Papers on bacterial viruses. Little Brown and Company, Boston, Massachusetts.
3. Kelln,R.A. and R.A.Warren. 1971. Isolation and properties of a bacteriophage lytic for a wide range of pseudomonads. Canadian Journal of Microbiology *17*:677–682.
4. Adams,M.D. 1959. Bacteriophages. Interscience Publishers, Inc., New York.
5. Delbrück,M. 1940. Adsorption of bacteriophages under various physiological conditions of the host. Journal of Physiological Chemistry *23*:631–642.
6. Kasman,L.M., A.Kasman, C.Westwater, J.Dolan, M.G.Schmidt, and J.S.Norris. 2002. Overcoming the phage replication threshold: a mathematical model with implications for phage therapy. Journal of Virology *76*: 5557–5564.
7. Haberer,K. and J.Maniloff. 1982. Adsorption of the tailed mycoplasma virus L3 to cell membranes. Journal of Virology *41*: 501–507.

Chapter 16

Measurement of the Bacteriophage Inactivation Kinetics with Purified Receptors

Andrew M. Kropinski

Abstract

Practical methods are described for studying the interaction between bacterial viruses and their surface receptors.

Key words: Adsorption, neutralization, inactivation, receptor, T4, M13, lipopolysaccharide, outer membrane protein, flagella, teichoic acids, pili.

1 Introduction

Bacteriophages bind to all possible cell surface receptors including pili [M13, D3112, F116] *(1, 2)*, flagella [χ, SP3, PBP1] *(3–5)*, lipopolysaccharide (LPS) [T7, P22], surface proteins [T1,T5, λ, AR1] *(6)*, teichoic acids [SP50, φ25] *(7)*, and capsules [K29, K1F, H4489A] *(8–11)*. In certain cases, such as T4, two receptors are used *(12)*. These observations make phages extremely useful tools for selecting receptor-deficient mutants, and for characterizing strains for specific receptors. An example of the latter would be the use of phages in typing systems.

Rather than using viable cells, receptor studies have been also carried out with cell extracts *(13)*, cell walls preparations *(14, 15)*, purified lipopolysaccharide *(16,17)*, and complexes of outer membrane proteins with LPS *(12, 15)*. In many cases the phages bind irreversibly to their isolated receptor resulting in inactivation. This can be tested in the following manner which is optimized from our studies of phage–LPS interactions *(17–20)*.

Martha R. J. Clokie, Andrew M. Kropinski (eds.), *Bacteriophages: Methods and Protocols, Volume 1: Isolation, Characterization, and Interactions, vol. 501*, © 2009 Humana Press, a part of Springer Science+Business Media
DOI 10.1007/978-1-60327-164-6_16 Springerprotocols.com

2 Materials

1. 1.0, 0.1, and 0.01 ml micropipetters such as Finnpipette or Eppendorf pipetters (Fisher Scientific). These should be periodically recalibrated.
2. Sterile tips.
3. Sterile capped 13 × 100 mm test tubes.
4. Two sterile 125 ml Erlenmeyer flasks.
5. Heating block or waterbath set at 48°C.
6. Waterbath shaker set at the growth temperature of your bacterium.
7. Visible spectrophotometer set at 650 nm.
8. Overnight culture of your bacterium grown in medium of choice supplemented with 1–10 mM $CaCl_2$, subcultured and grown to mid-log phase (**Note 1**).
9. Phage diluted in growth medium (plus Ca^{2+}) to a titer of 1–3 × 10^5. Prewarm to the assay temperature just prior to the experiment.
10. Bucket or styrofoam box containing crushed ice.
11. Agar plates and overlays (**Chapter** 7).

3 Methods

1. Set up a rack containing 12, 13 × 100 mm glass test tubes, and add carefully 1.6 ml of distilled water or buffer to the first tube (**Note 2**).
2. Add 0.9 ml of water or buffer to remainder of the tubes.
3. Number the tubes 1 through 11, and the last tube "C."
4. Add 0.2 ml of LPS to the first tube so as to achieve a final concentration of 200 μg/ml. The stock solution of LPS is therefore 1.7 mg/ml.
5. Mix, and using a new pipette tip transfer 0.9 ml from tube "1" to tube "2." Mix and continue to make doubling dilutions to tube number "11."
6. Discard 0.9 ml from tube "11."
7. Add 0.1 ml of phage preparation, diluted in broth or buffer, to each tube so as to achieve a final titer of 3 × 10^3 pfu/ml (**Note 3**).
8. Place the tubes in a waterbath or heating block at the desired temperature.
9. After an incubation period of 1 h remove 0.1 ml from each tube to molten overlay medium, seed with host cells and pour onto plates.

10. After appropriate incubation count and record the plague numbers, and calculate the percentage of phage neutralized at each concentration of LPS.
11. Plot the data on two or three cycle semilog paper (or in a software package using a log scale) with the final concentration of LPS on the log scale and the number of plaques on the linear scale. From this you can easily calculate the PhI_{50} i.e., the concentration of LPS which inactivates 50% of the phage.

4 Notes

1. The medium that which you use to grow your bacterium and propagate your phage. If you are starting a new project I recommend that you use the medium recommended for the propagating the host bacterium (*see* for example American Type Culture Collection (ATCC at http://www.atcc.org/) or Deutsche Sammlung von Mikroorganismen und Zellkulturen (DSMZ at http://www.dsmz.de/). N.B. many phages require 1–10 mM divalent ions (such as Ca^{2+} or Mg^{2+}) for optimal adsorption and the medium should be supplemented accordingly *(21, 22)*.
2. We noted that dilutions of LPS in broth had an inhibitory affect on the PhI_{50} value.
3. This assay is optimized for phages which produce small plaques in which accurate counts of 300 plaques per plate can be easily made. It is far more difficult to accurately count the number of phages, such as T7, which produce large plaques.

References

1. Pemberton,J.M. 1973. F116: a DNA bacteriophage specific for the pili of *Pseudomonas aeruginosa* strain PAO. Virology *55*: 558–560.
2. Roncero,C., A.Darzins, and M.J.Casadaban. 1990. *Pseudomonas aeruginosa* transposable bacteriophages D3112 and B3 require pili and surface growth for adsorption. Journal of Bacteriology *172*:1899–1904.
3. Shea,T.B. and E.Seaman. 1984. SP3: a flagellotropic bacteriophage of *Bacillus subtilis*. Journal of General Virology *65*: 2073–2076.
4. Samuel,A.D., T.P.Pitta, W.S.Ryu, P.N.Danese, E.C.Leung, and H.C.Berg. 1999. Flagellar determinants of bacterial sensitivity to chi-phage. Proceedings of the National Academy of Sciences of the United States of America *96*:9863–9866.
5. Lovett,P.S. 1972. PBPI: a flagella specific bacteriophage mediating transduction in *Bacillus pumilus*. Virology *47*:743–752.
6. Berrier,C., M.Bonhivers, L.Letellier, and A.Ghazi. 2000. High-conductance channel induced by the interaction of phage lambda with its receptor maltoporin. FEBS Letters *476*:129–133.
7. Givan,A.L., K.Glassey, R.S.Green, W.K.Lang, A.J.Anderson, and A.R.Archibald. 1982. Relation between wall teichoic acid content of *Bacillus subtilis* and efficiency of adsorption of bacteriophages SP 50 and ϕ25. Archives of Microbiology *133*:318–322.
8. Suthereland,I.W., K.A.Hughes, L.C.Skillman, and K.Tait. 2004. The interaction of phage and biofilms. FEMS Microbiology Letters *232*:1–6.

9. Stummeyer,K., A.Dickmanns, M.Muhlenhoff, R.Gerardy-Schahn, and R.Ficner. 2005. Crystal structure of the polysialic acid-degrading endosialidase of bacteriophage K1F. Nature Structural & Molecular Biology *12*:90–96.
10. Baker,J.R., S.Dong, and D.G.Pritchard. 2002. The hyaluronan lyase of *Streptococcus pyogenes* bacteriophage H4489A. Biochemical Journal *365*:317–322.
11. Bayer,M.E., H.Thurow, and M.H.Bayer. 1979. Penetration of the polysaccharide capsule of *Escherichia coli* (Bi161/42) by bacteriophage K29. Virology *94*:95–118.
12. Hantke,K. 1978. Major outer membrane proteins of *E. coli* K12 serve as receptors for the phages T2 (protein Ia) and 434 (protein Ib). Molecular & General Genetics *164*:131–135.
13. Tokunaga,T., T.Kataoka, K.Suda, and T.Yasuda. 1969. [Bacteriophage receptor of mycobacteria. 2. Inactivation of mycobacteriophages with the ethanol-ether extract from the cell wall fraction and electron microscopic studies]. [Japanese]. Igaku to Seibutsugaku – Medicine & Biology *78*:141–145.
14. Valyasevi,R., W.E.Sandine, and B.L.Geller. 1990. The bacteriophage kh receptor of *Lactococcus lactis* subsp. cremoris KH is the rhamnose of the extracellular wall polysaccharide. Applied & Environmental Microbiology *56*:1882–1889.
15. Yu,F. and S.Mizushima. 1982. Roles of lipopolysaccharide and outer membrane protein OmpC of *Escherichia coli* K-12 in the receptor function for bacteriophage T4. Journal of Bacteriology *151*:718–722.
16. Patel,I.R. and K.K.Rao. 1983. Studies on the *Pseudomonas aeruginosa* PAO1 bacteriophage receptors. Archives of Microbiology *135*: 155–157.
17. Jarrell,K. and A.M.Kropinski. 1977. Identification of the cell wall receptor for bacteriophage E79 in *Pseudomonas aeruginosa* strain PAO. Journal of Virology *23*:461–466.
18. Jarrell,K. and A.M.Kropinski. 1976. The isolation and characterization of a lipopolysaccharide-specific *Pseudomonas aeruginosa* bacteriophage. Journal of General Virology *33*:99–106.
19. Jarrell,K.F. and A.M.Kropinski. 1981. *Pseudomonas aeruginosa* bacteriophage phi PLS27-lipopolysaccharide interactions. Journal of Virology *40*:411–420.
20. Jarrell,K.F. and A.M.Kropinski. 1981. Isolation and characterization of a bacteriophage specific for the lipopolysaccharide of rough derivatives of *Pseudomonas aeruginosa* strain PAO. Journal of Virology *38*:529–538.
21. Haberer,K. and J.Maniloff. 1982. Adsorption of the tailed mycoplasma virus L3 to cell membranes. Journal of Virology *41*:501–507.
22. Adams,M.D. 1959. Bacteriophages. Interscience Publishers, Inc., New York.
23. Krueger,A.P. 1931. The sorption of bacteriophage by living and dead susceptible bacteria. Journal of General Physiology *14*: 493–503.

Chapter 17

Bacteriophage Plaques: Theory and Analysis

Stephen T. Abedon and John Yin

Abstract

Laboratory characterization of bacteriophage growth traditionally is done either in broth cultures or in semisolid agar media. These two environments may be distinguished in terms of their spatial structure, i.e., the degree to which they limit diffusion, motility, and environmental mixing. Well-mixed broth, for example, represents the microbiological ideal of a non-spatially structured environment. Agar, by contrast, imposes significant limitations on phage and bacterial movement and therefore gives rise to spatial structure.

The study of phage growth within spatially structured environments, such as that seen during phage plaque formation, is important for three reasons. First, a large fraction of environmental bacteria live within spatially structured environments such as within biofilms, within soil, or when growing in or on the tissues of plants and animals. Second, phage growth as plaques is a central technique to phage studies, yet appears to be under appreciated by phage workers in terms of its complexity. Third, selective pressures acting on phage during plaque growth differ from those seen during broth growth.

In this chapter we will discuss just what a plaque is, how one forms, and what can affect plaque size. We will describe methods, both experimental and theoretical, that have been employed to study plaque growth. As caveats we will discuss why plaque formation failure is not necessarily equivalent to virion inviability (**Note 1**). We also will consider problems with inferring phage broth growth fitness as a function of plaque characteristics (**Note 2**).

Key words: Phage plaques, plaque formation, plaque enlargement, mathematical modeling, phage virulence.

1 Introduction

A phage plaque is a clearing in a bacterial lawn. Plaques form via an outward diffusion of phage virions that is fed by bacterial infection. Anything that slows phage diffusion can impede plaque development and thereby plaque size. During plaque formation, the principal determinants of plaque size, therefore, are

Martha R. J. Clokie, Andrew M. Kropinski (eds.), *Bacteriophages: Methods and Protocols, Volume 1: Isolation, Characterization, and Interactions, vol. 501*, © 2009 Humana Press, a part of Springer Science+Business Media
DOI 10.1007/978-1-60327-164-6_17 Springerprotocols.com

the intrinsic characteristics of phage diffusion as observed within the medium, the effectiveness of porous barriers to phage movement as presented by cross-linked agar molecules, and the time phage spend infecting or otherwise interacting with immobile bacteria. Here, we review methods that have been employed to model and study plaque growth, including consideration of how various factors affect plaque development [*See (1)* for discussion of how selective pressures acting on phage can vary within plaques and *(2)* for discussion of plaque morphology].

2 Materials

The materials employed to produce phage plaques are presented elsewhere in this volume (**Chapter** 7). Analysis of resulting plaques can involve various probes employed for plaque sampling, sources of dark-field lighting (e.g., colony counters), digital cameras, and also appropriate software for image analysis. Recently plaque formation has also been followed using recombinant fluorescent phage capsid proteins *(3)*.

3 Methods

3.1 Soft-Agar Overlay

See **Chapter** 7 for methodology.

3.2 Observation of Growing Plaques

In this section we describe technologies that have been employed to observe both the kinetics of and variation in phage propagation within a growing plaque.

3.2.1 Whole Plaque Analysis

Various phages have been observed to establish a wave of "constant-velocity infection" during plaque formation *(4–9)*. Determination of this plaque growth rate may be achieved simply via periodic measurement of plaque diameters during incubation (ideally done within a hot room so that plaque development is not disrupted during measurement) or by employing time-lapse photography. Such measurement may be facilitated by employing lower densities of agar in both top and bottom agars (**Section 3.3.3.1**). For analysis of related phage populations, whole plaques may be cored—either in part or in full and either using Pasteur pipettes or larger bore glass (or metal) tubing. Coring allows subsequent analysis following resuspension in appropriate buffer. Populations of plaques may be sampled either by scraping the overlying top agar into appropriate buffer or by suspending top and bottom agar together in buffer.

3.2.2 Methods of Plaque Dissection

Comparison of phages present within different regions of a plaque may be accomplished either qualitatively or quantitatively. Qualitative determination of what phage types are present within different regions of a plaque may be accomplished via stab sampling. For example, different regions of plaques may be probed using sterile inoculating needles, tooth picks, or plastic pipette tips, with probes then incubated in appropriate buffer to release phage for subsequent examination. Note that heat can mutagenize as well as kill. Probes therefore should be cooled prior to contact with plaques, a process that may be speeded along by dipping flamed tools into sterile agar or water.

Quantitative comparison may be accomplished by careful removal of a bisecting slab of agar *(7)*. This slab may be incised using a sharp blade (ethanol-flame sterilized between cuts); creating two parallel slices that pass through the underlying bottom agar and which include lawn external to the plaque in question. To minimize contamination of regions of low phage density by regions of high phage density, slices are best made starting from outside of the plaque and continuing inward. Care should be taken that the plaque's center is included in the bisect and cuts may be guided using an ethanol-flame sterilized straight edge.

Removal of the slab is accomplished using a thin, flexible spatula, with removal, for example, to the inner surface of an empty, sterile Petri dish to minimize slab contamination. By placing the slab alongside a ruler (ideally a metal ruler so that it may be ethanol-flame sterilized), it is possible to cut individual, parallel blocks from the slab of approximately the same size. Effort should be made to specify block order as well as which blocks are removed both from a plaque's center and its periphery, such as via removal of the first block from what corresponds to the plaque's center. Blocks may be removed using a thin, flexible spatula, and then suspended in a constant volume of appropriate buffer in a well-labeled tube for extended virion diffusion out of the agar (e.g., as for 12 h at 37°C, adding a drop of chloroform or other metabolic inhibitor to inhibit concurrent phage replication). Disruption of the agar blocks using the spatula may also facilitate quantitative release of phage.

An alternative quantitative approach to plaque dissection involves removing a plaque bisect using straight-edge guided Pasteur pipettes (or equivalent coring device), with the caveat that with this latter method there exists less flexibility in determining "block" size, though with the advantage of providing less room for cross-contamination between blocks (assuming, of course, that pipettes are discarded after each use). As with bisect cutting, it is preferable to remove cores starting from outside of plaques and going inward in order to minimize contamination of low-phage density regions by high-phage density regions. Using either method, plaque dissection may be more easily achieved if phages are grown under conditions capable of supporting the formation of relatively large plaques (**Section 3.3.1**).

3.2.3 Tracking Phage Evolution as a Function of Plaque Perimeter Undulation

For phage T7, whose growth is not limited by lawn maturation *(7)*, it was found using a stab-sampling technique that considerable phage mutation and selection can occur during plaque growth *(10, 11)* (stab-sampling, using a plastic pipette tip, was done at equal spacings around the perimeter of individual plaques). Greater mutant densities were found to be associated with plaque peripheries following longer periods of plaque growth, and selection appeared to occur, not surprisingly, for phage variants displaying greater rates of plaque enlargement. Similarly, observation of sectored overgrowth in plaques of faster-growing phage variants has been known for some time *(12)*. Furthermore, plaque accumulation of overgrowing variants may be greatly minimized, such as for initiating broth stock cultures, by inoculating from only relatively young plaques, e.g., following 3.5–4 h of incubation versus 24 h at 37°C *(13)*.

Observation of phage evolution subsequently has been automated based on the inference that evolution, as a stochastic process, would not proceed uniformly across the perimeter of plaques. That is, there is an expectation that some regions of a plaque's perimeter will harbor phage that, as a consequence of variance in genotype, and thereby phenotype, will spread faster than others. The degree of variance in perimeter growth rate consequently may be quantified in terms of perimeter length, with greater undulation (deviation from circularity) indicating greater variance. Determination of perimeter shape in these studies was accomplished via the use of a digital camera, time-lapse digital imaging against a dark field, and the public-domain NIH Image (http://rsb.info.nih.gov/nih-image/) or ImageJ (http://rsbweb.nih.gov/ij/) programs using a custom-written image acquisition and analysis macro *(8, 9)*. Of interest, plaques initiated with wild-type phages displayed greater perimeter lengths (as a function of overall plaque size) than did plaques initiated with phage mutants already capable of displaying faster plaque enlargement than wild type *(8)*.

3.3 Plaque Growth Theory

We envisage plaque development as occurring in four phases [quoting from *(5)*, but *see* also *(14)*]: "(i) the primary adsorption event; (ii) the first round or first few rounds of viral multiplication; (iii) the enlargement phase; and (iv) the final phase, in which viral multiplication ceases." We in turn consider the impact each phase has on plaque size. Various authors explore the impact of plaque conditions on resulting plaque size *(4, 15–18)*.

3.3.1 The Primary Adsorption Event

To form a plaque there must be an initial adsorption of phage to bacterium. This adsorption can occur prior to phage and bacteria addition to top agar, within molten top agar, or following agar solidification. Delays in adsorption result in the formation of smaller plaques since, ultimately, plaque size is a function of the degree to which plaques can enlarge either prior to or in spite of bacterial lawn maturation. Three factors affect the timing of this initial adsorption: bacterial density, agar density (concentration), and the phage adsorption constant. The greater the bacterial density or the phage adsorption constant (**Chapters 15** and **18**) the earlier the phage adsorption. Alternatively, the greater the agar density, the slower phage diffusion, and the later the primary adsorption event occurs. Delays in adsorption not only can give rise to smaller phage plaques but also to greater variation in plaque size *(19)*. This occurs because the slower the adsorption, the greater the variance in the timing of adsorption by individual phage *(1, 20)*. This variance may be avoided by pre-adsorbing phage to bacteria for a number of minutes prior to plating *(19)*.

3.3.2 The First Round or First Few Rounds of Viral Multiplication

We limit discussion of this phase of plaque development to two considerations. First, if one employs non-log phase indicator bacteria then early plaque development may be affected by a less than ideal host physiology as these bacteria transit to log phase (*see*, however, **Section 3.3.4.2**). Second, while a number of authors indicate that plaque enlargement can occur at a constant rate *(4–9)*, in fact that constant rate is not necessarily achieved very early during plaque development, e.g., as during at least the first hour or two of plaque formation. In addition, it is not necessarily even straightforward to study very early plaque development—at least using traditional plaque-visualization techniques—given the absence of an easily observed bacterial lawn.

3.3.3 The Enlargement Phase

The bulk of plaque development occurs during the enlargement phase. Rates of plaque enlargement tend to vary as a function of phage growth characteristics and plaquing conditions, while the rate of enlargement of individual plaques tends to remain constant over the course of their development (**Section 3.3.2**).

3.3.3.1 Phage Diffusion

The faster the virions diffuse, the faster the plaques enlarge. Two factors impact phage diffusion: phage intrinsic properties and agar density. Faster intrinsic virion diffusion rates, such as may result from smaller phage particle size *(21)*, should contribute to greater plaque size. Too great agar densities, meanwhile, should slow phage diffusion and therefore reduce plaque size *(21, 22)*. Complicating factors such as differences in osmolarity between top and bottom agars should result in changes in soft-agar density. Excessive bottom agar drying, for example, will result in increases in solute densities (including agar densities) that can result in net water flow from top to bottom agar. This will increase the agar density of the soft-agar medium through which phage diffuse, resulting in a reduction in plaque size. Excessive drying of top agar also should result in smaller plaques, which perhaps contributes to the small size of plaques in the too thin overlay layers that result when top agar is allowed to solidify on a not-level surface.

It is a further complication that virions can enter bottom agar and diffuse. Diffusion in bottom agar should be slower than diffusion in top agar, owing to the higher agar density of the former. However, such movement could contribute to plaque enlargement particularly if lawn bacterial densities are high (which, by adsorbing phage, can lead to significant interruption of phage diffusion) and if one employs lower agar-density bottom agar (see following paragraph).

Changes in commercial processing of agar may also be relevant to the impact of agar on plaque enlargement. Higher molecular weight agar preparations are more effective in gelling at lower densities than standard agar preparations *(23)*. This observation has led to speculation that, as a consequence of modern recovery techniques, standard agar formulations may have become increasingly more effective at inhibiting phage diffusion. As a result, even though seemingly identical media formulations are employed, plaque sizes today may in fact be smaller than once observed, such as seen during the 1950s, because agar densities as specified from that era—e.g., 0.7% agar in top agar and 1.5% agar in bottom agar *(19)*—today may be excessive. For "traditional" plaque sizes, consider employing variations on Drake agar, which contains 0.65 and 1.0% agar for top and bottom agars, respectively. For specialized applications, top agar as low as 0.4% still provides sufficient gelling *(24)*, and it has been our experience that bottom agar even as low as 0.8% can mechanically support the overlay. In addition, employment of purposefully prepared higher molecular weight agar preparations can allow top agar densities of even 0.35%: "Agar solidification occurs normally but the virus particles presumably diffuse outwards more rapidly, and much larger plaques result" *(23)*.

3.3.3.2 Phage Adsorption

Because within most agar preparations virions can diffuse while bacteria cannot, it is particularly diffusing phage virions that contribute to plaque enlargement. As a consequence, a phage that adsorbs poorly to its host bacterium—if that poor adsorption is a consequence of phage–bacterium encounters that do not result in phage adsorption—will be freer to diffuse outward from a plaque's center than an equivalent but better adsorbing phage. Contrasting this tendency, however, less adsorption will result in fewer phage produced to feed plaque formation. Thus, larger plaques may form by reducing, from a probability of 1.0, the likelihood of virion adsorption given phage–bacterium encounter, but only if this reduction is not too great. *See (1)* for possible examples of poorer phage adsorption resulting in larger plaque diameters.

3.3.3.3 Phage Latent Period

Phage latent period impacts on plaque enlargement in a manner that is analogous to the impact of phage adsorption (**Section 3.3.3.2**): The less time phage spend infecting bacteria, the more time they can spend as diffusing virions, and therefore the potentially faster plaques can enlarge. *See (10)* and *(25)* for evidence consistent with this hypothesis that shorter phage latent periods can give rise to larger phage plaques and *(26, 27)* for consideration of the theory of phage latent period evolution in plaques.

3.3.3.4 Phage Burst Size

Like adsorption rates (**Section 3.3.3.2**) and latent periods (**Section 3.3.3.3**), phage burst size affects plaque size by modifying the size of the diffusing phage population. The impact of phage burst size on rates of plaque enlargement is not great, however *(5, 28)*. It is only with very low burst sizes, such as less than 10 *(29)*, that we expect burst size to greatly affect—in this case negatively—the rates of plaque enlargement. Not surprisingly, inactivation of virions during plaque development, as has been achieved via inclusion of anti-phage serum within agar, also reduces rates of plaque enlargement *(30)*. A similar effect may have been achieved by Bronfenbrenner and Korb *(22)* by mixing physiologically old (indeed, perhaps dead) bacteria with physiologically young indicator bacteria of the same strain (12 days versus 18 h, respectively). *See (26, 27)* for additional consideration of phage burst size evolution in plaques.

3.3.3.5 Lawn Density of Bacteria

Lower lawn densities delay primary phage adsorption, resulting in smaller plaques (**Section 3.3.1**). During plaque enlargement, however, such delays are equivalent to virions not rapidly adsorbing to bacteria, and therefore being freer to diffuse farther outward from the center of plaques prior to colliding with a bacterium. Thus, lower bacterial densities in lawns, like lower phage adsorption constants or shorter phage latent periods, ought to support more-extensive phage diffusion, though at a cost in virions contributing to the production of new phage. Given these various conflicts between phage freedom to diffuse, on the one hand, and phage ability to reproduce on the other, it should not be completely surprising that experiments performed by a number of authors indicate that rates of plaque enlargement remain constant even as densities of lawn bacteria rise during plaque formation *(5, 6, 9)*.

3.3.4 The Final Phase, in which Viral Multiplication Ceases

The impact of the final phase of plaque development on plaque size is a function of both when and if it occurs. For most phage systems (T7 is exceptional), plaque enlargement ceases upon lawn maturation to stationary phase. In that case, the sooner lawns reach stationary phase, the shorter the enlargement phase (**Section 3.3.3**) and therefore the smaller the resulting plaques. Three factors could affect when this final phase of plaque development occurs: The initial density of indicator bacteria, the rate of division of these bacteria, and the bacterial density at stationary phase. Plaque size as a function of initial bacterial density is explored by *(4, 5, 31–33)*. *See* also *(2)* for discussion of continued "plaque" enlargement via halo formation that is due to diffusion from plaques of low molecular weight extracellular polymer degrading "lysins."

3.3.4.1 Initial Indicator Bacteria

Greater initial densities of indicator bacteria give rise to earlier phage adsorption and therefore plaques that are larger and which display less variance in size (**Section 3.3.1**). Higher initial bacterial densities, however, also give the bacterial lawn a head start against the growing plaque, resulting in earlier lawn maturation and smaller plaques *(22)*. Too few indicator bacteria, by the same reasoning, could give rise to larger plaques, though one does not want to employ so few bacteria that the graininess of the lawn and sizes of individual microcolonies are greatly increased *(6)*.

3.3.4.2 Rate of Indicator Division

Phage growth parameters (**Chapter 18**) can vary as a function of bacterial division rates *(34)*. It is not a given, however, that phage growth will vary in a manner that is exactly compensated, vis-à-vis plaque enlargement, by changes in rates of bacterial growth. Therefore, it is uncertain whether reduced nutrient availability, for example, would have a greater impact on phage versus bacterial growth. Nutrient density, on the other hand, should have little impact on rates of phage diffusion. Consequently, slower phage and bacterial replication, especially resulting from poor nutrient quality, should result in larger plaques (though not faster-growing plaques), at least so long as all else is held constant, including peak bacterial densities at the point of lawn maturation. Perhaps similarly, some phages display larger plaques during anaerobic growth *(16)* or, for phages capable of growth on UV irradiated bacteria, "it may be useful to irradiate slightly with ultraviolet light the host cells to retard growth" *(35)*.

3.3.4.3 Bacterial Density at Maturation

Factors that result in bacteria cessation of growth at lower densities will also give rise to sooner termination of plaque enlargement. Greater nutrient richness, therefore, could result in greater plaque sizes by allowing bacterial growth to higher densities *(29)* (however, *see* **Section 3.3.4.2**). Alternatively, excessive bottom- and therefore top-agar dryness could result in solute densities that are greater than optimal. Excessive dryness also could lead to reduced rates of nutrient and waste-product diffusion into and out of the top agar due to higher agar densities, which impede diffusion rates. Both mechanisms could at least potentially lead to a premature maturation of the bacterial lawn and therefore to smaller plaques. Too thin bottom agar similarly may be inhibiting to optimal plaque development due to a reduced capacity for buffering of the top agar *(4)*. A reasonable upper limit of bottom agar volume per 9 cm plate, however, is about 35 ml *(24)*; much thicker and light flow may be impeded, thereby visually obscuring otherwise well-developed plaques. A lower limit is about 25 ml, and top agar volumes should range from about 2.5 to 3.5 ml, with ease of overlay delivery and oxygen penetration into poured overlays the primary considerations influencing top agar volume *(36)*.

3.4 Modeling Plaque Growth

We are aware of ten studies employing mathematical models to describe plaque development, those of Koch *(5)*, Kaplan *et al.*, *(6)*, Yin and colleagues *(28, 30, 37)*, Fort and colleagues *(38–40)*, and Abedon and Culler *(26, 27)*. *See* also *(14)*. From these models, along with supporting experimentation, we can better understand those factors affecting, in particular, rates of plaque enlargement during plaque development. Koch's model *(5)* uses heuristic arguments to suggest a form for the rate of plaque enlargement, r:

$$r = 10 \cdot \left(\frac{D}{L}\right)^{1/2} \tag{17.1}$$

where D is the virion diffusion rate and L is the phage latent period.

Yin and McCaskill *(31)*, using a more mechanistic approach than Koch, provide several expressions for the rate of plaque expansion under different conditions. Under conditions of equilibrated adsorption of the phage to its host, one attains the following expression for r:

$$\begin{aligned} r &= 2 \cdot \left[\frac{D \cdot k_2 \cdot (B-1) \cdot f \cdot K_{\max}}{(1 + f \cdot K_{\max})^2}\right]^{1/2} \\ &= 2 \cdot \left[\frac{D \cdot k_2 \cdot (B-1) \cdot N_o \cdot k_1/k_{-1}}{(1 + N_o \cdot k_1/k_{-1})^2}\right]^{1/2} \end{aligned} \tag{17.2}$$

where B is the phage burst size, N_o is a constant density of bacteria in the lawn, and k_1, k_{-1}, and k_2 are an association constant between bacteria and phage, a corresponding dissociation constant, and a rate of conversion of infected bacteria to virions, respectively. In addition, $f = N_o/N_{\max}$ and $K_{\max} = k_1 N_{\max}/k_{-1}$ where $N_{\max}$ is the "maximum attainable" lawn density of bacteria. Since k_2 is in inverse-time units, it is inversely proportional to the latent time (L). Expression (17.2) with the rate of plaque expansion proportional to $(D \cdot k_2)^{1/2}$ consequently is analogous to expression (17.1), where the rate of plaque expansion is proportional to $(D/L)^{1/2}$ *(26)*.

To derive the expression in equation (17.2), as well as the other limiting cases, several assumptions and approximations had to be made. An alternative approach is to numerically integrate the mechanistic equations for a given set of parameters *(28)*. Although the method does not provide analytic expressions for the rate of plaque enlargement, it does allow one to test the sensitivity of the rate to different parameters. For example, for a fixed set of parameters the numerical solutions predicts that a 2-fold increase in burst size will increase the enlargement rate more than a 2-fold increase in the adsorption rate.

Fort and Méndez *(38)* extend the plaque-growth model of Yin and McCaskill to explicitly incorporate time delays associated with phage infection, i.e., as observed between the point of adsorption and phage-induced bacterial lysis *(2)*. By incorporating an independent measurement of the time delay from one-step growth experiments (**Chapter 18**), they provide a closer approximation of experimentally observed rates of plaque development. Their inclusion of a time delay improves the predictive power of the model but, like You and Yin *(28)*, requires some numerical analysis. Subsequently, however, an exact solution has been attained by Ortega-Cejas *et al.* *(39)*:

$$r = 2 \cdot \left[D_{eff} \left(\frac{f}{L(f + f_o)} \right) \right]^{1/2} \quad \text{if } 0 \leq f \leq f_o \qquad (17.3a)$$

$$r = \left[\frac{2D_{eff}}{L} \right]^{1/2} \quad \text{if } f_o \leq f \leq 1 \qquad (17.3b)$$

where D_{eff} is a "hindered diffusion" constant defined as

$$D_{eff} = D\left(\frac{1-f}{1+(f/x)} \right) \qquad (17.3c)$$

and f_o is defined by

$$f_o \equiv 1/(Lk_1 N_{max}(B-1)) \qquad (17.3d)$$

The variable x in equation (17.3c), a description of bacterial shape, is set to 1.67 by Ortega-Cejas *et al.* As for equation (17.2), note that the rate of plaque expansion predicted by equations (17.3a) and (17.3b) is proportional to $(D/L)^{1/2}$ as seen in equation (17.1). *See (40)* for incorporation of rise time (**Chapter 18**).

Incorporating lawn growth into their model, Kaplan *et al.* *(6)* provide a prediction of final plaque radius, R:

$$R = (r/V) \cdot \ln(N_f/N_0) \qquad (17.4)$$

where r is the rate of increase of plaque radius, V is the rate of bacteria replication in a growing plaque, and N_f is the final density of bacteria in the mature lawn. Also from *(6)* is an estimation of the instantaneous rate of phage infection of bacteria,

$$dM/dt = N_0 \cdot e^{v \cdot t} \cdot 3 \cdot \pi \cdot (r \cdot t)^2 \qquad (17.5)$$

where M is the density of infected bacteria, t is time, e is the base of the natural logarithm, and π is from geometry (i.e., 3.14...). By integrating equation (17.5) with respect to t, as Kaplan *et al.*, present, one can determine the cumulative number of bacteria that have been infected (and subsequently lysed) at any point during plaque formation. *See (14, 27)* for further discussion of the Kaplan *et al.*, model.

4 Notes

1 **Plaque Formation Failure is not Necessarily Equivalent to Virion Inviability**

"Lack of plaque formation" in phage studies is commonly employed as a surrogate for "Phage inability to productively infect a bacterium." Significantly extended latent periods, as well as greatly reduced burst sizes, however, can both greatly impact plaque formation without actually preventing productive phage infection of individual bacteria. Consider, therefore, applying more stringent criteria than absence of plaque formation before declaring that a given phage under a given set of conditions is in fact inviable. Alternative approaches include (i) plating presumptive broth lysates using a permissive indicator (taking care to avoid carryover of virions that failed to adsorb the non permissive host; *see* **Chapter 15**, for various strategies), (ii) characterizing phage infection in broth by means other than via detection of infective phage progeny (e.g., by detecting physical, molecular, biochemical, or serological indicators, especially of late-stage phage infection), or (iii) determining a presumptive lysate's total virion count (*see* "Enumeration by fluorescent microscopy" and "Enumeration of phage using flow cytometry," **Chapters 8** and **9**) or, alternatively, a lysate's "Killing titer" *(29, 41, 42)*.

2 **Problems with Inferring Phage Fitness from Plaque Characteristics**

The rapidity with which phage populations grow is one measure of a phage's Darwinian fitness, and the negative impact that a phage population has on a bacterial population (a phage's "virulence") can also be a function of the rapidity of phage population growth *(43)*. Higher phage virulence often is a desirable property for phage-based antibacterial biocontrol, bioprocessing, or therapy *(44)*. A convenient surrogate for robust phage growth, i.e., high phage fitness or virulence, is phage plaque size *(33, 43)*. Nevertheless, not all phage characteristics positively affecting phage growth in broth will similarly positively affect plaque formation. *See (26, 27)* for discussion of how changing phage growth parameters may affect plaque growth and *(45–47)* for similar consideration especially in broth culture.

A primary example is phage adsorption rate. While rapid rates of phage diffusion should have an equivalent impact on phage fitness in both broth and soft agar, a lower likelihood of phage adsorption given phage–bacterium encounter will serve to lower a phage's growth rate in broth, but may actually support the formation of larger plaque sizes in agar-based media (**Section 3.3.3.2**). Phage latent period provides another example. For instance, T-even phages that are defective in the lysis-inhibition phenotype display larger plaques, but display reduced fitness relative to wild type during growth in broth culture *(48, 49)*. Also, phage growth in broth under conditions of relatively high bacterial density (e.g., in excess of 10^7 bacteria/ml), along with phage growth within plaques, may both result in selection for relatively short phage latent periods. However, this equivalence may be lost if one considers phage fitness in broth given lower bacterial densities *(25)*. We therefore caution, in general, against adapting phage lineages to a given host or set of growth conditions in order to improve phage fitness or bacterial killing under different or broader sets of conditions.

Acknowledgements

The authors thank Jan Drake for helpful comments on this chapter. The National Science Foundation provided support for JY.

References

1. Abedon, S.T. (2006) Phage ecology, in *The Bacteriophages* (Calendar, R. and Abedon, S T eds.), Oxford University Press, Oxford, pp. 37–46.
2. Abedon, S.T. and Yin, J. (2008) Impact of spatial structure on phage population growth, in *Bacteriophage Ecology* (Abedon, S.T. ed.), Cambridge University Press, Cambridge, UK, pp. 94–113.
3. Alvarez, L.J., Thomen, P., Makushok, T. and Chatenay, D. (2007) Propagation of fluorescent viruses in growing plaques. *Biotech. Bioeng.* **96**, 615–621.
4. Mayr-Harting, A. (1958) Die Entwicklung von Phagenloechern und der mechanismus der Phagenwirkung in festen Naehrboeden. *Zbl. f. Bakt. Paras. Infek. u. Hyg.* **171**, 380–392.

5. Koch, A.L. (1964) The growth of viral plaques during the enlargement phase. *J. Theor. Biol.* **6**, 413–431.
6. Kaplan, D.A., Naumovski, L., Rothschild, B. and Collier, R.J. (1981) Appendix: a model of plaque formation. *Gene* **13**, 221–225.
7. Yin, J. (1991) A quantifiable phenotype of viral propagation. *Biochem. Biophys. Res. Com.* **174**, 1009–1014.
8. Lee, Y. and Yin, J. (1996) Detection of evolving viruses. *Nat. Biotech.* **14**, 491–493.
9. Lee, Y. and Yin, J. (1996) Imaging the propagation of viruses. *Biotech. Bioeng.* **52**, 438–442.
10. Yin, J. (1993) Evolution of bacteriophage T7 in a growing plaque. *J. Bacteriol.* **175**, 1272–1277.
11. Yin, J. (1994) Spatially resolved evolution of viruses. *Ann. N. Y. Acad. Sci.* **745**, 399–408.
12. Hershey, A.D. (1946) Spontaneous mutations in bacterial viruses. *Cold Spring Harbor Symp. Quant. Biol.* **11**, 67–77.
13. Doermann, A.H., Carolyn, F.-R. and Dissosway, C. (1949) Intracellular growth and genetics of bacteriophage. *Year Book Carnegie Inst. Wash.* **48**, 170–176.
14. Krone, S.M. and Abedon, S.T. (2008) Modeling phage plaque growth, in *Bacteriophage Ecology* (Abedon, S.T. ed.), Cambridge University Press, Cambridge, UK, pp. 415–438.
15. Qanber, A.A. and Douglas, J. (1976) Enhancement of plaque size of a staphylococcal phage. *J. Appl. Bacteriol.* **40**, 109–110.
16. McConnell, M. and Wright, A. (1975) An anaerobic technique for increasing bacteriophage plaque size. *Virology* **65**, 588–590.
17. Carlson, K. and Miller, E.S. (1994) Enumerating phage: the plaque assay, in *Molecular Biology of Bacteriophage T4* (Karam, J.D. ed.), ASM Press, Washington, DC, pp. 427–429.
18. Lillehaug, D. (1997) An improved plaque assay for poor plaque-producing temperate lactococcal bacteriophages. *J. Appl. Microbiol.* **83**, 85–90.
19. Adams, M.H. (1959). Bacteriophages. Interscience, New York.
20. Abedon, S.T., Herschler, T.D. and Stopar, D. (2001) Bacteriophage latent-period evolution as a response to resource availability. *Appl. Environ. Microbiol.* **67**, 4233–4241.
21. Elford, W.J. and Andrews, C.H. (1932) The sizes of different bacteriophages. *Brit. J. Exp. Path.* **13**, 446–456S.
22. Bronfenbrenner, J.J. and Korb, C. (1925) Studies on the bacteriophage of d'Herelle. III. Some of the factors determining the number and size of plaques of bacterial lysis on agar. *J. Exp. Med.* **42**, 483–497.
23. Schnitzlein, C.F., Albrecht, I. and Drake, J.W. (1974) Is bacteriophage T4 DNA polymerase involved in the repair of ultraviolet damage? *Virology* **59**, 580–583.
24. Conkling, M.A. and Drake, J.W. (1984) Isolation and characterization of conditional alleles of bacteriophage T4. *Genetics* **107**, 505–523.
25. Abedon, S.T., Hyman, P. and Thomas, C. (2003) Experimental examination of bacteriophage latent-period evolution as a response to bacterial availability. *Appl. Environ. Microbiol.* **69**, 7499–7506.
26. Abedon, S.T. and Culler, R.R. (2007) Bacteriophage evolution given spatial constraint. *J. Theor. Biol.* **248**, 111–119.
27. Abedon, S.T. and Culler, R.R. (2007) Optimizing bacteriophage plaque fecundity. *J. Theor. Biol.* **249**, 582–592.
28. You, L. and Yin, J. (1999) Amplification and spread of viruses in a growing plaque. *J. Theor. Biol.* **200**, 365–373.
29. Carlson, K. and Miller, E.S. (1994) Enumerating phage, in *Molecular Biology of Bacteriophage T4* (Karam, J.D. ed.), ASM Press, Washington, DC, pp. 427–429.
30. Lee, Y., Eisner, S.D. and Yin, J. (1997) Antiserum inhibition of propagating viruses. *Biotech. Bioeng.* **55**, 542–546.
31. Hershey, A.D., Kalmanson, G.M. and Bronfenbrenner, J.J. (1944) Coordinate effects of electrolyte and antibody on the infectivity of bacteriophage. *J. Immunol.* **48**, 221–239.
32. Dennehy, J.J., Abedon, S.T. and Turner, P.E. (2007) Host density impacts relative fitness of bacteriophage ϕ6 genotypes in structured habitats. *Evolution* **61**, 2516–2527.
33. Burch, C.L. and Chao, L. (2004) Epistasis and Its relationship to canalization in the RNA virus ϕ6. *Genetics* **167**, 559–567.
34. Hadas, H., Einav, M., Fishov, I. and Zaritsky, A. (1997) Bacteriophage T4 development depends on the physiology of its host *Escherichia coli*. *Microbiology* **143**, 179–185.
35. Eisenstark, A. (1967) Bacteriophage techniques. *Meth. Virol.* **1**, 449–524.
36. Hershey, A.D., Kalmanson, G. and Bronfenbrenner, J.J. (1943) Quantitative methods in the study of the phage-antibody reaction. *J. Immunol.* **46**, 267–279.
37. Yin, J. and McCaskill, J.S. (1992) Replication of viruses in a growing plaque: A reaction-diffusion model. *Biophys. J.* **61**, 1540–1549.
38. Fort, J. and Méndez, V. (2002) Time-delayed spread of viruses in growing plaques. *Phys. Rev. Lett.* **89**, 178101.
39. Ortega-Cejas, V., Fort, J., Méndez, V. and Campos, D. (2004) Approximate solution to

the speed of spreading viruses. *Phys. Rev. E* **69**, 031909-1-031909-4.
40. Fort, J., Pérez, J., Ubeda, E. and García, F.J. (2006) Fronts with continuous waiting-time distributions: Theory and application to virus infections. *Phys. Rev. E* **73**, 021907-1-021907-8.
41. Stent, G.S. (1963). Molecular Biology of Bacterial Viruses. WH Freeman and Co., San Francisco, CA.
42. Carlson, K. (2005) Working with bacteriophages: common techniques and methodological approaches, in (Kutter, E. and Sulakvelidze, A eds.), CRC Press, Boca Raton, Florida, pp. 437–494.
43. d'Hérelle, F. (1922). The Bacteriophage: Its Role in Immunity. Williams and Wilkins Co./Waverly Press, Baltimore.
44. Goodridge, L. and Abedon, S.T. (2003) Bacteriophage biocontrol and bioprocessing: application of phage therapy to industry. *SIM News* **53**, 254–262.
45. Stopar, D. and Abedon, S.T. (2008) Modeling bacteriophage population growth, in *Bacteriophage Ecology* (Abedon, S.T. ed.), Bacteriophage Ecology, Cambridge University Press, Cambridge, UK, pp. 389–414.
46. Breitbart, M., Rohwer, F. and Abedon, S.T. (2005) Phage ecology and bacterial pathogenesis, in *Phages: Their Role in Bacterial Pathogenesis and Biotechnology* (Waldor, M.K., Friedman, D I and Adhya, S L eds.), ASM Press, Washington DC, pp. 66–91.
47. Bull, J.J., Millstein, J., Orcutt, J. and Wichman, H.A. (2006) Evolutionary feedback mediated through population density, illustrated with viruses in chemostats. *Am. Nat.* **167**, E39–E51.
48. Abedon, S.T. (1990) Selection for lysis inhibition in bacteriophage. *J. Theor. Biol.* **146**, 501–511.
49. Hershey, A.D. (1946) Mutation of bacteriophage with respect to type of plaque. *Genetics* **31**, 620–640.

Chapter 18

Practical Methods for Determining Phage Growth Parameters

Paul Hyman and Stephen T. Abedon

Abstract

Bacteriophage growth may be differentiated into sequential steps: (i) phage collision with an adsorption-susceptible bacterium, (ii) virion attachment, (iii) virion nucleic acid uptake, (iv) an eclipse period during which infections synthesize phage proteins and nucleic acid, (v) a "post-eclipse" period during which virions mature, (vi) a virion release step, and (vii) a diffusion-delimited period of virion extracellular search for bacteria to adsorb *(1)*. The latent period begins at the point of virion attachment (ii) and/or nucleic acid uptake (iii) and ends with infection termination, spanning both the eclipse (iv) and the post-eclipse maturation (v) periods. For lytic phages, latent-period termination occurs at lysis, i.e., at the point of phage-progeny release (vi). A second compound step is phage adsorption, which, depending upon one's perspective, can begin with virion release (vi), may include the virion extracellular search (vii), certainly involves virion collision with (i) and then attachment to (ii) a bacterium, and ends either with irreversible virion attachment to bacteria (ii) or with phage nucleic acid uptake into cytoplasm (iii). Thus, the phage life cycle, particularly for virulent phages, consists of an adsorption period, virion attachment/nucleic acid uptake, a latent period, and virion release (*(2)*, p. 13, citing d'Herelle). The duration of these steps together define the phage generation time and help to define rates of phage population growth. Also controlling rates of phage population growth is the number of phage progeny produced per infection: the phage burst size. In this chapter we present protocols for determining phage growth parameters, particularly phage rate of adsorption, latent period, eclipse period, and burst size.

Key words: Adsorption, adsorption constant, eclipse period, latent period, lysis timing, multiplicity of infection, MOI, rise period.

1 Introduction

Bacteriophagy always takes place in the same manner; the sequence of events is always the same. The bacteriophage corpuscle must invariably become fixed to the bacterium to

Martha R. J. Clokie, Andrew M. Kropinski (eds.), *Bacteriophages: Methods and Protocols, Volume 1: Isolation, Characterization, and Interactions, vol. 501*, © 2009 Humana Press, a part of Springer Science+Business Media
DOI 10.1007/978-1-60327-164-6_18 Springerprotocols.com

> exercise its action. Destruction of the bacterium is always accomplished by bursting. The bacteriophage corpuscles always multiply within the bacterial cell and are always liberated with the rupture of this cell. But the time required for the fixation to take place, the time necessary for the bacterium to undergo rupture, the number of young bacteriophage corpuscles developing within the bacterium to be liberated with its rupture, all vary in each particular case, according to a multitude of conditions which vary from one experiment to another. — Felix d'Herelle (*(3)*, p. 115)

In the study of phage life cycles, two break points bracket what we describe as the phage "extracellular search" for new bacteria *(4)*. These break points are the release of a virion from a phage-infected bacterium and the point of irreversible adsorption of the virion to a to-be-infected cell. The traditional study of phages as whole organisms, that is, as phage infective centers, both recognizes and helps define these break points: One key approach to phage whole-organismal characterization considers phage adsorption, which we will define as beginning some time during the phage extracellular search and ending with phage irreversible attachment to a bacterium. Another approach considers phage infection, which we will consider to begin at the point of phage irreversible adsorption and to continue until phages begin their extracellular search. In the (translated) words of d'Herelle *(5)*, phage are first "fixed" to bacteria, "each . . . penetrates to the interior," "there multiplies," and then "liberates" the phage "that have been formed in the bacterial protoplasm" when the infected bacterium "bursts" (pp. 60–61).

Here we describe methods involved in measurement of the phage life cycle: the phage adsorption curve and the phage one-step growth experiment. Especially with the latter *(6)*, Ellis and Delbrück in 1939 established that phage infection is amenable to quantitative dissection, with subsequent experimentation along this conceptual framework leading to our modern understanding of the molecular basis of life *(7)*. In considering one-step growth, we will also describe both eclipse period estimation and stand-alone burst size determination, plus provide alternative methods for determining phage latent period.

All the presented protocols may be performed without employing any molecular techniques. Indeed, the primary technique involved, other than various manipulations of broth culture (pipetting, dilution, centrifugation, etc.), is the plaque assay (*see* also **Chapters 7, 14, 16** and **17**). Although these methods remain mostly unchanged from those presented by Adams *(2)* for adsorption constant determination and Ellis and Delbrück *(8)* for one-step growth, we will provide both refinements and, where applicable, technical variations that may apply to particular bacteria or phage. As a caveat, we describe why it is important to consider time dependence when calculating phage multiplicities, which are ratios of phage to bacteria (**Notes Section 4.5**).

2 Materials

While the particular materials used in the various protocols will be determined by the specific phage–bacteria system being studied, in general the materials for phage whole-organismal characterization *(9, 10)* include: (i) phage and bacterial growth media, (ii) diluent for serial dilution, (iii) bottom and top agars for plating, (iv) pipetting devices for measuring and moving about different volumes of liquid, (v) water baths, shakers, and incubators for maintaining constant conditions, and (vi) chloroform or other lysing agents for eclipse period or adsorption constant determination.

3 Methods

3.1 Adsorption Constant Determination

If we consider adsorption as a reaction in which the substrates are free phage and bacteria, then the product would be the phage-adsorbed cell. The adsorption reaction thereby may be followed by looking at the disappearance of either substrate or the appearance of product. Adsorption rates are presented as adsorption constants (k) and are specific for a given phage, host, and physical and chemical adsorption conditions. An adsorption constant is presented as a unit volume per time, typically ml/min, and is a function of bacterium size, phage particle effective radius, rate of phage diffusion, and the likelihood of phage attachment given collision. Adsorption, in principle, may be differentiated into phases of diffusion and attachment *(11)*. Historically, however, it is the undifferentiated adsorption constant that most phage workers have determined.

Adsorption measurements may be used to identify phage and host receptor mutants *(12, 13)*, bacterial membrane stability (as indicated by continued ability to adsorb phage; 14), organic and inorganic cofactors for adsorption *(15–19)*, and specific environmental niches for phage infections *(20–22)*. The same basic protocol for determining phage adsorption to bacteria has also been used to determine phage adsorption to fragments of bacteria *(23, 24)* as well as to abiotic substances such as clays *(25–27)*.

Phage adsorption constant determination begins by mixing phage with bacteria within an appropriate medium. This is followed by assessment, as a function of time, of free phage loss

(**Section 3.1.3** and **Notes Section 4.1**), infected-bacteria gain (**Notes Section 4.2**), or uninfected-bacteria loss *(2)*. The latter is employed especially if dealing with infection-competent but otherwise nonviable phage and essentially involves repeated determination of a phage's "killing titer" *(9, 11, 28)*. Here, we present practical considerations for determining adsorption rates and then an annotated sample protocol that considers phage adsorption as a function of free phage loss (**Section 3.1.3**). *See* **Fig. 18.1** and **Table 18.1** for examples of adsorption curves and adsorption constant calculations.

3.1.1 Adsorption Conditions

Ideally phage adsorption determinations are done within media that approximates—in terms of adsorption cofactors, osmolarity, pH, temperature, etc.—the environment in which the phage under study would normally adsorb. Bacteria size and/or physiology can also be a concern *(29, 30)*, in one system affecting adsorption rates over 60-fold *(31)*, and may be especially important with bacteria that have multiple life phases *(32, 33)* or that can produce capsule layers *(34, 35)*. Furthermore, not all bacteria express phage receptors constitutively nor at constant levels *(36, 37)*. Another factor affecting rates of phage adsorption is motion within or of the adsorption medium, where too lit-

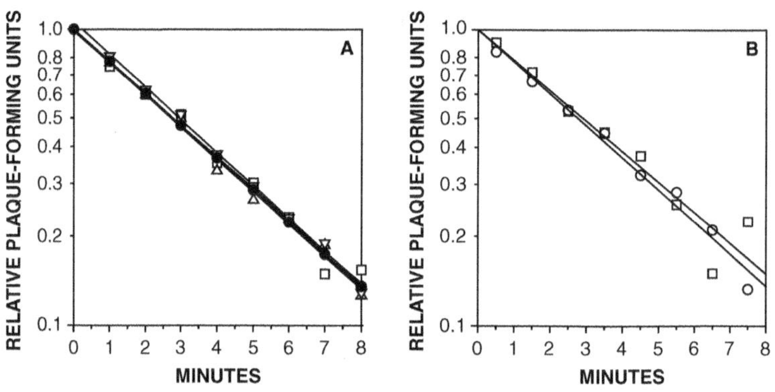

Fig. 18.1. Comparison of theoretical and actual adsorption experiments. For panel A data is either from **Table 18.1** (circles and squares) or generated similarly, with *y*-axis adjusted by dividing by 1000. "Experimental" titers are derived as from a single plating per time point with randomly generated error as for **Table 18.1**. Linear regression lines for each curve are as shown but the zero point is indicated (by a closed circle) only for the theoretical curve. Phage titers have *not* been divided by a calculated *y* intercept. Panel B is from Abedon *et al.* *(66)* and represents an actual adsorption experiment, one comparing phage RB69 wild type (circles, slope $= -0.249$, $r = -0.991$, $k = 9.03 \times 10^{-10}$ ml/min) and an RB69 mutant, sta5, which displays a shorter latent period than wild type, but apparently identical or nearly identical adsorption constant (squares, slope $= -0.237$, $r = -0.956$, $k = 8.60 \times 10^{-10}$ ml/min). Note that time points for this experiment were taken on the half minute (i.e., 0.5, 1.5,..., 7.5) and that increased error can be seen with lower plate counts (6.5 and 7.5 min time points). Panel B is reprinted from *(66)* with permission from the American Society for Microbiology.

Table 18.1
Theoretical and Hypothetical Adsorption Experiments

Min (*t*)	Theoretical titer (*P*)[a]	ln (*P*)	"Experimenttal" titer (*P*)[b]	ln (*P*)	% Difference
0	1000.00	6.91	—	—	—
1	778.80	6.66	748.44	6.62	–3.9%
2	606.53	6.41	601.48	6.40	–0.8%
3	472.37	6.16	513.75	6.24	8.8%
4	367.88	5.91	353.10	5.87	–4.0%
5	286.50	5.66	303.03	5.71	5.8%
6	223.13	5.41	232.12	5.45	4.0%
7	173.77	5.16	149.10	5.00	–14.2%
8	135.34	4.91	153.84	5.04	13.7%
Slope		–0.2500		–0.2451	
Corr (*r*)		–1.000		–0.989	
k		2.50×10^{-9}		2.45×10^{-9}	

[a] Theoretical titers are calculated assuming an adsorption constant (k) of 2.5×10^{-9} ml/min and a bacterial density (N) of 1×10^{8} bacteria/ml such that the resulting free phage titer (P) is calculated as $P = P_0 e^{-kNt}$ (equation (18.1)) where t is in minutes as indicated in the table and P_0 is the initial phage density (at $t = 0$).
[b] "Experimental" titers were calculated as above except that titers were varied using a random number generator that increased or decreased theoretical titers up to 2 times the square root of the expected titer. Percent differences between theoretical and "experimental" values are shown in the last column.

tle motion (i.e., lack of mixing or agitation) or too much motion (e.g., placing phage and bacteria into a running blender) can both result in reduced rates of phage adsorption *(31, 38, 39)*. Consequently, it is important to indicate both adsorption conditions and bacterial preparation conditions when reporting adsorption constants. Because of the potential for differences in bacterial strains, physiology, or even techniques, it is preferable to compare adsorption constants of more than one phage or condition by making the measurements oneself rather than comparing values obtained from different sources.

One approach to assuring similarity between adsorption and growth environments is to employ identical conditions for both. This places limitations on adsorption protocols, however, since within complete growth media bacteria can grow and phage can produce virion progeny. It is possible to slow or stop bacterial and phage metabolisms (**Notes Section 4.3**). Care must be taken, however, that the chosen inhibitors do not also modify rates of phage adsorption, change bacterium size, or distort the phage receptor molecules found on the surface of bacteria.

3.1.2 Phage, Bacterial, and Experimental Dilution

Phage densities generally should be low enough that multiplicities are less than one. These low multiplicities serve to prevent multiple adsorption, "lysis from without" *(10, 30, 40, 41)*, and other means of interference with subsequent phage adsorption (e.g., limits to a bacterium's adsorption capacity; 2,30,40). Phage densities also should be chosen to minimize diluting steps for plating. Phage dilutions during experimental set up, if at all possible, should be made into the adsorption media being employed so as to minimize dilution of this media upon phage mixture with bacteria and/or to avoid carry over of ingredients found in phage diluent but not in adsorption media.

Bacterial densities should be chosen with phage multiplicity in mind, but especially as a determinant of experimental duration. Generally the more bacteria present, the faster phage will adsorb, the faster data points must be collected, and the sooner experiments will be over (**Notes Section 4.4**). Faster determination is also preferable particularly if bacteria are allowed to metabolize over the course of experiments *(8,31)*. Adams *(2)* suggests adjusting conditions so that between 20 and 90% of phage adsorb over the course of adsorption determination. Plating error limits the precision of measurements when free phage titers are less than a few percent of total infective centers. Potential phage-stock inhomogeneities or inefficiencies in free phage separation limit the accuracy when free phage titers are comparable to phage-infected cell titers.

We prefer to employ phage and bacterial densities, as well as experiment durations, such that plating during experiments results in approximately 100 to 700 plaques per plate. For example, to 900 μl of adsorption medium containing an appropriate density of bacteria we might add 100 μl of 5×10^6 phage/ml (the latter equals 10^3 phage/plate $\times$ 5 ml of chloroform-containing broth $\times 10 \times 10 \times 10$ which, respectively, represent the inverse of three serial removals of 100 μl, i.e., 0.1 ml, each, one to the adsorption mixture, one to the chloroform-containing broth, and one to the plate). Given some expectation of what rates of adsorption a phage will display, one can also estimate what bacterial density (N) should be employed during adsorption rate determinations:

$$N = -\ln(P/P_0)/kt \qquad (18.1)$$

where P and P_0 are ending and starting phage densities, respectively, k is the phage adsorption constant, and t is the time over which one desires to have phage adsorption to take place. Ending phage density (P) can be expressed as a percentage, with value, for example, between 80 and 10%, as suggested by Adams (i.e., as indicated two paragraphs above in terms of percentage of phage adsorbing rather than the numbers or fractions of unadsorbed phage remaining that define P). P_0 consequently would be set equal simply to 100%. Thus, for example, an order of magnitude reduction in free phage density may be achieved over an 8-min period (with $k = 2.5 \times 10^{-9}$ ml/min) by employing $N = -\ln(0.1)/(2.5 \times 10^{-9} \times 8) \approx 10^8$ bacteria/ml. Note that equation (18.1) is simply a rearrangement of

$$P = P_0 e^{-kNt} \quad (18.2)$$

which calculates rates of free phage loss to adsorption as a function of time (**Notes Section 4.4**).

A wider range of adsorption may be obtained by employing greater initial phage densities and incorporating additional phage dilutions. A simple approach to accomplishing this is to double phage densities and then initially plate 50 μl, e.g., for the first five or six time points, and then 100 μl for the last five or six time points (**Section 3.1.3**). To save on materials one may employ spot tests of 5 or 10 μl *(9, 42)* for pilot experiments, particularly if one follows adsorption in terms of free phage loss (**Notes Section 4.1**).

3.1.3 Quantitative Determination of Phage Adsorption

Well prior to the actual experiment one should determine the titer of any phage stock or stocks which will be characterized (*see* **Section C, this volume**). Obtaining a highly accurate titer (e.g., within 10%) is not crucial since the zero point is not employed in the adsorption constant calculation. Determination of a phage adsorption constant by measuring the decline of free phage may then be accomplished as follows:

(i) Obtain a bacterial culture of appropriate physiology *(9)*. This can be a growing culture or, more conveniently, one for which bacterial growth has been halted (**Notes Section 4.3**).

(ii) Determine bacterial density by some combination of total count, viable count, or standardized estimation *(9)*. Accurate determination is crucial for accurate adsorption-constant calculation and is used to adjust experimental bacterial densities prior to phage addition (*see* **Section 3.1.2** to determine what bacterial density one should employ).

(iii) Mix phage with bacteria by swirling or gentle vortexing within a suitable adsorption medium that has been pre-equilibrated to the temperature at which the adsorption experiment is being performed. Time of initiation of this mixing represents the zero time point.

(iv) Though there exist many strategies for distinguishing free phage from phage-infected bacteria (**Notes Sections 4.1** and **4.2**), we prefer the approach of Séchaud and Kellenberg *(43)*, which is the chloroform-mediated inactivation of bacteria *(6, 43, 44)*. To do this, remove 4.5 ml of the phage–bacterial mixture to 4.9 ml of experimental- or room-temperature broth that has been saturated with chloroform (i.e., by adding a few drops). Vortex this mixture and then let it stand at experimental or room temperature until enumeration is convenient, e.g., no more than a few hours, or shorter if phage are demonstrably labile under these conditions.

(v) Generally one obtains eight data points per adsorption curve, at times 1, 2, 3, 4, 5, 6, 7, and 8 min. Two curves may be easily done simultaneously with the second one done on the half minute. For phage with very short eclipse periods, or bacteria with rapid doubling times, curves instead may be done by removing volumes at 0.5, 1.0, 1.5, 2.0, 2.5, 3.0, 3.5, and 4.0 min., though close time points necessitate greater timing precision *(31)*. Precise control of the timing of the zero point and the number of phage is not crucial, though ideally six or more usable counts may be found among the eight data points taken. Precise control of bacterial densities also is not crucial but, as noted, accurate determination of bacterial density *is* important.

Slopes of adsorption curves are determined using natural-log (i.e., ln) transformed free phage determinations graphed as a function of time (**Table 18.1** and **Fig. 18.1**). The adsorption constant (k) is then equal to the opposite of the resulting slope divided by the density of bacteria (N) present in the adsorption mixture (that is, $k = -\text{slope}/N$). The correlation coefficient (r) of an adsorption curve provides an easily obtained measure of curve quality, though not of curve accuracy. For example, one might retain only those curves falling above a given cut-off such as $r \geq -0.90$ or -0.95. It is preferable, also, that one visually inspect adsorption curves to identify consistent deviations from linearity. For graphical presentation of data, one can anchor graphs at an initial phage density of 1.0 by dividing free phage densities by the calculated y-intercept. However, do not represent this calculated zero point as a data point on graphs nor in adsorption constant calculations (**Fig. 18.1B**).

To minimize error in adsorption constant determination we recommend multiple experimental repeats with single platings per time point rather than multiple platings per data point (replating if necessary is OK, though for consistency one should replate experiments in whole if replatings have been greatly delayed). Curves done under different conditions or involving different phages should then be compared in terms of calculated adsorption constants. To produce publication-quality figures of individual experiments one can reduce per-experiment noise by making multiple titer determinations per time point. However, note that technically these individual time points should not be presented with error bars *(45)*. For graphical comparison of multiple experiments it is especially important to keep bacterial densities consistent.

3.1.4 End point Determination of Phage Adsorption

End point determinations supply a more qualitative indication of phage adsorption *(46–48)* since they can miss biphasic (*(8)*, especially due to a phage "residual fraction"; 30) or other non linear adsorption kinetics *(49)*. They may be performed similarly to the above-described kinetic determination (and with similar caveats) except, of course, by taking fewer time points. This is often reported as a percentage of adsorbed (or unadsorbed) phage without any calculation of adsorption rate *(46–48)*. Even simpler, a qualitative indication of adsorption may be obtained simply by spotting free phage onto a nascent bacterial lawn *(9, 42)*, with spot formation indicative of successful phage adsorption. The converse is not also true, however, since phage failure to form spots could be due to reasons other than phage failure to adsorb to a bacterium. In general we feel that actual adsorption curves should be employed whenever quantitative indication of phage adsorption properties is desired, that is, when reaching any conclusion other than whether adsorption did or did not occur.

3.2 Latent Period Determination

The latent period is the delay between phage adsorption of a bacterium and subsequent phage-progeny release as observed for a given phage infecting a given bacterial strain under a given set of growth conditions (which is a "problem of three bodies" as described by d'Herelle *(5)*, p. 6). Measurement especially of a phage's latent-period duration may be accomplished either by detecting the liberation of phage virions (**Section 3.2.1**) or by detecting the destruction of bacterial infections (**Section 3.2.2**). It is also possible to follow phage lysis by microscopic observation *(2, 4, 5, 11, 40, 50)*.

The minimum latent period also may be described as a *constant* period *(30)* because plaque-forming units (pfus) do not appreciably change in number until culture lysis has begun *(8)*. This constant period is followed by a "rise," referring to the "rise" in pfu numbers observed upon lysis during one-step growth *(8)* (**Fig. 18.2A**). The rise is the finite time over which lysis of a bacterial population occurs *(30)*. For simplicity we limit our protocols to lytic phage. Latent period (as well as eclipse period and burst size) may be determined for temperate phages following lysogen induction, which is equivalent to initiation of infection via phage absorption.

3.2.1 One-Step Growth

One-step growth experiments allow "one to determine very simply the effect of changes in the physical and chemical environment on the duration of the infectious cycle and on the yield of virus per infected host cell" (*(2)*, p. 15). One-step (a.k.a., single-step) growth may also be employed to determine the duration

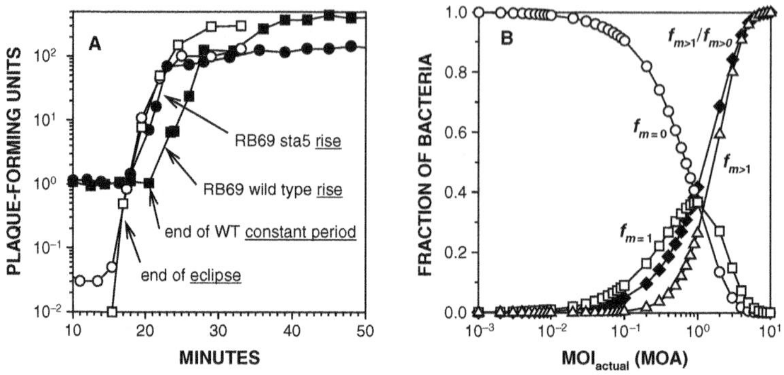

Fig. 18.2. One-step growth experiment (panel A, closed symbols) with eclipse period determination (panel A, open symbols). Squares represent phage RB69 wild type (WT) while circles represent the shorter latent-period phages RB69 mutant, sta5. Note the similarity of eclipse periods between the two phage but the differences in latent periods and burst sizes. Indicated are the end of the eclipse period for each phage, the end of the constant period for phage RB69 WT, and a portion of the rise period for each phage (the latter is for solid-symbol curves only). Curves were normalized to an initial pfu count of 1.0. Panel B explores fractions of bacteria which have been adsorbed to a various degrees as functions of MOI_{actual}. Shown are $f_{m=0}$ (open circles) which is the fraction of bacteria that are uninfected, $f_{m=1}$ (open squares) which is the fraction of bacteria that are adsorbed/infected by a single phage, $f_{m>0}$ (open triangles) which is the fraction of bacteria that are adsorbed by one or more phage, and $f_{m>1}/f_{m>0}$ (closed diamonds) which is the fraction of *infected* bacteria that are infected/adsorbed by more than one phage. In all cases, multiplicity of infection assumes 100% phage adsorption (**Notes Section 4.5**). Panel A is reprinted from *(66)* with permission from the American Society for Microbiology.

of the phage eclipse period (**Section 3.2.1.7**). Since latent period typically is measured as a bulk property of phage-infected cultures, precise measurement requires some degree of metabolic synchronization of the start of phage infections (**Section 3.2.1.1**). A sudden increase in pfus signals the end of the phage constant period and the beginning of the phage rise.

In **Section 3.2.1.6** we provide protocols for one-step growth characterization. First, however, we present an overview of one-step growth theory and practice. This we do so that individuals may effectively adapt methods to the peculiarities of individual laboratories and phage–host systems, plus avoid common pitfalls during inevitable protocol tinkering.

3.2.1.1 Synchronizing Phage Infections

One-step growth typically begins with bacteria grown to a suitable log-phase density and then, ideally, entails only minimal manipulation so as to preserve an optimal physiology for phage replication. Subsequent synchronization of the phage infections, however, represents the "essential feature" *(30)* of one-step growth experiments, allowing greater precision in determining the length and timing of the constant, rise, and eclipse periods. A number of different approaches can be used to synchronize phage infections, ranging from short, rapid adsorption periods followed by culture dilution to terminate phage adsorption (**Notes Section 4.3** and **4.6**) to halting bacterial metabolism during phage adsorption (**Note Section 4.3**) and/or post-adsorption virion inactivation (**Notes Section 4.1**) done in conjunction with culture dilution. The particular choice depends on the phage/bacteria system being studied as well as the desired precision of timing determination. Note that, wherever possible, phage dilutions prior to phage addition to bacteria should be made into the same type of media that bacteria will be suspended in over the course of phage adsorption.

3.2.1.2 Using Phage Multiplicities of Less Than One

One-step growth usually assumes that a large majority of phage-infected bacteria are infected with only a single phage, even though it is typically assumed that phage one-step characteristics are not necessarily affected by phage multiplicity, other than by lysis from without *(2, 30)*. To assure a reasonable approximation of singly infected bacteria it is important to initiate phage infections using a phage multiplicity that is considerably less than one. One assumes a Poisson distribution to describe the likelihood of bacteria adsorption by only a single phage for a given phage multiplicity (M) *(2, 11, 51)*, at least for phage multiplicities of less than 2 *(2, 52)* (*see* **Notes Section 4.5** for discussion of the concept of phage multiplicity). More generally, the likelihood of bacterial adsorption by a total of m phage (f_{m}), where m is a non negative integer, is described by

$$f_{\mathrm{m}} = \mathrm{e}^{-M} M^{m}/m! \tag{18.3}$$

which for $m = 0$ and $m = 1$ reduces to $f_{m=0} = \mathrm{e}^{-M}$ and $f_{m=1} = \mathrm{e}^{-M}M$: the fraction of bacteria infected with no phage and one phage, respectively (**Fig. 18.2B**). The fraction of bacteria infected by more than one phage therefore is described by $f_{m>1} = 1 - \mathrm{e}^{-M}(1 + M)$. Of the total bacteria infected by at least one phage, the fraction of bacteria infected by more than one phage is described by

$$f_{m>1}/f_{m>0} = (1 - \mathrm{e}^{-M}(1 + M))/(1 - \mathrm{e}^{-\mathrm{M}}) \tag{18.4}$$

Thus, for a multiplicity of $M = 1$ we find that 42% of *infected* bacteria are infected by more than one phage whereas for a multiplicity of $M = 0.1$, as suggested by *(2, 51)* for one-step growth, this fraction reduces to 5%.

3.2.1.3 Post-Adsorption Dilution

Following the initial period of synchronized phage adsorption (**Section 3.2.1.1**), one must prevent subsequent phage adsorption to uninfected bacteria *(30)*, which could skew burst size and rise measurements, or to infected bacteria, which can inactivate virions or, for some phages, induce lysis inhibition *(41)*. Inhibition of subsequent phage adsorption is complicated, however, by three factors: (i) a need to maintain optimum cell physiology (which is not necessarily consistent with deployment, for example, of chemical inhibitors of phage adsorption or removal of phage adsorption cofactors; *(53)*); (ii) a requirement for subsequent titering of liberated free phage (which means that, at best, virion inactivation can be employed only transiently to inhibit subsequent phage–bacterial interaction); and (iii) the fact that both phage and bacterial densities will tend to rise over the course of one-step growth, thereby increasing the likelihood of phage adsorption to bacteria (**Notes Section 4.4**). The inhibition of subsequent phage adsorption consequently is typically accomplished via culture dilution *(2, 6, 31)*. In some cases dilution may be done in conjunction with means of reducing the phage adsorption constant *(40)*, such as via the removal of adsorption cofactors necessary for subsequent phage adsorption *(54)* or by adding excess salts (as cited by 40).

3.2.1.4 Retaining Sufficient pfus

Since one-step experiments are employed to determine bulk properties of phage-infected bacteria, it is important to retain statistically reasonable numbers of infected bacteria while simultaneously diluting cultures to inhibit subsequent phage adsorption. To determine what dilutions to employ to accomplish these goals it is best to work backwards. In the following protocol (**Section 3.2.1.6**), for example, we employ a maximum post-dilution pfu density of 4000/ml. With a phage burst size of 100 this pfu density will produce a total of 4×10^5 phage/ml (= 4000 × 100), which is sufficiently low that phage adsorption to bacteria over a given time interval will be minimal (**Notes Section 4.4**). If one employs a phage multiplicity of 0.1 (to minimize multiple adsorptions) and an initial bacterial density of 10^8/ml, then a 2,500-fold culture dilution is required to produce a pfu density of 4000/ml ($2,500 = 10^8 \times 0.1/4000$). This will result in a bacterial density of 4×10^4/ml($= 10^8/2,500$), which is also sufficiently low, given relatively short phage latent periods, that post-dilution interaction between free phage and bacteria will be minimal.

One can check the likelihood of phage adsorption to bacteria by multiplying bacterial density, phage density, adsorption constant, and latent period. For a phage with an adsorption constant of 5×10^{-9} ml/min, burst size of 100, and latent period of 30 minutes, this would result in 5×10^{-9} ml/min $\times 30$ min $\times 4 \times 10^5$ phage/ml (as present post-rise) $\times 4 \times 10^4$ bacteria/ml (which, given low initial phage multiplicities, will be mostly intact) ×3 (which accounts for roughly one and one-half 20-minute bacterial doublings) = 7200 phage-bacteria/ml, which is just 1.8% of total phage present following lysis (4×10^5/ml). By titering phage post-lysis from a 10-fold or 100-fold dilution of the already diluted culture, this reduces the number of adsorptions by 100-fold or 10,000-fold, to 72 or 0.72, respectively. This is just 0.18 or 0.018% (again, respectively) of the phage present in this diluted culture, post-rise (i.e., of 4×10^4 or 4×10^3 phage/ml, respectively). To further minimize any contribution to the overall post-lysis titer by these later-adsorbing phage, we recommend that post-rise titering for burst size determination be completed within about 2 latent periods of the initial point of phage-induced bacterial lysis, unless phage are found to display an unusually long rise.

Note that if one employs a higher initial phage density due to use of higher cell densities, such as to effect more rapid phage adsorption (**Section 3.2.1.1**), and/or one employs a higher phage multiplicity, then this only changes the size of the dilution necessary to result in a final concentration of 4000 pfu/ml. Use of higher phage multiplicities and therefore greater dilution is desirable given determination of very long latent periods because of the potential for uninfected bacteria to grow and thereby repopulate broth cultures *(2)*.

We caution against using combinations of infected-bacteria concentrations and culture volumes that, following maximum dilution of cultures (**Section 3.2.1.6** step (iii)), result in fewer than about ~ 100 infected bacteria per tube. To address this issue, we present a protocol (**Section 3.2.1.6**) in which no fewer than 40 infected bacteria are present per milliliter of a 10 ml culture and suggest continued culture mixing as well as sufficient culture volumes such that at least a few milliliters are retained per tube. For phages displaying very large burst sizes, even greater culture dilution may be desirable, which can be compensated for by concomitantly increasing volumes at maximal culture dilutions (e.g., by a 10-fold increase in culture volume for each additional 10-fold increase in culture dilution beyond those recommended during step (iii) in **Section 3.2.1.6**). We additionally recommend avoiding initial phage multiplicities that are lower than ~ 0.01 due to resulting conflicts between sufficiently diluting bacterial populations and not excessively diluting phage populations.

3.2.1.5 Assaying for Unadsorbed Phage

During experiments, and prior to the onset of lysis, it is desirable to assay for unadsorbed phage *(30)*. This can be accomplished prior to the end of the eclipse period, for example, by removing three 1.0 ml aliquots of culture to sterile tubes, adding a few drops of chloroform to each tube, vortexing, and then letting tubes sit at room or experimental temperature (for alternative approaches to removal of infected bacteria, *see* **Notes Section 4.1**). When plating is convenient, such as following completion of one-step growth, one should plate 500 μl from these chloroform-treated tubes for infective centers, taking care to avoid removing undissolved chloroform. Assaying for unadsorbed phage allows an end-point measure of phage adsorption ability (**Section 3.1.4**) in conjunction with one-step growth, though for precise determination of phage adsorption constants a kinetic analysis is much preferred (**Section 3.1.3**). Just one sampling for unadsorbed phage is needed, however, if steps have been taken to remove these phage from cultures (**Notes Section 4.2**) and a phage eclipse period determination is not also being made (**Section 3.2.1.7**).

3.2.1.6 One-step Growth Protocol

To prepare the infective centers necessary for a one-step growth experiment one needs to first synchronize the initiation of phage infections (**Section 3.2.1.1**), employing a phage multiplicity of 0.1 (**Section 3.2.1.2**), then dilute phage and bacteria into pre-warmed growth media (**Sections 3.2.1.3** and **3.2.1.4**) and, if necessary, remove unadsorbed phage (**Notes Section 4.2**). Our preference is to design experiments such that 50 μ l of the diluted culture will contain approximately 200 pfu (i.e., 4000 pfu/ml). If one begins with 10^8 bacteria/ml and employs a phage multiplicity of 0.1 then this entails a 2,500-fold dilution (**Section 3.2.1.4**). Bacterial densities will be sufficiently low as to make post-dilution culture aeration unnecessary *(9,51)*. Nevertheless, we suggest that cultures still be shaken, or at least periodically gently vortexed or swirled by hand, so as to promote culture mixing over of the course of one-step growth. The remainder of the experiment is then performed as follows:

i. For burst size determination, pfu enumeration prior to the onset of lysis is necessary, and it is important to do sufficient replicate platings (e.g., at least three, ideally more) since for subsequent burst size determination (below) the average of these pre-lysis titers will be found in the denominator of the ratio of liberated phage to originally infected bacteria (= burst size). At this point one should also assay for unadsorbed phage (**Section 3.2.1.5**). For constant period determination, without simultaneous burst size or eclipse-period measurement, these initial enumeration steps may be skipped.
ii. To precisely determine latent period it is important to achieve as many platings as possible just before and during lysis since it is the increase in pfus that defines the end of the latent period *(6, 11)*. For phage and experimental conditions in which lysis occurs over relatively short periods, it can be best to record "on the fly" the timing of sequential platings sampled as rapidly as possible rather than attempting to plate at a rapid but constant, pre-set rate (e.g., plating on 30 s intervals). Trial and error will be necessary to determine just when this lysis is expected to occur.
iii. For rise as well as burst size determination it is necessary to follow cultures well past the initiation of lysis and, ideally, post-rise, which is when phage titers stabilize. To capture the

rise portion of one-step growth do the following: Just prior to the initiation of lysis begin interspersing 50 μ l samplings of a 10-fold dilution with 50 μ l samples of the original culture. Subsequently, as the rise progresses, one can replace sampling of the original culture with 50 μ l sampling of a 100-fold dilution. At approximately the end of the rise one then continues sampling from just this latter dilution, unless phage burst sizes are in excess of approximately 250, thereby necessitating further dilution prior to plating. To minimize the impact of dilution errors, consider generating (subsequent to the initial culture dilution step) ~five 10-fold dilutions and ~ten 100-fold dilutions, both in 10-ml volumes. These dilutions will respectively contain 400 and 40 of the original pfus per ml so, in 10-ml volumes, should represent an adequate sampling of the population. Maintain these dilutions at experimental temperatures. At appropriate times (**Table 18.2**), swirl or vortex dilutions and then plate 50 μ l, plating from each tube only once. For phages displaying relatively small burst sizes, consider sampling, from 100-fold dilutions, volumes that are in excess of 50 μ l. As high as a 20-fold greater volume (1000 μl) is usually easily plated.

Taking dilutions into account, we prefer to present one-step growth data employing log-transformed pfu determinations. The beginning of the rise minus the time of initiation of infection

Table 18.2
Plating Recommendations for One-Step Growth[a]

Period	0.1–fold[b]	0-fold[c]	10-fold	100-fold
Eclipse	1 or 3	—	—	—
Constant	—	3 or more	—	—
early rise	—	up to 3[d]	up to 3[d]	—
middle rise	—	—	up to 3[d]	up to 2[d]
late rise	—	—	—	up to 2
early post-rise	—	—	—	up to 2
later post-rise	—	—	—	4 or more

[a] Shown are the number of recommended platings with each plating representing an independent dilution.
[b] Remove 1.0 ml to a sterile tube, add a few drops of chloroform (recording time), incubate at room or experimental temperature, then plate 500 μ l. Do at least once if unadsorbed phage had been actively removed from cultures (**Notes Section 4.2**) or at least three times if they were not. All other recommended platings are of 50 μ l.
[c] 0-, 10-, and 100- fold refer to different degrees of dilution of cultures from which pfus are enumerated.
[d] Platings at multiple dilutions within a given row should be alternated.

metabolism operationally defines the phage constant period, often equated with latent period, which is the minimum lysis timing observed within a population of individual phage infections *(2)*.

One can assess whether post-rise platings truly have been plated post-rise by noting the timing of each of plating and then examining the resulting titers. If titers consistently rise from one supposedly post-rise data point to the next, then the titers may not have been gathered after population lysis was complete. Alternatively, one does not want to wait too long to determine post-rise phage titers since with time, even given culture dilution, there is an expectation of at least some progeny phage infecting bacteria and then bursting. One way to avoid this latter problem is to remove adsorption cofactors such as cations during the incubation period *(54)*, at least so long as this demonstrably has no effect on phage replication and maturation.

Note that when interpreting the time points taken during the phage rise, phage bursts which by chance are confined to a single plating will inappropriately suggest that a more rapid rise in phage titer is occurring than is actually the case. As Adams *(2)* points out, "...any point along the *rise* portion of the single step curve may lie well above the curve. Since there is no compensating error which may lead to correspondingly low counts, these high points must be disregarded in drawing the curve" (p. 481). The potential for plating these "confined" bursts serves as good justification for both keeping cultures well mixed over the course of one-step determination (and especially immediately pre-sampling) and to achieve rapid sampling and plating, especially during the rise. Similarly, if supposedly post-rise samplings are somewhat high, particularly if the very earliest are, then this may represent the plating of a confined burst associated with the tail end of the rise.

See also Adams *(2)*, Carlson *(9, 51)*, and Eisenstark *(6)* for one-step growth protocols plus discussion of methodology. *See* Carlson *(9)* for the protocol of a pilot analysis of phage growth kinetics for use prior to formal one-step growth determination.

3.2.1.7 Eclipse Period Determination

The end of eclipse period (or, simply, the "eclipse"; 44)—particularly among phages that lyse their host bacteria to effect phage release—occurs when the first infectious virion is found within the bacterial cytoplasm, a state that may be detected only by artificially lysing the bacterial host to release what virion particles may be present. Bacteria can be lysed in the same way as for phage adsorption rate determination (**Section 3.1.3** step (iv) and **Notes Section 4.1**). Eclipse periods may be determined in conjunction with latent-period determination since chloroform treatment allows one to delay plating. We recommend the same approach to sampling as described in **Section 3.2.1.5** for determination of unadsorbed phage: removal of 1.0 ml of culture followed by addition of a few drops of chloroform, incubation, and then subsequent plating of either 50μl or 500μ l volumes, depending on tube titers (which can be determined initially via spot testing). Alternatively, Doermann *(44)* suggests rapid chilling of cultures, such as by removal of 1.0 ml of culture to pre-cooled (6°C or lower) test tubes as a means of hindering phage progeny maturation, a delaying strategy requiring even less manipulation than immediate chloroform treatment. Cell lysis may then be effected at leisure, though it is important to determine that lysis efficiency has not been compromised. Post-eclipse, change to removing 1.0 ml samples from 10- and 100-fold diluted culture tubes.

We recommend recording the timing of each sampling rather than taking samples on a set schedule. In **Fig. 18.2A** we provide an experiment where two one-step growth curves—including two eclipse-period determinations—were acquired by a single individual (S.T.A.), in parallel for two relatively short latent-period phages. Note that it is at the point where the lysed cultures first possess pfus, which are not simply unadsorbed phage contaminants, that defines the end of the phage eclipse period. Note also in the same experiment that the initial time points were taken well prior to the beginning of the phage rise (i.e., well before the end of the phage constant period), and indeed well prior to the end of the phage eclipse period. Note also that later time points, post-rise, were taken but are not presented.

3.2.1.8 Post-Eclipse and Pre-rise

The historical importance of the discovery of the phage eclipse period somehow "eclipsed" the next and at least equally important intracellular period during which phage progeny mature within infected bacteria. This latter period may be called, for example, a period of "intracellular phage growth" *(55)*, a reproductive *(1)* or adult period *(56)*, a post-eclipse period *(57)*, or, as we prefer, a period of phage-progeny maturation or accumulation *(58)*. Technically this period is not equivalent to Delbrück's "rise" *(30)*, which is a term he affixed to the phage-induced bacterial-lysis period that follows synchronized phage population growth (which, in **Fig. 18.2A**, occurs after 20 min for wild type RB69; solid squares).

The minimum duration of the maturation period is equal to the constant period minus the eclipse period. The rate of phage-progeny maturation during this time is explicitly characterized during eclipse period determination or may be estimated by dividing burst size by the constant period minus the eclipse period. In ecological terms we can describe the phage period of maturation as a sole reproductive period within an otherwise pre-reproductive lytic-phage life cycle *(1)*.

3.2.2 Turbidometric Determination of Phage Latent Period

Determination of phage latent period duration employing culture turbidity, rather than via virion liberation or by microscopic observation, can be traced at least back to Delbrück *(40)*, who compared culture turbidities by eye. The "modern" era of turbidometric determination appears to have begun a short time later with Underwood and Doermann *(59)*, who describe a "photoelectric nephelometer." Doermann *(60)* subsequently employed this device to characterize the extended latent periods associated with the T-even phage *(61)* lysis inhibition phenotype *(41)*. More recently, Young and colleagues (as reviewed in 62,63) have employed turbidometric measures to characterize phage λ lysis. We, too, have extensively employed this technique to study phage T4 lysis inhibition *(4,64,66,67)* as well as phage RB69 lysis-timing evolution *(66)* (**Fig. 18.3**).

Modern turbidometric analysis of phage growth generally employs one of two wavelengths, that employed by Klett colorimeters (660 nm) and A_{550} at 550 nm *(62)*. For phage λ and other temperate phages, turbidometric analysis of latent period typically begins with lysogen induction whereas for virulent phages, such as phage T4, latent periods instead are initiated with phage adsorption. For observation of culture turbidity, densities of infected bacteria must be relatively high, e.g., $\sim 10^8$ bacteria/ml *(2)*. As a consequence, culture manipulation to achieve adsorption synchronization (**Notes Section 4.3**) is not nearly as necessary as with one-step growth experiments (**Section 3.2.1.1**). Since one's goal is to infect a majority of bacteria so that significant lysis may be observed, multiplicities well in excess of 1 are routinely employed (**Fig. 18.2B**). Maintenance of consistent

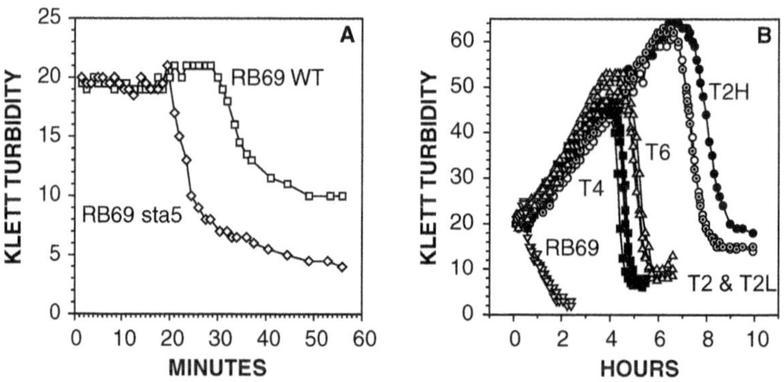

Fig. 18.3. Latent period determination using turbidity measurements. Experiments with phages not displaying lysis inhibition are shown in panel A (MOI = 5, added at time = 0) and experiments with phages displaying lysis inhibition (phage RB69 is exceptional) are shown in panel B (MOI = 10, added at time = 0). Different phage types are as indicated. Figures are reprinted from *(66)* with permission from the American Society for Microbiology.

bacterial physiology, especially via constant and robust aeration is crucial given the ongoing high densities of bacteria required for turbidometric determination.

Typical "lysis profiles" are presented in **Fig. 18.3** where the steep drops in culture turbidity correspond to phage-induced bacterial lysis. Note that though not technically one-step growth curves, in **Fig. 18.3** effectively only single rounds of phage infection and lysis are observed. This is due to the use of relatively high starting multiplicities *(5–10)*. With lower initial phage multiplicities it is possible to observe multiple rounds of infection by turbidity *(4, 66, 67)*, and one can automate the determination of lysis profiles by employing a shaking, incubating, and kinetic microtiter plate reader *(67)*. By using various "tricks" it is also possible to perform lysis profiles in which phage secondary adsorption of already-infected bacteria does not occur. Such experiments can begin with mechanisms of synchronized adsorption (though tailored to involve only minimal dilution; **Notes Section 4.3**), but additionally require—given the high bacteria and infection densities—mechanisms that interfere with subsequent phage adsorption. For instance, one can employ conditional phage mutants that produce adsorption-incompetent virions under non-suppressing conditions, or chase phage infections with anti-phage serum *(4, 65)*. One additionally can augment infections by adding a dosage of secondary phage *(4, 65)*.

3.3 Stand-Alone Burst Size Determination

Burst size determinations have a long history in phage research. We note, for example, that Felix d'Herelle provides an estimated burst size of 18 for a phage of "Shiga bacillus" (pp. 59–60 in the English translation of his 1921 monograph *(5)*). Burst size determination typically is done in a manner similar to that employed for one-step growth determination, except with emphasis placed on the beginning and the end of such curves rather than the middle (i.e., especially steps (i) and (iii) of **Section 3.2.1.6** but not step (ii)). To do these experiments one can take a minimalist approach and just take one pre-lysis data point and one post-lysis data point, and then perform numerous experimental repetitions. However, as per step (i) of **Section 3.2.1.6**, we recommend up to three platings for enumeration of unadsorbed phage per burst size determination. Likewise, we also recommend at least three pre- and also at least three post-rise platings to determine burst size, taking care with the latter that platings really are done post-rise (keeping in mind, again, that presentation of error values is not appropriate if describing individual experiments; *(45)*). Multiple platings yield more robust data with only minimal additional effort. To increase the independence of individual data points, we also highly recommend that generally one employ no more than one plating per dilution for any dilution series employed (step (iii), **Section 3.2.1.6**). Note that with sufficient dilution prior to lysis it is possible to determine burst sizes associated with individual bacteria *(6, 8, 11, 30)*.

4 Notes

4.1 Removing Infected Bacteria from Cultures

Adsorption may be followed as a function of free phage loss. Such protocols require some means of elimination of infected bacteria. This may be accomplished using a number of approaches:

(i) Free phage loss may be determined by employing phage–bacteria combinations that result in phage inactivation upon adsorption. For example, conditionally lethal phage mutants may be adsorbed to non suppressor bacteria *(10)*, phage may be adsorbed to non permissive restriction-modification types, and even dead cells may be employed (provided that the method of killing does not greatly reduce phage absorptive ability) *(6, 31)*. Free phage loss in these approaches is followed by plating using permissive indicator bacteria. Consistent with the use of virion-inactivating agents in general, it is important to sufficiently dilute non-permissive bacteria prior to plating.

(ii) Infected bacteria may be removed prior to phage enumeration. Bacteria removal may be accomplished by physically separating infected bacteria from phage, such as by employing low-speed centrifugation (**Notes Section 4.2**) or filtration *(68)*. Note that separation must be accomplished prior to phage completing their latent period since the resulting phage-induced lysis from within would add to the free phage pool.

(iii) Another approach is to inactivate infected bacteria. This may be accomplished by addition of chloroform *(2, 6, 69)*, though only for phage that are stable in its presence *(70)*. KCN or other energy poisons can activate phage holins *(62)*, thereby prematurely lysing infected bacteria, but may be less effective than chloroform for these purposes *(44, 69)*. High multiplicities of superinfecting phage that are capable of displaying lysis from without, such as phage T4 or especially phage T6, can also lyse phage-infected bacteria, particularly in conjunction with metabolic inhibition *(6, 44, 58)*. Additional possibilities for selectively eliminating or lysing infected bacteria, potentially without harming phage virions, include inducing osmotic lysis using lysozyme (*(44)*, citing 71) plus EDTA *(6)*, employing lysostaphin for *Staphylococcus aureus* phage *(72)*, or even disrupting infected bacteria via sonication *(6)*, at least for phage with sonication-resistant virions *(44)*.

One should expect differences among phage–host combinations with regard to the efficacy of these various approaches.

Regardless of the method employed, infected bacteria must be destroyed in a manner that does not significantly distort free phage numbers, either by damaging free phage or by releasing free phage from post-eclipse period bacteria *(2)*.

4.2 Removing Free Phage from Cultures

The premise of determining phage adsorption as a function of rates of phage infection is that each phage adsorption gives rise to an infective center, an entity capable of giving rise to a single plaque *(2)*. To accomplish this, one must be careful to employ phage multiplicities of much less than one so that each phage adsorption may be registered as an individual infective center (**Section 3.2.1.2**). Subsequent phage adsorption often can be inhibited by diluting the bacteria/phage mixture (**Section Number 4.6**). Precise enumeration of infected bacteria, however, requires separation of free phage from bacteria and/or selective inactivation of free phage. Fortunately, there exist a variety of methods for removing free phage from cultures:

(i) Bacteria may be separated from free phage by low-speed centrifugation *(2, 5, 30)*. Depending on the extent of washing involved, however, separation by centrifugation could involve some carryover of free phage. In addition, phage adsorption could continue throughout the centrifugation step. Care must also be taken to avoid phage-induced bacterial lysis from within since this can simultaneously decrease numbers of infected bacteria while increasing apparent free phage carry over. Such avoidance may be accomplished by inhibiting bacterial metabolism, such as by centrifuging in the cold *(2)*.

(ii) Filtration with a 0.45 μ filter to separate unadsorbed phage from bacteria, followed by resuspension of the bacteria in growth media *(73, 74)* has also been used to separate free phage from bacteria.

(iii) Free phage may be inactivated by exposure to anti-phage serum *(2)*, a method that is commonly used *(60)*. Care must be taken (a) to allow sufficient time for virion inactivation, (b) to reduce the activity of antiserum prior to plating by dilution, and (c) to avoid clumping infected bacteria (since individual plaques could then be formed from multiple infected bacteria). Recently, less specific viricides have been employed to remove free phage from cultures *(75)*.

(iv) In principle, free phage may be inactivated by exposure to heat-killed bacteria *(10)*. As with antiserum-mediated virion inactivation, rapidity of inactivation and a requirement for dilution before plating are concerns. Furthermore, in at least some circumstances boiling bacteria presumably can modify phage adsorption rates or ability *(2)*.

One advantage to inactivating or otherwise removing free phage is that it allows one to initiate experiments using excess phage numbers in cases where phage adsorption is particularly slow *(75)* or where experiments could otherwise benefit from a shorter adsorption period *(2)*, since these excess unadsorbed phage will be mostly removed prior to plating for infective centers. In such cases one can estimate the actual phage multiplicity (ratio of *adsorbed* phage to bacteria; *see* **Notes Section 4.5**) by separately enumerating, post free phage removal, both infective centers (as plaques) and unadsorbed bacteria (as colonies).

4.3 Suppressing Bacterial Metabolism

Phage maturation may be inhibited or delayed by metabolically suppressing bacterial metabolism. Generally the goal of metabolic suppression is some degree of metabolic synchronization of phage cultures particularly over the course of phage adsorption. There exist various strategies aimed at achieving this end.

(i) The simplest approach towards adsorption synchronization is to allow for an only short period of phage adsorption, though any duration of this period will produce an extension of the phage rise *(2)*. Rapid phage adsorption is especially useful for this approach, requiring high bacterial densities and a reasonably large phage adsorption constant. Adams *(2)* suggests employing bacteria (such as *Escherichia coli*) grown to 5×10^7/ml whereas Carlson *(9, 51)* recommends growing bacteria (*E. coli* in at least the first instance) to 3×10^8 and 2×10^8/ml, respectively. Eisenstark *(6)* similarly suggest a culture density of 2×10^8/ml for *Salmonella typhimurium*. Ellis and Delbrück grew *E. coli* also to 2×10^8/ml in their initial studies on phage growth *(8)*, diluting the bacteria-phage mixture to abruptly end adsorption. This dilution should be at least 100-fold *(2)* and is best done into prewarmed growth media to allow an uninterrupted infection process. Additional methods—particularly addition of anti-phage serum (**Notes Section 4.2**)—can be used prior to dilution to terminate free phage adsorption.

(ii) To truly metabolically synchronize phage infections one must inhibit bacterial metabolism prior to phage addition. At the end of the adsorption period the inhibitor is removed, often in conjunction with unadsorbed phage, and the infection cycle is then allowed to begin. A commonly used metabolic inhibitor is KCN *(2, 6, 29, 58, 76–79)*, the use of which for adsorption synchronization is attributed to Benzer and Jacob *(80)*.

(iii) Another method of metabolic inhibition is to use centrifugation to remove the bacteria from the growth medium, washing, and then resuspending bacteria in non nutritive salt buffer, i.e., one containing necessary adsorption cofactors but otherwise lacking in carbon or energy sources *(2)*. Alternatively, filtration may be employed to wash bacteria *(6)*. Adsorption takes place in the salt buffer and then the infection is initiated by again centrifuging down the bacteria *(2, 5)*, leaving the unadsorbed phage in the supernatant, and then resuspending the infected bacteria into prewarmed, complete growth media. An advantage of employing washed cells is that higher bacterial densities may be employed, thereby either hastening or allowing more complete phage adsorption. However, starvation or the presence of metabolic inhibitors can increase bacteria susceptibility to lysis from without *(44)*, especially if one employs higher phage multiplicities. Starving or metabolic inhibition also can potentially impact a bacterium's physiology post resuspension into growth medium.

(iv) Bacteria metabolism may be halted by chilling in an ice bath, then warming bacteria prior to infection, or by employing a protein-synthesis inhibitor such as chloramphenicol. Carlson *(10)*, however, cautions that these approaches may not always yield reproducible results. More generally, it is advisable to test different approaches to suppression of bacterial metabolism to compare their relative impact, if any, on phage growth parameters.

4.4 Adsorption Theory

Theory of phage adsorption to bacteria is discussed by Schlesinger *(81)* and reviewed by Stent *(11)*. Consider the simplest case where a unit volume of homogeneous adsorption medium contains a single phage virion and a single phage-susceptible bacterium. The probability of phage adsorption within this volume over one unit of time is the phage adsorption constant (k), e.g., $2.5 \times 10^{-9}\,\mathrm{ml\,min^{-1}}$. Note that the likelihood that phage will adsorb in this system will increase linearly with bacterial number such that with two bacteria the probability that a single phage will become adsorbed is approximately $2k$, which over one minute would be $5.0 \times 10^{-9}\,\mathrm{ml\,min^{-1}}$ (which actually equals $1 - \mathrm{e}^{-kNt} = 1 - \mathrm{e}^{-2.5\times10(-9)\times2\times1} \approx kNt$ where N is bacterial density and t is the duration of phage adsorption). The probability that a given bacterium will become phage adsorbed also increases approximately linearly with phage density, at least when phage multiplicities are much less than 1.0. Thus, four phage present at time, $t = 0$, will result in a probability that the single bacterium will be adsorbed, over one minute, of $1.0 \times 10^{-8} = 4 \times 2.5 \times 10^{-9}$. *See (82)* for theory of phage adsorption to abiotic surfaces.

4.5 Phage Multiplicity of "Adsorption" (MOA)

Rates of phage adsorption to bacteria are a function of bacterial density (**Section 3.1.2** and **Notes Section 4.4**). In practice, except if bacterial densities are very high or adsorption intervals are very long, this means that not all phage added to a bacterial culture will adsorb. The practical consequence of this observation is that phage multiplicity of infection (MOI)—often defined as the "number of phage particles *added* per cell" *(10)* (p. 423; emphasis added)—typically will be greater than phage multiplicity of adsorption (MOA) or MOI_{actual} *(16,17)*. In other words, MOI has an often-ignored time component. The difference between MOI and MOI_{actual} is of some concern either when MOI precision is called for or if one dilutes a mixture of phage and bacteria expecting constancy in the number of phage adsorbing. Abedon *(64)* considered the time-dependence of phage multiplicities in terms of phage secondary adsorption to phage-infected bacteria, which can only occur prior to phage-induced bacterial lysis. That is, MOI (*M*) should be defined as

$$M = \mathrm{MOI}_{\mathrm{actual}} = \mathrm{MOA} = P(1 - \mathrm{e}^{-kNt})/N \qquad (18.5)$$

where *P* is the free phage density, *t* is the interval of time over which adsorption occurs, and N is the bacterial density. Note that this equation only holds if free phage density is assumed to remain constant, which may be approximated over relatively short intervals or if *N* is small. Otherwise equation (18.5) will overestimate MOI_{actual}.

MOI as often defined will also overestimate the actual MOI:

$$M = \mathrm{MOI} = P/N \qquad (18.6)$$

By contrast, Adams *(2)* explicitly defines MOI as the "ratio of *adsorbed* phage particles to bacteria in a culture" (p. 441, emphasis added) and elsewhere describes how to rigorously derive MOI as the ratio of phage to bacteria once unadsorbed phage have been removed or otherwise accounted for (*(9)*, *see* also 51). *See* Kasman *et al., (83)* for rederivation as well as rigorous testing of equation (18.5). For additional discussion of MOI and MOA or MOI_{actual}, *see (84, 85)*.

For $k = 2.5 \times 10^{-9}$ ml/min and $t = 10$ min, MOI as defined by equation (18.6) is reduced to MOI_{actual} as defined by equation (18.5) as follows: Assuming a starting ratio of phage to bacteria of 10, for $N = 10^9, 10^8, 10^7, 10^6$, or 10^5 bacteria/ml, then MOI_{actual} equals 10.0, 9.18, 2.21, 0.25, or 0.025 phage adsorbed per bacterium, respectively. Thus, for assuring equivalence between added and adsorbed ratios of phage and bacteria it is crucial that high concentrations of bacteria (e.g., 10^8 or even 10^9/ml) and/or long periods of adsorption be employed. Furthermore, not all phage display adsorption constants are as high as that assumed above. For example, the adsorption constant for phage M13 is approximately 100-fold lower or 3×10^{-11} ml/min *(83)*. Plugging that adsorption constant into the above calculations for even $N = 10^9$/ml and $t = 10$ min yields an MOA of 2.59 rather than the "expected" MOI of 10.

4.6 Dilution Inhibits Adsorption

The probability of adsorption of a given virion is a function of the phage adsorption constant, bacterial density, and time. The probability of adsorption is not a function of phage-to-bacteria ratios. Indeed phage-to-bacteria ratios are relevant only when considering likelihoods of multiple phage adsorptions per bacterium (**Section 3.2.1.2**; **Fig. 18.2B**), and even then such likelihoods are more a function of adsorbed virions (MOI_{actual}) than they are of starting ratios of free virions and bacteria (MOI as defined by equation (18.6)). As a consequence, diluting mixtures of phage and bacteria, while having no impact on ratios of phage to bacteria, in fact will greatly reduce likelihoods of phage–bacteria encounter. This can be inconvenient should one want to study phage infections initiated at different bacterial densities, or convenient since it allows an effective termination of phage adsorption simply by diluting mixtures of phage and bacteria (**Section 3.2.1.3**).

Acknowledgement

We would like to dedicate this chapter to Harris Bernstein, who, serving as Ph.D. advisor to both of us, introduced us to both the power and the joy of phage whole-organismal analysis.

References

1. Abedon, S.T. (2006) Phage ecology, in *The Bacteriophages* (Calendar, R. and Abedon, S T eds.), Oxford University Press, Oxford, pp. 37–46.
2. Adams, M.H. (1959). Bacteriophages. Interscience, New York.
3. d'Hérelle, F. and Smith, G.H. (1926). The Bacteriophage and Its Behavior [English translation]. The Williams &Wilkins Co., Baltimore.
4. Abedon, S.T. (1992) Lysis of lysis inhibited bacteriophage T4 infected cells. *J. Bacteriol.* **174**, 8073–8080.
5. d'Herelle, F. (1922). The Bacteriophage: Its Role in Immunity. Williams and Wilkins Co./Waverly Press, Baltimore.
6. Eisenstark, A. (1967) Bacteriophage techniques. *Meth. Virol.* **1**, 449–524.
7. Cairns, J., Stent, G. and Watson, J.D. (1966). Phage and the Origins of Molecular Biology. Cold Spring Harbor Laboratory Press, Cold Spring Harbor, NY.
8. Ellis, E.L. and Delbrück, M. (1939) The growth of bacteriophage. *J. Gen. Physiol.* **22**, 365–384.
9. Carlson, K. (2005) Working with bacteriophages: common techniques and methodological approaches, in *Bacteriophages: Biology and Application* (Kutter, E. and Sulakvelidze, A eds.), CRC Press, Boca Raton, Florida, pp. 437–494.
10. Carlson, K. and Miller, E.S. (1994) Working with T4, in *Molecular Biology of Bacteriophage T4* (Karam, J.D. ed.), ASM Press, Washington, DC, pp. 421–426.
11. Stent, G.S. (1963). Molecular Biology of Bacterial Viruses. WH Freeman and Co., San Francisco, CA.
12. Heller, K.J. and Bryniok, D. (1984) O antigen-dependent mutant of bacteriophage T5. *J. Virol.* **49**, 20–25.
13. Prehm, P., Jann, B., Schmidt, G. and Stirm, S. (1976) On a bacteriophage T3 and T4 receptor region within the cell wall lipopolysaccharide of Escherichia coli B. *J. Mol. Biol.* **101**, 277–281.
14. Brumfitt, W. (1960) Studies on the mechanism of adsorption and penetration by bacteriophage. *J. Pathol. Bacteriol.* **79**, 1–9.

15. Fildes, P. and Kay, D. (1957) Tryptophan as a bacteriophage adsorption factor. *Brit. J. Exp. Path.* **38**, 563–572.
16. Dandekar, A.M. and Modi, V.V. (1978) Interaction between Rhizobium japonicum phage M-1 and its receptor. *Can. J. Microbiol.* **24**, 685–688.
17. Landry, E.F. and Zsigray, R.M. (1980) Effects of calcium on the lytic cycle of Bacillus subtilis phage 41c. *J. Gen. Virol.* **51**, 125–135.
18. Garen, A. and Puck, T.T. (1951) The first two steps of the invasion of host cells by bacterial viruses. II. *J. Exp. Med.* **94**, 177–189.
19. Rountree, P.M. (1951) The role of certain electrolytes in the adsorption of staphylococcal bacteriophages. *J. Gen. Microbiol.* **5**, 673–680.
20. Daniels, L.L. and Wais, A.C. (1990) Ecophysiology of bacteriophage S5100 infecting *Halobacterium cutirubrum*. *Appl. Environ. Microbiol.* **56**, 3605–3608.
21. Conley, M.P. and Wood, W.B. (1975) Bacteriophage T4 whiskers: A rudimentary environment-sensing device. *Proc. Natl. Acad. Sci. USA* **72**, 3701–3705.
22. Abedon, S.T. The Ecology of Bacteriophage T4. 1990. University of Arizona. Ph.D. dissertation
23. Takumi, K., Takeoka, A., Kinouchi, T. and Kawata, T. (1985) Solubilization and partial properties of receptor substance for bacteriophage alpha 2 induced from Clostridium botulinum type A 190L. *Microbiol. Immunol.* **29**, 1185–1195.
24. Watanabe, K., Takesue, S. and Ishibashi, K. (1977) Reversibility of the adsorption of bacteriophage PL-1 to the cell walls isolated from Lactobacillus casei. *J. Gen. Virol.* **34**, 189–194.
25. Rossi, P. and Aragno, M. (1999) A fast method for assessing rapid inactivation and adsorption kinetics of bacteriophages using batch agitation experiments and colloidal clay particles. *Can. J. Microbiol.* **45**, 9–17.
26. Lipson, S.M. and Stotzky, G. (1985) Specificity of virus adsorption to clay minerals. *Can. J. Microbiol.* **31**, 50–53.
27. Fildes, P. and Kay, D. (1963) The conditions which govern the adsorption of a tryptophan-dependent bacteriophage to kaolin and bacteria. *J. Gen. Microbiol.* **30**, 183–191.
28. Carlson, K. and Miller, E.S. (1994) Enumerating phage, in *Molecular Biology of Bacteriophage T4* (Karam, J.D. ed.), ASM Press, Washington, DC, pp. 427–429.
29. Hadas, H., Einav, M., Fishov, I. and Zaritsky, A. (1997) Bacteriophage T4 development depends on the physiology of its host *Escherichia coli*. *Microbiology* **143**, 179–185.
30. Delbrück, M. (1942) Bacterial viruses (bacteriophages). *Adv. Enzymol.* **2**, 1–32.
31. Delbrück, M. (1940) Adsorption of bacteriophage under various physiological conditions of the host. *J. Gen. Physiol.* **23**, 631–642.
32. Rosner, A. and Gutstein, R. (1981) Adsorption of actinophage Pal 6 to developing mycelium of Streptomyces albus. *Can. J. Microbiol.* **27**, 254–257.
33. Preissner, W.C., Maier, S., Volker, H. and Hirsch, P. (1988) Isolation and partial characterization of a bacteriophage active on Hyphomicrobium sp. WI-926. *Can. J. Microbiol.* **34**, 101–106.
34. Ohshima, Y., Schumacher-Perdreau, F., Peters, G. and Pulverer, G. (1988) The role of capsule as a barrier to bacteriophage adsorption in an encapsulated Staphylococcus simulans strain. *Med. Microbiol. Immunol.* **177**, 229–233.
35. Wilkinson, B.J. and Holmes, K.M. (1979) Staphylococcus aureus cell surface: capsule as a barrier to bacteriophage adsorption. *Infect. Immun.* **23**, 549–552.
36. Schwartz, M. (1976) The adsorption of coliphage lambda to its host: Effect of variation in the surface density of the receptor and in phage-receptor affinity. *J. Mol. Biol.* **103**, 521–536.
37. Chai, T.-J. (1983) Characteristics of *Escherichia coli* grown in bay waters as compared with rich medium. *Appl. Environ. Microbiol.* **45**, 1316–1323.
38. Koch, A.L. (1960) Encounter efficiency of coliphage-bacterium interaction. *Biochim. Biophys. Acta* **39**, 311–318.
39. Anderson, T.F. (1949) On the mechanism of adsorption of bacteriophages on host cells, in *The Nature of the Bacterial Surface* (Miles, A.A. and Pirie, N W eds.), Blackwell Scientific Publications, Oxford, pp. 76–95.
40. Delbrück, M. (1940) The growth of bacteriophage and lysis of the host. *J. Gen. Physiol.* **23**, 643–660.
41. Abedon, S.T. (1994) Lysis and the interaction between free phages and infected cells, in *The Molecular Biology of Bacteriophage T4* (Karam, J.D. ed.), ASM Press, Washington, DC, pp. 397–405.
42. Carlson, K. and Miller, E.S. (1994) General procedures, in *Molecular Biology of Bacteriophage T4* (Karam, J.D. ed.), ASM Press, Washington, DC, pp. 427–437.
43. Séchaud, J. and Kellenberger, E. (1956) Lyse prêcoce, provoquée par le chloroforme, chez les bactéries infectés par du bactériophage. *Ann. Inst. Pasteur* **90**, 102–106.
44. Doermann, A.H. (1992) The eclipse in the bacteriophage life cycle, in *Phage and the*

Origins of Molecular Biology (expanded edition) (Cairns, J., Stent, G S and Watson, J D eds.), Cold Spring Harbor Laboratory Press, Cold Spring Harbor, NY, pp. 79–87.
45. Cumming, G., Fidler, F. and Vaux, D.L. (2007) Error bars in experimental biology. *J. Cell Biol.* **177**, 7–11.
46. Jarvis, A.W. (1993) Analysis of phage resistance mechanisms encoded by lactococcal plasmid pAJ2074. *Can. J. Microbiol.* **39**, 252–258.
47. Gupta, B.M., Chandra, K. and Clifford, I. (1963) Human intestinal bacteriophage. V. Effect of yeast adenylic acid on virus adsorption and plaque formation by coli-dysentery phage CVX-5 on purine-deficient *Escherichia coli*. *Ann. Biochem. Exp. Med.* **23**, 395–400.
48. Sanders, M.E. and Klaenhammer, T.R. (1983) Characterization of phage-sensitive mutants from a phage-insensitive strain of *Streptococcus lactis*: Evidence for a plasmid determinant that prevents phage adsorption. *Appl. Environ. Microbiol.* **46**, 1125–1133.
49. Chakrabarti, B.K., Si, K. and Chattopadhyay, D. (1996) Characterization of Vibrio cholerae EIT or typing phage D10. *J. Gen. Virol.* 77, 2881–2884.
50. Gründling, A., Manson, M.D. and Young, R. (2001) Holins kill without warning. *Proc. Natl. Acad. Sci. USA* **98**, 9348–9352.
51. Carlson, K. (1994) Single-step growth, in *Molecular Biology of Bacteriophage T4* (Karam, J.D. ed.), ASM Press, Washington, pp. 434–437.
52. Dulbecco, R. (1949) On the reliability of the Poisson distribution as a distribution of the number of phage particles infecting individual bacteria in a population. *Genetics* **34**, 122–125.
53. Gots, J.S. (1959) Chemical interference with phage growth, in *Bacteriophages* (Hershey, A.D., Hotchkiss, R D, Pappenheimer, A M, Jr. and Racker, E eds.), Interscience Publishers, Inc., New York, pp. 265–287.
54. Smart, K.M. and Steinberg, B.M. (1977) Simultaneous presence of antiviral activity and its degrader in Bacillus extracts. *Can. J. Microbiol.* **23**, 726–732.
55. Doermann, A.H. and Dissosway, C. (1948) Intracellular growth and genetics of bacteriophage. *Year Book Carnegie Inst. Wash.* **48**, 170–176.
56. Kerr, B., West, J. and Bohannan, B.J.M. (2008) Bacteriophage: models for exploring basic principles of ecology, in *Bacteriophage Ecology* (Abedon, S.T. ed.), Cambridge University Press, Cambridge, UK, pp. 31–63.
57. Heineman, R.H. and Bull, J.J. (2007) Testing optimality with experimental evolution: lysis time in a bacteriophage. *Evolution* **61**, 1695–1709.
58. Doermann, A.H. (1952) The intracellular growth of bacteriophages. I. Liberation of intracellular bacteriophage T4 by premature lysis with another phage or with cyanide. *J. Gen. Physiol.* **35**, 645–656.
59. Underwood, N. and Doermann, A.H. (1947) A photoelectric nephelometer. *Rev. Scient. Instr.* **18**, 665–672.
60. Doermann, A.H. (1948) Lysis and lysis inhibition with *Escherichia coli* bacteriophage. *J. Bacteriol.* **55**, 257–275.
61. Abedon, S.T. (2000) The murky origin of Snow White and her T-even dwarfs. *Genetics* **155**, 481–486.
62. Young, R. (1992) Bacteriophage Lysis: Mechanisms and regulation. *Microbiol. Rev.* **56**, 430–481.
63. Young, R. and Wang, I.-N. (2006) Phage Lysis, in *The Bacteriophages* (Calendar, R. and Abedon, S T eds.), Oxford University Press, Oxford, pp. 104–125.
64. Abedon, S.T. (1990) Selection for lysis inhibition in bacteriophage. *J. Theor. Biol.* **146**, 501–511.
65. Abedon, S.T. (1999) Bacteriophage T4 resistance to lysis-inhibition collapse. *Genet. Res.* **74**, 1–11.
66. Abedon, S.T., Hyman, P. and Thomas, C. (2003) Experimental examination of bacteriophage latent-period evolution as a response to bacterial availability. *Appl. Environ. Microbiol.* **69**, 7499–7506.
67. Paddison, P., Abedon, S.T., Dressman, H.K., Gailbreath, K., Tracy, J., Mosser, E., Neitzel, J., Guttman, B. and Kutter, E. (1998) Lysis inhibition and fine-structure genetics in bacteriophage T4. *Genetics* **148**, 1539–1550.
68. Sarimo, S.S., Hartiala, M. and Aaltonen, L. (1976) Preparation and partial characterization of a *Lactobacillus lactis* bacteriophage. *Arch. Microbiol.* **107**, 193–197.
69. Brown, A. (1956) A study of lysis in bacteriophage-infected *Escherichia coli*. *J. Bacteriol.* **71**, 482–490.
70. Ackermann, H.-W., Roy, R., Martin, M., Murthy, M.R.V. and Smirnoff, W.A. (1978) Partial characterization of a cubic *Bacillus* phage. *Can. J. Microbiol.* **24**, 986–993.
71. Wollman, E. and Wollman, E. (1937) Les "phases" des bactériophages (facteurs lysogènes). *Compt. Rend. Soc. Biol.* **124**, 931–934.
72. Rees, P.J. and Fry, B.A. (1981) The morphology of staphylococcal bacteriophage K and DNA metabolism in infected Staphylococcus aureus. *J. Gen. Virol.* **53**, 293–307.

73. Hooper, I., Woods, W.H. and Egan, B. (1981) Coliphage 186 Replication is delayed when the host cell is UV irradiated before infection. *J. Virol.* **40**, 341–349.
74. Zachary, A. (1976) Physiology and ecology of bacteriophages of the marine bacterium *Beneckea natriegens*: salinity. *Appl. Environ. Microbiol.* **31**, 415–422.
75. de Siqueira, R.S., Dodd, C.E.R. and Rees, C.E.D. (2006) Evaluation of the natural virucidal activity of teas for use in the phage amplification assay. *Int. J. Food Microbiol.* **111**, 259–262.
76. Josslin, R. (1971) Physiological studies on the *t* gene defect in T4-infected *Escherichia coli*. *Virology* **44**, 101–107.
77. Josslin, R. (1970) The lysis mechanism of phage T4: Mutants affecting lysis. *Virology* **40**, 719–726.
78. Reddy, A.B. and Gopinathan, K.P. (1987) Characterization of mycobacteriophage I8 and its unrelatedness to mycobacteriophages I1, I3 and I5. *J. Gen. Virol.* **68**, 949–956.
79. Stevens, R.H., Hammond, B.F. and Lai, C.H. (1982) Characterization of an inducible bacteriophage from a leukotoxic strain of *Actinobacillus actinomycetemcomitans*. *Infect. Immun.* **35**, 343–349.
80. Benzer, S. and Jacob, F. (1953) Étude du développement du bactériophage au moyen d'irradiations par la lumière ultra-violette. *Ann. Inst. Pasteur* **84**, 186–204.
81. Schlesinger, M. (1932) Absorption [*sic?*] of bacteriophages to homologous bacteria [translation], in *Bacterial Viruses*, Little, Brown and Co., Boston, pp. 26–36.
82. Gerba, C.P. (1984) Applied and theoretical aspects of virus adsorption to surfaces. *Adv. Appl. Microbiol.* **30**, 133–168.
83. Kasman, L.M., Kasman, A., Westwater, C., Dolan, J., Schmidt, M.G. and Norris, J.S. (2002) Overcoming the phage replication threshold: a mathematical model with implications for phage therapy. *J. Virol.* **76**, 5557–5564.
84. Abedon, S.T. (2008) Phage population growth: constraints, games, adaptations, in *Bacteriophage Ecology* (Abedon, S.T. ed.), Cambridge University Press, Cambridge, UK, pp. 64–93.
85. Goodridge, L.D. (2008) Phages, bacteria, and food, in *Bacteriophage Ecology* (Abedon, S.T. ed.), Cambridge University Press, Cambridge, UK, pp. 302–331.

Chapter 19

Phage Production and Maintenance of Stocks, Including Expected Stock Lifetimes

Louis-Charles Fortier and Sylvain Moineau

Abstract

In microbiology, preservation of an archival stock or a "master stock" of a given microorganism is essential for many reasons including scientific research, conservation of the genetic resources and providing the foundation for several biotechnological processes. The objective is to preserve the initial characteristics of the microorganism and to avoid the genetic drift that occurs when the organism is maintained indefinitely in an actively growing state. The same holds true in phage biology and it is of particular interest when a collection of phages is to be maintained. The aim of this chapter is to provide phage biologists with general procedures to prepare and maintain bacteriophage stocks on a long-term basis. The protocols described below should be considered as general guidelines because although many phages and bacterial strains can be propagated and stored in these conditions, specific media and/or growth and storage conditions must be evaluated for each phage and bacterium. Since it was not the scope of this chapter to provide an exhaustive list of these particular conditions, we instead highlighted the main factors affecting phage amplification and storage. We hope this will help phage biologists to develop their own strategies for their preferred phages.

Key words: Bacteriophage, storage, amplification, stock, collection.

1 Introduction

1.1 Phage Collections

The last decade has seen a heightened awareness of the value of collections of microorganisms both in the conservation of genetic resources and biodiversity, in providing the foundation for emerging biotechnologically based industries, and for the training of future generations of researchers *(1)*. Moreover, most scientific journals also recommend depositing the described isolates in public repositories, and ideally in more than one collection in different countries, to ensure access and allow scientific

Martha R. J. Clokie, Andrew M. Kropinski (eds.), *Bacteriophages: Methods and Protocols, Volume 1: Isolation, Characterization, and Interactions, vol. 501*, © 2009 Humana Press, a part of Springer Science+Business Media
DOI 10.1007/978-1-60327-164-6_19 Springerprotocols.com

reproducibility. Too often, biological samples are lost due to the retirement or death of principal investigators. Collections can provide the perennial key microbes. The knowledge covered in this chapter should be helpful to any scientist working with phages who wishes to properly safeguard this biological material for an extended period of time. It is worth mentioning at this point that there are a few public collections that maintain phage stocks namely, the American Type Culture Collection (www.atcc.org) and the German Collection of Microorganisms and Cell Cultures (www.dsmz.de). To our knowledge, the only public collection devoted entirely to bacteriophages is the Félix d'Hérelle Reference Center for Bacterial Viruses (www.phage.ulaval.ca) located at the Université Laval (Québec, Canada). The mission of the Félix d'Hérelle Center is to collect, preserve and distribute reference bacteriophages of importance in taxonomy as well as with interesting applications or properties to foster research and education. The collection now contains more than 450 reference phages infecting over 120 bacterial species. Thus, our expertise in manipulating various bacteriophages gives us the opportunity to draw general guidelines for phage amplification, manipulation, and storage.

1.2 Stabilization of Biological Material

In phage biology, preparation and conservation of phage stocks is obviously a vital issue. When a particular phage is first isolated, some basic characterization must be conducted in order to properly identify the phage and to determine its affiliation regarding other known phages. Electron microscopy (**Chapter 10**), host range analysis (**Chapter 12**), and genome analysis (**Chapters 23–25** and **29**) are among the most common methods for phage identification. Once this preliminary characterization is made, the isolated phage must be stored adequately in order to preserve its integrity and to make sure there will be no alterations during prolonged storage. It is even more important these days as more and more phage genomes are being sequenced and therefore, it is critical to maintain the blueprint of the original phage. Archiving a stock of such a phage isolate is essential to avoid the genetic drift that will inevitably occur if the phage is maintained over a prolonged period of time by repeated amplification cycles. The notion of genetic drift is best exemplified when taking into account the number of phage particles that are produced during infection of a bacterial cell. For example, if 100 virions are released per infected cell (referred to as the burst size; **Chapter 14**), it also means that its genome must be replicated at least 100 times. The rates of spontaneous mutations per genome per replication are similar in double-stranded DNA (dsDNA) containing phages and bacteria but the rates of mutations per base pair per replication can be as much as 1000-fold greater in an *E. coli* phage genome compared to its bacterial host *(2)*.

Moreover, RNA phages such as f2, MS2, and Qβ, are known to have higher rates of mutations, mainly due to their shorter generation times and the error-prone nature of viral RNA polymerases compared to DNA polymerases *(3–6)*. Thus, taking into account that high titer lysates can be reached (as much as 10^{10} pfu/ml for a single amplification) it becomes obvious that mutants will naturally occur in a relatively short time frame if a given bacteriophage is maintained by serial consecutive amplification cycles. Under certain circumstances, these phage mutants may even become dominant in a lysate. In recent years, with the advent of whole genome studies, it has been shown that phages evolve by exchanging clusters of genes or functional modules. The exchanges that occur between infecting phage and prophages that are integrated in the bacterial chromosome are responsible for subsequent gene rearrangements and thus represent an important mechanism of phage evolution *(7, 8)*.

These examples strengthen the need to perform several "quality controls" on the isolated phage, as mentioned above and to store phages in the most stable form that is possible.

1.3 Storage of Bacteriophages

Previous investigations on phage stability and survival under various storage conditions were generally conducted on a period of time ranging from a few weeks to 1–3 years *(9–12)*. Other authors reported stability data over 3–5 years *(13, 14)*. All things considered, these may be viewed as short time studies. To our knowledge, the only available reports on long-term storage (≥ 5years) and viability of bacteriophages is from Ackermann et al., *(15)* and Zierdt *(16)*. Accumulation of data over two decades, including electron microscopy analysis and host range determination brought new information regarding the long-term stability of several phages and to the effectiveness of the storage conditions. Ackermann et al., drew general trends regarding viability of hundreds of different phages from the collection that were stored for as much as 20 years at 4°C, −80°C, −196°C and freeze-dried *(15)*. They reported that cleared lysates of model phages such as the T series, λ group and Φ29 were stable for 10–12 years and phages of the T4 and T7 groups were extremely resistant *(15)*.

1.3.1 A Comparison of Different Approaches to Storing Phage Stocks

For the long-term preservation of phages, several methods have been evaluated in the last few decades *(9, 10, 12–14, 16–19)*. Among these different methods, lyophilization or freeze-drying is probably the most interesting for the following reasons: (1) it has been proven highly effective for the long-term preservation of bacterial cells, (2) once lyophilized, the biological material is generally stable at room temperature, thus eliminating the need for refrigerators and freezers or liquid nitrogen tanks, (3) it reduces the room space needed to preserve the vials, (4) the lyophilized material can be easily shipped to other labs without wet or dry-ice. Thus, it was tempting to apply this technology for the storage of bacteriophages. Clark and collaborators from the ATCC conducted some analysis with over 60 different phages infecting strains of *Bacillus*, *Escherichia*, *Mycobacterium*, *Pasteurella*, *Pseudomonas*, *Rhizobium*, *Serratia*, *Shigella*, *Staphylococcus* and *Vibrio* *(10, 13, 17, 18)*. These studies showed that freeze-drying was particularly harmful (up to almost 3-log loss in titer) to phages of the *Myoviridae* family with large particle sizes such as T2 and T6 whereas *Siphoviridae* phages with small particle sizes like T1 or the *Microviridae* ΦX174 were more resistant (e.g., less than 1-log loss in titer) *(10, 18)*. Some phages were even shown to be more resistant to freeze-drying than to slow-cooling followed by deep-freezing (liquid nitrogen). These include the single-stranded RNA (ssRNA) phage f2 and the well known single-stranded DNA (ssDNA) phage M13 *(18)*. As a general trend, either deep-freezing or storage at 4°C of a lysate was the most effective in preserving phage viability. Conversely, Carne and Greaves reported high stability of 14 corynebacteriophages after freeze-drying and storage up to 30 months *(9)*. Engel et al., observed a higher viability with 44 mycobacteriophages (out of 53) that were freeze-dried in sodium glutamate and gelatin followed by storage in the dark at room temperature. The overall loss in titer was less than 1-log pfu/ml over a 2.5 years period *(14)*. The mycobacteriophage D35 was the most sensitive in the conditions reported by Engel (2-log loss in titer after 12 months) *(14)*.

Similar conclusions were obtained by Zierdt with 27 phages infecting *Staphylococcus aureus*: 12 phages retained their original titer whereas 14 phages showed a 1-log drop in titer directly post-lyophilization, although titers remained stable thereafter during storage at −20°C for up to 8 months. Only one phage (type 7) had a decrease in titer greater than 1-log but less than 2-log for the same period *(12)*. Zierdt also reported in 1988 a long-term study where the 25 *S. aureus* phages lyophilized in 1959 were still highly infectious after storage for 12–18 years at −20°C, showing no more than 1-log drop in the phage titer *(16)*. Recently, Ackermann et al., reported the use of lyophilization for the storage of various phages from the Félix d'Hérelle collection and also frequently observed a 1-log drop in titer for most phages after only one month of storage *(15)*. This general trend was also observed with phage lysates stored at 4°C or deep-frozen at −80°C or −196°C. After 1 year at room temperature, most lyophilized vials had lost their vacuum and no viable phages could be detected. Over 20 years later, the remaining ampoules that had been stored in a cold room were analyzed again and all ampoules with intact vacuums contained active phages *(15)*. Thus, it turned out from this study that the quality of the vacuum in the ampoules is most likely the decisive parameter for long-term storage of phages.

1.4 General Considerations Regarding Phage Infection and Amplification

Amplification and storage of most bacteria is relatively straightforward and the same general principle holds true for most bacteriophages, albeit the latter needs a bacterial host as an intermediate to propagate, which can sometimes bring some difficulties. Preparation of a phage stock can be made either in a liquid broth or on soft agar overlays. The choice between these two methods depends mostly on the phage to be amplified and sometimes on the preference of the investigator. Whenever possible, the authors of this chapter prefer to amplify phages in liquid broth as it is simpler.

Most of the phages and their hosts (~ 75%) maintained at the Félix d'Hérelle Reference Center for Bacterial Viruses can be readily amplified on Trypticase Soy Broth (TSB) or Brain Heart Infusion (BHI). These include, among others, several phages infecting bacterial species of *Bacillus*, *Escherichia*, *Pseudomonas*, *Salmonella*, and *Yersinia*. In addition, most dairy phages (*Lactococcus*, *Lactobacillus*, *Leuconostoc*, *Streptococcus thermophilus*) are amplified in M17 or MRS media. Thus, the protocols described below are based on these media. Obviously, it should be pointed out that specific growth conditions are likely to be needed for other phages and they should be developed on a case by case basis.

2 Materials

2.1 Isolation of a Bacterial Indicator Strain

1. Ready-to-use powdered TSB or BHI. You can also prepare your own media by combining separate components listed in **Table 19.1** (**Note 1**).
2. For agar plates, add 15 g/l of Bacto agar to the liquid formulation, or purchase ready-to-use Trypticase soy agar (TSA) or BHI media. For top agar (double agar overlay method), add 7.5 g/l agar or less depending on your needs to the liquid formula (**Note 2**).
3. Sterile Petri dishes (9 cm diameter).

2.2 Preparation of a Glycerol Stock of the Bacterial Indicator Strain

1. Glycerol can be sterilized by autoclaving and it is stable at room temperature. You can aliquot 150 μl glycerol into 2.0 ml cryogenic vials and sterilize by autoclaving with the cap fitted loosely. After autoclaving, screw the caps tightly and store at room temperature. These tubes can be readily used to prepare glycerol stocks of bacteria by adding 850μl of the cell suspension to the glycerol tube (**Notes 3** and **4**).
2. Ethanol-, isopropanol- or ethylene glycol-dry ice slush.

Table 19.1
Composition of common growth media used for routine bacterial and phage propagation (Note 1)

Trypticase soy broth (TSB) Bacto cat # 211 825	g/l
Casein peptone (Pancreatic digest of casein)	17.0
Soy peptone (Papaic digest of soybean meal)	3.0
Dipotassium phosphate	2.5
Dextrose	2.5
Sodium chloride	5.0
pH 7.3 ± 0.2 @ 25°C	
Brain Heart Infusion (BHI) Bacto cat # 237 500	**g/l**
Calf brains (infusion from)	200.0
Beef heart (infusion from)	250.0
Proteose peptone	10.0
Dextrose	2.0
Sodium chloride	5.0
Disodium phosphate	2.5
pH 7.4 ± 0.2 @ 25°C	

2.3 Lyophilization of the Bacterial Indicator Strain

1. Reconstituted non fat dried milk (NFDM) (**Note 5**). Make a 10% (w/v) solution and sterilize by autoclaving. The solution should be stored at 4°C and is stable for a few weeks. Sterility of the milk solution should be monitored prior to use.
2. Lyophilization ampoules (Wheaton Science Products; Millville, NJ; http://www.wheatonsci.com/; cat # 651 502). Introduce a cotton plug to close the end of the ampoule before autoclaving.

2.4 Phage Isolation and Amplification

1. Glass tubes (18 mm) containing 10 ml TSB or BHI. Aliquot the liquid broth into the tubes and then sterilize by autoclaving. Also sterilize by autoclaving 13 mm glass tubes that will be used to prepare molten soft agar overlays.
2. Prepare one liter of 10× phage buffer stock solution (200 mM Tris–HCl pH 7.4, 1 M NaCl, 100 mM $MgSO_4$) and sterilize by autoclaving. Store the stock solution at room temperature. To prepare a 1× working solution, simply dilute the desired volume of 10× buffer with sterile water in a pre-sterilized bottle. Alternatively, you can also dilute the 10× buffer with non-sterile distilled water and then autoclave the buffer again (**Note 6**).

3. Prepare a 2 M $CaCl_2$ stock solution and sterilize by autoclaving (**Note 6**).

2.5 Phage Storage

1. Sterile 2 ml screw cap cryogenic vials, sterile glycerol, and lyophilization ampoules (*see* **Sections 2.2** and **2.3** above).

3 Methods

3.1 Isolation of a Bacterial Indicator Strain

1. It is quite important to make sure that the bacterial strain that will serve as a host indicator to propagate the phage is a pure culture (**Note 7**). We recommend to isolate a single sensitive colony and to make a master stock that will be stored deep-frozen at $\leq -70°C$ and lyophilized.
2. On day one, streak the bacterial host on a TSA plate and incubate the plate upside down under optimal growth conditions. The objective is to obtain isolated colonies that will be assayed for phage sensitivity (make sure the colonies are large enough so that enough bacteria can be picked up for subsequent streaking, see below).
3. On day two, apply a phage sample (0.1 ml from a previous phage lysate) on a TSA agar plate (**Notes 8** and **9**). Start from one side of the Petri dish and streak the phage aliquot across the plate trying to make a uniform and straight line. Then, pickup a single bacterial colony with a sterile toothpick and streak the bacteria across the plate, making a straight perpendicular line crossing the phage sample. You can easily test up to 8–10 colonies per plate. Incubate the plates upside down overnight under optimal growth conditions for the bacterial host.
4. On day three, identify the isolates that were lysed by the phage as shown by the absence of growth when the cells have come in contact with the phage. This means that this isolate is sensitive to the phage and will be a suitable host. Bacterial isolates which have grown across the entire streak are not sensitive to the phage and should be discarded.

3.2 Preparation of a Glycerol Stock of the Bacterial Indicator

1. Pickup the sensitive indicator bacteria with a sterile loop and inoculate a tube containing 5–10 ml TSB and grow the cells under optimal conditions until late exponential or the beginning of the stationary growth phase (**Note 3**). Mix 0.85 ml of the cell culture with 0.15 ml sterile glycerol (100%) into a 2.0 ml screw cap cryogenic vial so as to get a final glycerol concentration of 15% (v/v; **Note 4**). Most investigators will prepare a few more vials as precautionary measure. For example, one vial can be used as your working stock while the others are kept for long-term storage. Mix the cell suspension with

the glycerol and snap-freeze the tubes in an ethanol dry-ice slush (**Note 10**). Transfer half the vials into a freezer held at $\leq -70°C$ and if available, the other half in a liquid nitrogen storage container.

2. Streak five TSA plates with the same bacteria so as to inoculate the entire surface of the agar. The objective is to get enough biomass to proceed to the lyophilization of this bacterial strain (*see* next section and **Note 11**).

3.3 Lyophilization of the Bacterial Indicator Strain

1. Collect the cells from the five agar plates that have been streaked with the sensitive bacteria using 2.5 ml per plate of 10% reconstituted NFDM (**Note 5**). Use a sterile glass rod to recover the cells. Pool the suspension from all four plates in a sterile tube and homogenize the suspension.
2. Dispense 0.5 ml aliquots of the cell suspension in sterile lyophilization ampoules. Transfer all the ampoules in a lyophilization bottle and fill with water up to the level of cell suspension in the ampoules. Cover the bottle with perforated paraffin paper. Freeze the bottle containing the ampoules at $-80°C$ for at least 1 h. It can be stored frozen overnight for lyophilization on the next day.
3. Lyophilize the ampoules overnight at a temperature $\leq -50°C$ following the recommendations of the manufacturer of your freeze-dryer unit. Ampoules are then sealed, verified using a vacuum tester and stored at 4°C in the dark.

3.4 Phage Isolation and Amplification in Liquid Broth (Note 12)

1. Inoculate 10 ml of TSB with the sensitive bacterial strain isolated previously (**Section 3.1**) and grow overnight at the optimal temperature.
2. The next day, prepare 10-fold serial dilutions of the phage lysate, archival stock or enriched sample from which you want to isolate single plaques. For example, this can be done in sterile Eppendorf tubes by transferring 100 μl of the initial phage sample into 900 μl of 1× phage buffer (**Note 6**). Then, proceed to 10-fold serial dilutions as described above up to the desired dilution.
3. Add 0.1–0.5 ml of the sensitive bacterial culture (**Note 13**) into a tube filled with 3 ml molten (45°C) soft agar (**Note 2**) containing the proper cofactor if needed (for example 10 mM $CaCl_2$) (**Note 9**). Quickly add 0.1 ml of the phage dilution and pour onto a bottom agar plate (TSA + cofactors). Spread uniformly by gently rocking the plate and let stand at room temperature for 10–15 min to allow the agar to solidify. Transfer the plates upside down at the optimal temperature and incubate until plaques are visible and large enough (usually overnight).

4. Inoculate the indicator strain (1%) in seven tubes containing 10 ml of TSB with cofactors, if needed. With a sterile 1 ml truncated pipette tip or Pasteur pipette, pick up six individual phage plaques (**Note 14**). Carefully insert the pipette tip around the plaque and push through the agar to the bottom of the Petri dish. Twist the pipette and gently rub the pipette against the bottom of the dish to dislodge the agar plug. Remove the pipette and blow each phage-containing plug into one of the inoculated tube (**Note 15**). Keep one tube without phage as a control for bacterial growth. Incubate 4–8 h or more, depending on the phage–host pair, and regularly monitor cell lysis (**Notes 9** and **16**). This can be done by simply looking at the phage-containing tubes and compare them with the control or if a more rigorous monitoring is needed, cell growth (or lysis) can be monitor by optical density with a spectrophotometer. For the latter, you will need an 8th non inoculated TSB tube for blank purposes.
5. After cell lysis, filter the phage lysate through a sterile 0.45 μ m filter adapted to a 10 ml syringe (**Note 17**). You may need to centrifuge the lysate (10 min at 8000 $\times g$) prior to the filtration if incomplete lysis is observed or to remove cell debris.
6. Most of the time, it is advantageous to perform a second amplification cycle in order to reach a higher titer. Grow the indicator strain into 10 ml TSB until an $OD600_{nm} \sim 0.1$ is reached and then, inoculate the culture with 50 μl of the cleared lysate from the first amplification (**Note 13**). Incubated for 4–8 h until a cleared lysate is obtained. Filter as in step 5 and store at 4°C.

3.5 Phage Quality Control

1. Scientists usually obtain research material from recognized collections because such material has undergone quality control and authentication testing as part of the routine procedures of the collection. Use of the wrong organism in investigations is time wasting, expensive, and leads to invalid published results. Moreover, without proper authentication, noxious organisms could be inadvertently supplied. So, it is important to maintain high standards of quality control when you store phages and hosts. Consequently, personnel should be trained accordingly.
2. Before archiving a phage stock, it is highly recommended to confirm the identity of the phage lysates prepared in **Section 3.4**. Electron microscopy analysis (**Chapter 10**) host range and/or genetic profiling are usually performed for quality control purposes. For example, if studying dsDNA phages, you can isolate the phage DNA and digest it with endonucleases. Starting with 1.5 ml from the six phage lysates obtained in step 6 from **Section 3.4**, purify the phage DNA (Volume 2 **Section I**) and perform a restriction digest with

2–3 different endonucleases. Compare the restriction profiles of the six phages. If all profiles are identical, pool the lysates and make an archival stock with this mixture (see below).

3.6 Phage Storage

3.6.1 Conservation at 4°C

1. Keep a phage lysate master stock in a screw cap glass tube at 4°C. These lysates are generally stable for several months, even years, without significant loss of infectivity (**Section 1**). Add a few drops of chloroform to avoid microbial contamination if the phage is known to tolerate chloroform.

3.6.2 Conservation at ≤ −70°C and in Liquid Nitrogen (−196°C)

1. In a 2 ml screw cap cryogenic vial, mix 0.5 ml of phage lysate with 0.5 ml sterile glycerol to obtain a final concentration of 50%. Make several tubes per phage.
2. Place the cryogenic vials into an ethanol dry-ice slush to snap freeze the phages. After 5–10 min, transfer half the vials into deep freezers at ≤ −70°C for long-term storage. Transfer the other half vials into liquid nitrogen tanks (−196°C). Store duplicates in two different locations (preferably in different buildings for maximum security).

3.6.3 Lyophilization of Phages

3. Mix 2.5 ml of phage lysate with 2.5 ml of sterile glycerol to obtain a final concentration of 50%. Transfer 0.5 ml aliquots into lyophilization ampoules (**Note 18**).
4. Transfer all the ampoules to a lyophilization bottle and fill with water up to the level of phage suspension in the ampoules. Cover the bottle with perforated paraffin paper. Freeze the bottle containing the ampoules at −80°C for at least 1 h. It can be stored frozen overnight for lyophilization on the next day.
5. Lyophilize the ampoules overnight at a temperature ≤ −50°C following the recommendations of the manufacturer of your freeze-dryer unit. Ampoules are then sealed, verified using a vacuum tester and stored at 4°C in the dark.

3.6.4 Databases

1. Ideally, you should develop a documentation system for your collection of phages and hosts. For example, you should have a form (ideally on a computer) to be completed by the depositor for each phage that is stored for a long period of time. The forms should include all available information regarding the phage such as name of isolator, date/time/geographic location of isolation, taxonomic identification (if known), phenotypic/genotypic strain properties, and references. This information is important for providing maximum scientific data to future users. Well-developed databases are crucial for this knowledge transfer.
2. When cultures and phages are recovered from stock during maintenance, routine preservation work, or when they

are being dispatched, care should also be taken to ensure they conform to the original deposit by carrying out appropriate tests such as electron microscopy analysis and DNA restriction profiles for phages and a biochemical analysis e.g., using API strips (bioMérieux INDUSTRY; Hazelwood, MO; http://industry.biomerieux-usa.com/), pulsed-field gel electrophoresis (PFGE), and phage sensitivity assay for the bacterial hosts.

3. Detailed records of users of the cultures should also be available. In the case of unsatisfactory performance (such as contamination), or if it is necessary to supply subsequent information, users can then be notified.
4. Finally, it should be reminded that the management of a culture collection is necessarily labor-intensive due to the various tasks that include culture supply, preservation, maintenance, documentation and viability checking and technical support.

4 Notes

1. A clear advantage of using premixed formulas is that you just have to add water and sterilize by autoclaving. We recommend using premixed dehydrated formulas of the highest quality available to your laboratory because they give reproducible results from batch to batch and minimize the possible variation in the composition that could occur if one prepares the medium from separate components. Over the years, we have noticed that the media supplier can greatly affect the phage yield and purity as well as the quality of the isolated phage DNA. For example, cheaper media formulations may sometimes give lower phage yields and low quality DNA. It is also noteworthy that some bacterial strains will grow well in different media, but the efficacy of phage infection is sometimes highly variable depending on the medium used. This is mainly due to the speed at which the bacterial host grows, which affects the overall phage yield. If the host grows too fast, a number of uninfected cells will interfere with the phage purification. On the other hand, if the host grows too slowly, the phage yield will be low. This could partially be due to microelements that may vary from one supplier to another and to the degree of purity of the formula. Thus, the best medium for amplification should be determined for each phage. We found that in certain cases, preparing a half strength medium (0.5×) gives excellent results and at the same time reduces the cost of phage amplification (especially in large batch amplifications). Sterilized liquid media are stable for several weeks at room temperature. Agar plates should be kept for a few days on a

bench top, or left opened for 10 min under a laminar flow hood to eliminate excess humidity before storing. Stack the plates back into their original packaging bag and store at room temperature or at 4°C if space is not limiting. Plates should be stable for several weeks if good microbiological techniques have been used during pour plating.

2. Some phages are difficult to amplify and it can be sometimes very hard to detect plaques on agar plates, especially when the size of the plaques is very small (≤ 0.1 mm diameter) and turbid. Lillehaug *(20)* reported that some modifications in the top agar composition can greatly increase the size of the plaques formed by some bacteriophages infecting lactic acid bacteria *(20)*. Such modifications include decreasing the top agar concentration to $\leq 0.3\%$ and the addition of glycine at a final concentration ranging between 0.25 and 1.25%, which corresponds to the concentration that increases the bacterial doubling time by ~2-fold. The softer agar helps the phages to diffuse and infect adjacent cells more easily while the glycine, which likely destabilizes the peptidoglycan cell wall of bacteria *(21–23)*, helps cell lysis. The resulting effect is a dramatic increase in plaque size for certain phages, going from pinpoint plaques on regular top agar to 1–2 mm plaques on the modified top agar.
3. We found that for most strains, an overnight culture is adequate to prepare a glycerol stock. However, for best results and long storage, we suggest growing the cells until mid-exponential phase ($OD_{600\,\mathrm{nm}} \sim 0.5 - 0.8$) and preparing a glycerol stock as described in **Note 4**.
4. We recommend collecting 1 ml of cells by centrifugation for 1 min at full speed in a sterile Eppendorf tube using a table-top microcentrifuge and then recovering the bacteria in 1 ml of TSB supplemented with 15% glycerol (alternatively, you can recover the cells in 850 μl of TSB and then mix with a sterile 150 μl glycerol aliquot as mentioned in **Section 2.2**). This eliminates toxic end products present in the final growth medium that may decrease the viability of the cells upon storage.
5. Lyophilization of bacteria can be performed in different media and you may need to find one that will be suitable for your needs and strain of interest. As a general rule, rich media are usually preferred because they provide a better cryoprotection. NFDM (10%) is widely used because it contains natural cryoprotectants such as milk proteins and sugars. Bovine serum albumin (10%) has also been used with several bacterial species *(15)*.
6. Phage contamination from aerosols (e.g., often generated in micropipettes) and lab environment (dust and clothes) is frequent and special care must be taken to reduce the risks of

contamination. For example, we recommend changing frequently the working solutions as well as using filter-containing pipette tips. It is also a good laboratory practice to regularly clean lab coats and rigorously clean all working areas, including bench tops, shelves and other lab equipments. Use a freshly prepared solution of sodium hypochlorite (500 ppm) or 70% ethanol to disinfect and clean the surfaces. Bleach solutions should be freshly prepared each 2–3 days since there is a rapid loss of disinfecting activity after dilution.

7. When several closely related strains of a same bacterial species are routinely used along with different phages, one strain may get contaminated with another one. The contamination might not always be noticed if the phenotypes are identical. However, this may result in a turbid phage lysate or colonies that start to grow inside the lysed zone obtained after phage spotting on a bacterial lawn. It is thus essential to make sure that a pure culture is used.
8. If a phage lysate is not readily available, you may apply a sample from a frozen phage stock. This is done by scraping the top of a frozen phage stock with the tip of a sterile serological 1 ml pipette and then streaking a line on the plate. If the phage titer is too low, the cross streaks method may not give definite results and you might confound a sensitive colony with a resistant one. If so, try to amplify the phage using the host at hand.
9. Many phages will need cofactors such as Ca^{2+}, Mg^{2+}, or other cations to properly complete their lytic cycle *(24)*. It may be necessary to add 5–10 mM $CaCl_2$ or $MgSO_4$ to the growth medium (liquid or solid) to obtain an efficient phage infection. Note that these solutions of cations (autoclaved or filtered) must be added under sterile conditions to autoclaved medium, otherwise precipitates may form during sterilization. You may still notice the formation of a precipitate with certain rich media after the addition of $CaCl_2$. The presence of such a precipitate may sometimes be mistaken with residual bacterial growth in the final lysate but has generally no negative impact on phage amplification.
10. The methods used to freeze the cells vary from one lab to another. Some prefer to snap-freeze the cell suspension in liquid nitrogen or in an ethanol dry-ice or ethylene glycol dry-ice slush, whereas others simply transfer the cryogenic vials from room temperature to $\leq -70°C$ without any pre-cooling steps. Others slowly cool ($-1°C/min$) the vials down to $-20°C$ and then transfer the vials to $\leq -70°C$. In our laboratory, we routinely use an isopropanol dry-ice slush stored in $-80°C$ freezer. Cryogenic vials are placed in the slush for a few minutes to quickly freeze the cells and then the vials are transferred to their respective cryogenic boxes for storage. We also

recommend keeping a series of vials in liquid nitrogen in addition to $\leq -70°C$. Replicates should be stored in both conditions in different freezers or liquid nitrogen tanks (ideally in different buildings) for maximum stock security. The same holds true for ampoules of lyophilized bacteria where replicates should be stored in at least two separate locations.

11. We also highly recommend the metabolic and molecular characterization of the strains to be archived. For example, API strips, sequencing of the 16S rDNA, and genomic restriction profiles separated by PFGE are often used to identify a strain as well as its genus/species.
12. Some phages grow poorly in liquid broth. For example, some mutant phages from *Streptococcus thermophilus* isolated in our lab could only be amplified on plates. If this is the case with your phage, you may need to use one of the following alternative methods:

Soft-Agar Method: (1) add 0.2 ml of a mid-log phase culture of the indicator strain in a series of 10 mm glass tubes (the amount of tubes is determined according to your needs); (2) add $10^3 - 10^5$ pfu of phages to each tube; (3) add 3 ml of molten soft-agar to each tube, mix gently (do not vortex) and pour on top of bottom TSA plates; (4) incubate 6–8 h or until complete lysis; (5) scrape the top agar with a glass spreader or spatula and transfer to a sterile tube; (6) rinse the bottom plates with a few ml of TSB or phage buffer (~ 0.5 ml/plate) to recover residual phages and transfer into the tube; (7) add a few drops of chloroform (if your phage is not sensitive) and let stand for 30 min at room temperature to allow phages to elute from the soft-agar; (8) transfer the agar and the liquid into a centrifuge tube, taking care not to transfer chloroform as it will dislodge the pellet after centrifugation and spin at $4,000 \times g$ for 10 min; (9) recover the supernatant and centrifuge again (you may need to filter your phage lysate through a $0.45\ \mu m$ filter to eliminate residual agar); (10) transfer the supernatant to a sterile screw cap glass tube, add a few drops of chloroform (if your phage is insensitive) and store at 4°C or directly proceed to phage titration. Some phages may be unstable using this method. If this is the case, you may prefer the method reported by Zierdt *(12)* and used to amplify *S. aureus* phages. The author said it is superior to broth amplification as it gives more reliable and uniform results, needs less attention and is easy to perform *(12)*. However, the phage yield is generally lower.

Bottom Agar Method: (1) pour about 500 µl of a mixture of a mid-log phase indicator strain and a proper dilution of your phage onto a bottom TSA plate (make several plates according to your needs); (2) spread evenly with a sterile glass spreader and incubate at the appropriate temperature on a leveled surface until complete lysis (3–6 h); (3) add 0.5 ml of TSA or phage buffer to each plate and spread evenly with a glass spreader; (4) slant the plates and recover the liquid containing the phages, transfer to a screw cap glass tube, add a few drops of chloroform (if your phage is tolerant) and store at 4°C or proceed to phage titration. If the titer is adequate, prepare stocks as described in **Section 3.6**.

13. Depending on the phage–host interactions, you may need to adjust the volume (concentration) of host and phage (multiplicity of infection; MOI) to be added to the top agar in order to get a proper balance between cell growth and phage amplification. Phages that have a long latent period or small burst size take more time to amplify and thus, cells may reach the stationary phase too early, especially for fast growing species. As most phages will amplify mainly on growing cells, this physiological state of the cell will affect the capacity of the phage to infect its host. The same rule applies for liquid broth amplification. If cells grow too fast, the culture will not be completely lysed and the phage yield (titer) will likely be low. These phages may benefit from a low cell density (higher MOI) at the time of infection (either in liquid broth or in top agar). Inversely, if not enough cells are added to the top agar, a non uniform bacterial lawn will result, which may limit plaque formation, identification, and numbering. A high MOI for liquid broth amplification will result in fast lysis of the initial inoculum, limiting further rounds of infection by the progeny phages and thus reducing the final yield. Finally, we also recommend adding to the top agar an exponential growing culture ($OD_{600\,nm}0.5 - 0.8$) as the resulting bacterial lawn will be generally more uniform. Note that you may still get an acceptable bacterial lawn using an overnight culture or even a culture that was stored at 4°C for a day or two, but we don't recommend this as a general practice.
14. As mentioned in **Section 1.2**, the frequency of occurrence of phage mutants is high as well as the possibility of contamination if working with several phages. Therefore, to avoid picking up the wrong phages or mutants from a single plaque, it is advisable to characterize and compare phages from 5 to 6 plaques in order to confirm they are all the same. If all isolated plaques are identical, then depending on the type of work, you may wish to pool the phage lysates obtained from each plaque and make a phage stock.
15. Inoculation of a phage plug directly into a culture may not be advisable when bacteriophage insensitive mutants (BIMs) are easily obtained. This is particularly true for some strains of the Gram-negative bacterium *E. coli* and Gram-positive bacterium

Lactococcus lactis. Thus, it is not unusual to find bacterial colonies growing inside a lysed zone in a spot test or if too much phage is added into a top agar. On a plate containing isolated plaques, these BIMs may not be readily visible but after transfer of a phage plug containing these mutants into fresh medium, they may rapidly outgrow the initial sensitive bacteria leading to a trouble lysate, as well as a low phage titer. Thus, it may be necessary to elute the phages from the agar plug by soaking at least 30 min in 0.5 ml of phage buffer containing a few drops of chloroform. This will kill any residual bacteria or BIMs but again, this might not be advisable for lipid-containing phages.

16. Sometimes, a turbid lysate after 4–8 h incubation may be left overnight at 4°C for residual cells to be lysed. Some phages (e.g., T4-like phages) will exhibit lysis-inhibition *(25)* and prolonged incubation and/or addition of a few drops of chloroform to the lysate will increase the phage yield.
17. It should be noted that filtration may not be suitable for large phages (>450 nm), which may clog the membrane. Alternatively, you may add a few drops of chloroform to the phage lysate to kill and lyse residual bacteria. This may even increase the phage yield in the presence of a lysis-inhibition mechanism (e.g., T4-like phages) *(25)*. However, lipid-containing phages will be inactivated by chloroform and other phages may be sensitive to chloroform, even if they do not contain lipids. It is advisable to test the sensitivity of your phage before using chloroform to prepare stocks.
18. Different media can be used for lyophilization of phages. Clark et al., at the ATCC and Zierdt reported the use of double strength skim milk (20%) as a routine practice *(12, 18)*. Ackermann et al., used 50% glycerol for all phages stored at the Félix d'Hérelle Reference Center *(15)*. Carne and Greaves used 10% peptone, 5% sucrose, and 1% sodium glutamate as preservative *(9)*. Engel et al., reported the use of 5% sodium glutamate combined with 0.5% gelatin for the long-term storage of mycobacteriophages *(14)*. We thus recommend to test different media for your particular phages and assay for infectivity directly post-lyophilization and at different time intervals afterwards.

References

1. Sigler, L. (2004) Culture collections in Canada: perspectives and problems. *Can. J. Plant. Pathol.* **26,** 39–47.
2. Drake, J. W. (1991) A constant rate of spontaneous mutation in DNA-based microbes. *Proc. Natl. Acad. Sci. USA* **88,** 7160–7164.
3. Domingo, E., Sabo, D., Taniguchi, T., and Weissmann, C. (1978) Nucleotide sequence heterogeneity of an RNA phage population. *Cell* **13,** 735–744.
4. Holland, J., Spindler, K., Horodyski, F., Grabau, E., Nichol, S., and VandePol, S.

(1982) Rapid evolution of RNA genomes. *Science* **215,** 1577–1585.
5. Steinhauer, D.A., and Holland, J.J. (1987) Rapid evolution of RNA viruses. *Annu. Rev. Microbiol.* **41,** 409–433.
6. Drake, J. W., and Holland, J. J. (1999) Mutation rates among RNA viruses. *Proc. Natl. Acad. Sci. USA* **96,** 13910–13913.
7. Casjens, S. (2003) Prophages and bacterial genomics: what have we learned so far? *Mol. Microbiol.* **49,** 277–300.
8. Brussow, H., and Desiere, F. (2001) Comparative phage genomics and the evolution of *Siphoviridae*: insights from dairy phages *Mol. Microbiol.* **39,** 213–222.
9. Carne, H. R., and Greaves, R. I. (1974) Preservation of corynebacteriophages by freeze-drying *J. Hyg. (Lond)* **72,** 467–470.
10. Clark, W. A. (1962) Comparison of several methods for preserving bacteriophages. *Appl. Microbiol.* **10,** 466–471.
11. Mendez, J., Jofre, J., Lucena, F., Contreras, N., Mooijman, K., and Araujo, R. (2002) Conservation of phage reference materials and water samples containing bacteriophages of enteric bacteria. *J. Virol. Methods* **106,** 215–224.
12. Zierdt, C. H. (1959) Preservation of staphylococcal bacteriophage by means of lyophilization *Am. J. Clin. Pathol.* **31,** 326–331.
13. Clark, W. A., and Klein, A. (1966) The stability of bacteriophages in long term storage at liquid nitrogen temperatures. *Cryobiology* **3,** 68–75.
14. Engel, H. W., Smith, L., and Berwald, L. G. (1974) The preservation of mycobacteriophages by means of freeze drying. *Am. Rev. Respir. Dis.* **109,** 561–566.
15. Ackermann, H.-W., Tremblay, D., and Moineau, S. (2004) Long-term bacteriophage preservation. *World Federation for Culture Collections Newsletter.* **38,** 35–40.
16. Zierdt, C. H. (1988) Stabilities of lyophilized *Staphylococcus aureus* typing bacteriophages *Appl. Environ. Microbiol.* **54,** 2590.
17. Clark, W. A., Horneland, W., and Klein, A. G. (1962) Attempts to freeze some bacteriophages to ultralow temperatures. *Appl. Microbiol.* **10,** 463–465.
18. Clark, W. A., and Geary, D. (1973) Preservation of bacteriophages by freezing and freeze-drying. *Cryobiology* **10,** 351–360.
19. Davies, J. D., and Kelly, M. J. (1969) The preservation of bacteriophage H1 of *Corynebacterium ulcerans* U103 by freeze-drying. *J. Hyg. (Lond)* **67,** 573–583.
20. Lillehaug, D. (1997) An improved plaque assay for poor plaque-producing temperate lactococcal bacteriophages. *J. Appl. Microbiol.* **83,** 85–90.
21. Hammes, W., Schleifer, K. H., and Kandler, O. (1973) Mode of action of glycine on the biosynthesis of peptidoglycan. *J. Bacteriol.* **116,** 1029–1053.
22. Holo, H., and Nes, I. F. (1989) High-frequency transformation, by electroporation, of *Lactococcus lactis* subsp. *cremoris* grown with glycine in osmotically stabilized media. *Appl. Environ. Microbiol.* **55,** 3119–3123.
23. Cruz-Rodz, A. L., and Gilmore, M. S. (1990) High efficiency introduction of plasmid DNA into glycine treated *Enterococcus faecalis* by electroporation. *Mol. Gen. Genet.* **224,** 152–154.
24. Guttman, B., Raya, R., and Kutter, E. (2005) Basic phage biology, *in* "Bacteriophages: biology and applications" (Kutter, E., and Sulakvelidze, A., Eds.), CRC Press, Boca Raton, pp. 29–66.
25. Kutter, E., Raya, R., and Carlson, K. (2005) Molecular mechanisms of phage infection, *in* "Bacteriophages: biology and applications" (Kutter, E., and Sulakvelidze, A., Eds.), CRC Press, Boca Raton, pp. 165–222.

Section 3

Bacteriophage-Host Interactions

Chapter 20

Construction of Phage Mutants

Robert Villafane

Abstract

Recent studies have established that the most abundant life form, that of phages, has had major influence on the biosphere, bacterial evolution, bacterial genome, and lateral gene transmission. Importantly the phages have served and continue to serve as valuable model systems. Such studies have led to a renewed interest and activity in the study of phages and their genomes. In order to determine the details of the involvement of phages in these important processes and activities, it is critical to assign specific functions to the phage gene products. The initial functional and gene assignments can be made by general mutagenesis of the phage genomes and of these specific gene products. A very informative mutagenic protocol that has found renewed interest is that using hydroxylamine. This mutagenic protocol has been used to obtain gene mutations involved in the lysogenic cycle of the *Salmonella enterica serovar* Anatum var. 15+ phage ε^{34} (hereafter phage ε^{34}) and to isolate conditional lethal mutants of phage ε^{34}. A similar protocol using plasmid is also described. A plate complementation method is presented to determine quickly the number of genes which are present in the population of mutations isolated from hydroxylamine mutagenesis.

Key words: Complementation, ε^{34}, hydroxylamine (HA), mutagenesis, phage, amber (am) mutants, temperature-sensitive (ts) mutants, suppression, permissive temperature, restrictive temperature, *Salmonella*.

1 Introduction

The use of hydroxylamine (HA) has a long and rich tradition in gene identification, analysis of gene product, and genetic regulation. The reason for this tradition is the nature of the mutation which causes a nonreversible single nucleotide substitution from a Cytosine → Thymine *(1)*. The action of hydroxylamine has a two step process *(2)*. A classic report showed that the first step of mutagen treatment rapidly inactivated the T4 phage yet this inactivation was not mutagenic. The first step is dependent

Martha R. J. Clokie, Andrew M. Kropinski (eds.), *Bacteriophages: Methods and Protocols, Volume 1: Isolation, Characterization, and Interactions, vol. 501*, © 2009 Humana Press, a part of Springer Science+Business Media
DOI 10.1007/978-1-60327-164-6_20 Springerprotocols.com

on a metal-inactivating process. The second step mutagenizes the phage but is rather slow. Inclusion of a chelating agent eliminates the first step and makes mutagenesis a first order reaction and an extremely potent tool. This procedure has been used to construct several different types of ε^{34} mutants (unpublished and described herein) and host range mutants of the *Salmonella* phage P22 *(3)*. An extremely useful offshoot of this HA protocol on phages was developed by Humphreys et al., for use on multi-copy plasmids *(4)*. The plasmid version is a regularly used protocol for both phage and other genes or groups of genes that can be cloned (e.g., *(5–7)*) and unpublished data from the author's laboratory).

2 Materials

2.1 Bacterial and Phage Growth

2.1.1 Bacterial Strains (Note 1)

There is now a consensus nomenclature for many of the *Salmonella* strains described in this chapter but for brevity the original lab nomenclature will be followed here. This is outlined below. The suppressor-negative strain *Salmonella newington* or $A1\varepsilon^{15}$ (BV7001 lab designation) is more correctly designated as *Salmonella enterica serovar* Anatum var, 15+. The strain, $37A_2su+\varepsilon^{15}$, (or BV7004, *(8)*) is a derivative of $A1\varepsilon^{15}$ which contains a glutamine suppressor mutation. The phage predominantly described in this chapter is known as ε^{34} the full designation is *Salmonella enterica serovar* Anatum var, 15+ phage ε^{34}. This phage is a clear plaque mutant in gene cI (ε^{34}C16; *(9)*).

The plasmid DNA, pJS28 (originally from Dr. Peter Berget – Carnegie Mellon University), was derived from our stock strain (BV1300). This plasmid contains a cloned version of the wild-type tailspike gene from phage P22 *(7)*. The pJS28 plasmid concentration for the study described in this Protocol was 0. 1 μg/μl. BV1300 is the source of pJS28 plasmid which contains a cloned copy of the WT P22 tailspike gene *(7)*. BV1300 is equivalent to KK2186 (pJS28) where KK2186 is an *E. coli* strain, JM103 which is also P1−.

2.1.2 CsCl Stock Solutions (for phage particle purification)

The CsCl formula weight is 168.4. For a CsCl density gradient centrifugation, four densities are used which are: density (ρ) of 1.3, 1.4, 1.5, and 1.7 g/ml. These densities can be achieved by making concentrated solutions of CsCl: ρ of 1.3 is 2.4 M CsCl (40.41 g of CsCl in 100 ml of Phage Buffer), ρ of 1.4 is 3.2 M CsCl (53.88 g of CsCl in 100 ml of Phage Buffer), ρ of 1.5 is 4.0 M CsCl (67.48 g of CsCl in 100 ml of Phage Buffer) and ρ of 1.7 is 5.6 M CsCl (94.29 g of CsCl in 100 ml of Phage Buffer; **Note 2**).

2.1.3 Dialysis casettes

Pierce Slide-A-Lyzer Dialysis Cassette (Cat No. 66810; Pierce, Rockford, IL; http://www.piercenet.com/) with a 10,000 molecular weight cut-off (MWCO) and a 3–12 ml capacity.

2.1.4 Luria Broth (LB)

Ten grams of Bacto Tryptone (Baxter), 5 g of Bacto Yeast Extract, 5 g of NaCl, and 1 ml of 1.0 N NaOH per liter of H_2O and sterilized by autoclaving (25 min at 15 psi [103.4 kPa] and 121°C).

2.1.5 LB Agar Plates

LB broth plus 15 g of Bacto agar per liter and sterilized by autoclaving.

2.1.6 LB Top Agar (TA)

The TA consists 6 g of Bacto agar per liter of H_2O plus the components of LB broth and is sterilized by autoclaving. The melted TA is distributed into sterile glass bottles and allowed to solidify.

2.1.7 Phage Buffer

Phage Buffer consists of 50 mM Tris–HCl (pH 7.4), 100 mM $MgCl_2$, 10 mM NaCl which is sterilized by autoclaving *(9)*.

2.1.8 Sterile Toothpicks

For picking and patching phage plaques.

2.1.9 Tubes: Small Tubes

Small tubes with caps for plating out phage. Those used here are disposable culture tubes: Fisher brand 13 × 100 mm borosilicate glass tubes. (Cat. No. 14-961-27). Dilution tubes are used for dilution of phage and bacteria. Those used here are disposable culture tubes: Fisher brand 16 × 150 mm borosilicate glass tubes (Cat. No. 14-961-31).

2.2 Hydroxylamine Mutagenesis of Phage Stocks

2.2.1 Hydroxylamine / NaOH Stock Solution (1M pH6)

To a sterile small tube is added 0.175 g of hydroxylamine (NH_2OH, Cat. No. # 255580; Sigma-Aldrich St. Louis, MO; http://www.sigmaaldrich.com) and 0.28 ml of 4M NaOH and this is brought up to a total of 2.5 ml total volume.

2.2.2 LBSE (10)

0.2 ml of 0.5M EDTA and 5.85 g NaCl is added to 100 ml of LB broth and sterilized by autoclaving for 20 min.

2.2.3 Phosphate–EDTA Buffer (for use in hydroxylamine mutagenesis)

70 ml of sterile water is slowly added to 6.8 g of KH_2PO_4. Once dissolved, the pH is adjusted to pH 6.0 with 1M KOH. This is then brought up to 99 ml with sterile water. Finally 1 ml of 0.5 M EDTA is added and the resultant solution is sterilized by autoclaving for 20 min.

2.3 Hydroxylamine Mutagenesis of Plasmid DNA

2.3.1 Competent Cells

Electrocompetent cells were from Novagen (EMD Biosciences; San Diego, CA; http://www.emdbiosciences.com/home.asp). They are designated NovaBlue GigaSingles competent cells (Catalogue number 71227-3).

2.3.2 DNA Dialysis Buffer (TE)

10 mM Tris–HCl, 1 mM EDTA, pH 8

2.3.3 Plasmid DNA

Plasmid DNA was prepared by the Qiagen Midi Plasmid protocol essentially as described in their instruction sheet (QIAGEN Inc.; Valencia, CA; http://www1.qiagen.com/).

3 Methods

3.1 Bacterial Growth Conditions

1. Overnight cultures were prepared by the inoculation of 5 ml of sterile LB broth in a sterile test tube with a colony of BV7001 or BV7004 from a LB Petri plate bacterial stock followed by incubation at 37°C with shaking in a New Brunswick Scientific (NBS) gyratory water bath for at least 12 h *(11)*.
2. This saturated culture was used for determining titer of the phage stocks. Saturated cultures were stored at 4°C for a maximum of 2 weeks until needed.
3. The number of bacterial cells in a culture was determined by measuring the optical density (OD) of a culture either in a Klett colorimeter or a spectrophotometer (at 590 nm). The number of viable cell counts was also measured. These parameters were measured by diluting a fresh overnight culture of bacteria 1:20 into LB in a 125 ml sterile Erlenmeyer flask and allowing the cells to grow in LB broth in a shaking water bath at 37°C. At specific times aliquots of cells were withdrawn and the OD was determined and a portion of the aliquot is diluted and plated on LB plates to determine colony forming units (cfu) for a particular OD value. Generally a concentration of 2×10^8 cells per ml corresponds to a Klett reading of 70. This correspondence can be done any number of ways including using a Petroff-Houser counting glass slide instead of plating out for viable cell on Petri plates (**Note 3**).
4. Bacterial cultures were maintained on LB agar plates and stored at 4°C for no longer than one month.

3.2 Preparation of a Phage Stock

1. Phage stocks of ε^{34} were prepared by diluting an overnight culture of BV 7001 cells, 0.5 ml in 30 ml of LB in a 125 ml sterile flask.

2. A phage plaque plug ($\sim 10^6$ pfu) was obtained by using a large bore Pasteur pipette and capturing the plaque in the middle of the Pasteur pipette and adding the plug to the flask containing cells and incubated with shaking at 30°C (**Note 4**) until lysis was complete as indicated by the transformation of the cloudy, turbid cell suspension into a clear solution (about 4 h).
3. Whether lysis was complete or not, after 5 h, lysis was completed by addition of about 1 ml of chloroform to the flask containing the infected culture and the culture was allowed to shake for 5 additional minutes (**Note 4**).
4. The unwanted bacterial debris is separated from the phage particles by carefully decanting the lysed cells (care taken not include the chloroform) into 50 ml sterile plastic Oakridge tubes and centrifuging in a Sorval SS34 rotor (or equivalent) for 15 min at 7000 rpm [$\sim 11,000\times$ g] and 4°C.
5. Following centrifugation, the phage supernatant is decanted into another sterile Oakridge tube and the phage is pelleted by centrifugation in the SS34 rotor for 90 min at 15,000 rpm [$\sim 23,000\times$ g]. The pellet is resuspended in about 1–2 ml of Phage Buffer. This is the normal working phage stock.
6. Typically, a 30 ml infected culture will yield a phage stock with a titer of 10^{11} pfu in 1–2 ml. The use of phage mutants can significantly enhance the yield of phages *(12)*.
7. If the phage being studied is known to be structurally fragile then the hard spin can be replaced by a gentler phase separation *(13)*. The resulting supernatant solution was brought to a final concentration of 0.5 M NaCl and 6% polyethylene glycol (PEG) (empirically determined for the ε^{34} phage) and allowed to stir slowly overnight and maintained at 4°C *(8, 13)*.
8. The PEG/NaCl–phage complex was obtained after centrifugation for 20 min at 6000 rpm [$\sim 9,000 \times g$] and 4°C as the pellet.
9. The pellet was inverted to allow residual PEG to decant (**Note 5**). The pellet was resuspended in Phage Buffer. The phage was further dialyzed in Phage Buffer.
10. Phage purification by CsCl. For routine studies small phage stocks prepared as described above were adequate but for mutagenesis or other studies, it is advisable to prepare larger phage stocks and purify them as described as follows. One liter of infected cells is treated with PEG as described above and the pellet was resuspended in 10 ml of Phage Buffer and purified in a CsCl gradient *(12)*. The CsCl gradient was formed by placing 2 ml of 1.7 ρCsCl at the bottom of a 13 ml polyallomer centrifuge tube (Beckman, for SW41Ti rotor), then carefully layering, (one on top of the other), 3 ml of 1.5 ρCsCl, 3 ml of 1.4ρ CsCl, 1 ml of 1.3 ρCsCl,

with 3 ml of the phage sample placed on top. The gradient is then centrifuged in a SW41Ti rotor for 90 min at 28,000 rpm [$\sim 141{,}000\times$ g] and 20°C. The resulting viral band, which appears bluish, is carefully removed by a 20 gauge syringe and placed in a dialysis cassette with a 10,000 MWCO and dialyzed with three changes of 1000-fold excess of Phage Buffer every 8 h at 4°C.

3.3 Determination of the Concentration of a Phage Stock (Titering) See Chapters 7 and 12, 13

The relative phage concentration (the phage titer) is determined by the plate overlay (**Note 6**). The phages were titered on host cells (plating bacteria), as follows:

1. Sterile solidified TA is melted in a microwave until liquid. The TA is placed in a small standing water bath at 55°C and allowed to equilibrate to this temperature for at least 30 min.
2. At this point, 3 ml of the TA is distributed into small tubes and either placed in a test tube warmer or a water bath at 55°C.
3. 5 ml of sterile Phage Buffer was added sterilely to five sterile capped dilution tubes. This will serve as the source tube into which the phage will be diluted to determine its concentration.
4. 50 μl of the phage stock to be titered is added to the first dilution tube. This constitutes a 1:100 dilution (our "−2" dilution) of the phage stock. Similarly, dilutions of $10^{-4}, 10^{-6}, 10^{-7}, 10^{-8}$ are made (**Note 8**).
5. 1 drop of an overnight culture of bacteria is added using a sterile transfer pipette to a small tube containing 3 ml of TA and 0.1 ml of a phage dilution. This is quickly mixed and poured over the surface of a room temperature LB plate. This is allowed to solidify for approximately 20 min and is inverted (to avoid condensation droplets onto the surface of the Petri plate) and incubated at 30°C for at least 16 h. This temperature is necessary for some phages which are partially temperature sensitive at 37°C such as Salmonella phages P22 and ε^{34}.
6. Under these conditions one finds several hundred plaques on the plate onto which one has added 0.1 ml from the 10^{-7} dilution of the phage stock (−7 dilution). The wild-type phages appear as turbid areas of lysis, i.e., turbid plaques (due to the ability of these temperate phages to lysogenize their hosts). Concentrations are in units of the number of plaque forming units per milliliter (pfu/ ml). If 200 pfu are found on the −7 plate, the titer is 2×10^{10} pfu / ml (from $200/0.1\ \mathrm{ml}/10^{-7}$).
7. Typically a 30 ml infected culture will yield a phage stock with a titer of 10^{10}–10^{11} pfu in 1–2 ml.
8. Although one could proceed to the mutagenesis protocol at this stage it is preferable to use purified phage for the mutagenesis (by CsCl as was described in **Section 3.2.10**) to avoid the possibility of contamination with bacterial debris.

3.4 Hydroxylamine Mutagenesis of Phage ε^{34}C16 (2, 10, 14)

1. 0.4 ml of the Phosphate–EDTA Buffer is added to a small tube along with 0.4 ml sterile double-distilled water, 0.8 ml of HA, 0.2 ml of 100 mM $MgSO_4$ and 0.2 ml of ε^{34} (or the relevant phage) at 10^9–10^{11} pfu/ml. A control reaction is also done in which the hydroxylamine is replaced by Phage Buffer.
2. The small tubes, containing the control reaction and the mutagenic phage reaction are capped and incubated at 37°C, generally for about 48 h.
3. The mutagenic action on the phage particle will cause a decrease in the viable phage as measured by the plaque overlay technique. Aliquots are withdrawn and phages are titered. **Figure** 20.1 shows the survival of the phage as a function of no mutagen, 0.4 M HA and 0.8 M HA. Mutagenesis was generally carried out using 0.4 M HA. Treatment of phage with mutagen for 48 h results in a decrease in phage titer to as low as 0.2–2% of the original phage concentration. These surviving phage normally yield phage mutants with single nucleotide substitutions. Another indicator of mutagenesis for temperate phage which form turbid plaques is the appearance of an increasing number of clear plaques with increasing time of incubation with mutagen.
4. The mutagenized phage is dialyzed at 4°C in Phage Buffer with three changes of buffer at 12 h intervals with Pierce Slide-A-Lyzer Cassette (**Note** 9). The sample is now ready for isolation of various types of mutants.

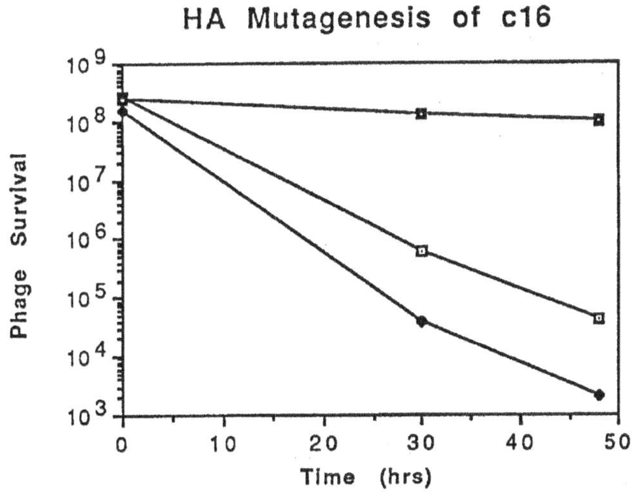

Fig. 20.1. Phage Survival in the Presence of Hydroxylamine (HA). In this figure filled squares represent the control phage reaction in which no mutagen was added while the open squares and the filled diamond symbols represent reaction in which the phage was incubated with 0.4M and 0.8M HA, respectively. Phage survival was measured at 0, 30, and 48 h of incubation with HA.

3.5 Isolation of Various Mutant Phage ε^{34} Types

3.5.1 Isolation of Amber Phage Mutants

1. Isolation of amber mutants in the phage ε^{34}c16. The bacterial strains BV7001 (su−) and BV7004 (su+) are used. These strains are suppressor negative and gln-suppressor respectively. The operational definition of a phage amber mutant in this system is a phage which can form plaques on the BV7004 (suppressor strain) but not on BV7001 (suppressor-negative strain).
2. To amplify the pool of mutants the strain, BV7004, is infected with $\sim 10^6$ pfu of the mutagenized ε^{34}C16 phage. Among the phages that grow will be mutants which contain amber mutations which are suppressed by the inserted glutamine (**Note 10**).
3. The phage stock is prepared as described in **Section 3.2**, and the phage titer is determined as described in **Section 3.3** on the strain BV7004.
4. The amber mutants are then selected from the phage plaques that appear on the phage titer plates.
5. LB plates are separately seeded with BV7001 and BV7004 by adding one drop of the bacteria to 3 ml of TA and pouring over the surface of the LB plate. The Petri plates are allowed to solidify at room temperature for at least 30 min.
6. From one of the phage titer plates, gently poke a plaque through its center with a sterile toothpick and touch the surface of the plate seeded with BV7001 (su−) and then touch the surface of the plate seeded with BV7004 (su+) in an analogous position. This step is repeated for as many of the plaques that can be picked and patched on the surfaces of the two seeded plates.
7. The sets of seeded plates are incubated at 30°C for at least 16 h.
8. Putative amber mutants are those phages which are found only on the BV7004 plates but not on the BV7001. Stocks of these putative amber mutants are prepared on BV7004.
9. Once the amber phage stocks are prepared, it is possible to determine if more than one amber mutation is present on each phage mutant by a reversion analysis. One phage mutation reverts in bacterial systems at a frequency of about 10^{-6}–10^{-8}. The frequency for the simultaneous reversion of two amber mutations would therefore be about $(10^{-6}–10^{-8}) \times (10^{-6}–10^{-8})$, or 10^{-12}–10^{-16}– a number which would preclude its detection.
10. Reversion frequency is determined as the titer of the phage under non permissive conditions (such as a wild-type host strain for amber mutants) divided by the titer under permissive conditions. Table 20.1 displays the reversion frequencies obtained from an amber phage mutant isolation from a hydroxylamine treatment of ε^{34}c16 phage.

Table 20.1
Reversion frequencies of amber derivatives of ε^{34}C16

Phage	Titers A1ε^{15}	37A_2su$^+$ ε^{15}	*Reversion frequency
ε^{34}C16am3.1	1.0×10^3	7.7×10^9	1.2×10^{-6}
ε^{34}C16am4.1	1.1×10^4	3.4×10^{10}	2.7×10^{-7}
ε^{34}C16am5.1	3.7×10^4	7.7×10^{10}	4.0×10^{-7}
ε^{4}C16am6.1	3.0×10^4	6.1×10^9	4.9×10^{-6}
ε^{34}C16am9.2	3.0×10^3	2.4×10^9	1.3×10^{-5}
ε^{34}C16am10.1	4.0×10^4	4.8×10^9	8.3×10^{-6}
ε^{34}C16am12.1	5.3×10^5	1.4×10^{10}	3.8×10^{-5}
ε^{34}C16am13.1	1.0×10^4	5.4×10^9	1.8×10^{-6}
ε^{34}C16am15.1	6.0×10^5	1.1×10^{10}	4.3×10^{-6}
ε^{34}C16am16.1	9.0×10^4	7.3×10^{10}	1.2×10^{-6}
ε^{34}C16am18.1	4.9×10^4	1.6×10^9	3.1×10^{-5}
ε^{34}C16am19.1	1.1×10^4	3.2×10^8	3.5×10^{-5}
ε^{34}C16am22.1	1.0×10^6	1.0×10^{11}	1.0×10^{-5}
ε^{34}C16am23.1	5.0×10^5	1.0×10^{11}	5.0×10^{-6}
ε^{34}C16am24.1	1.3×10^5	1.5×10^{10}	8.7×10^{-6}

*Reversion frequency determined as the titer on A1ε^{15} / titer on 37A_2su$^+\varepsilon^{15}$

3.5.2 Isolation of Temperature-Sensitive (*ts*) Phage Mutants

1. For the isolation of *ts* mutants in the phage ε^{34}c16, strain BV7001 (su−) is used. The operational definition of a ts phage mutant in this system is a phage which can form plaques on the BV7001 at 30°C but forms plaques with an efficiency of 0.01 or below or not at all at 39°C or above.
2. BV7001 is infected with ~ 10^6 pfu of the mutagenized ε^{34}c16 phage. This phage stock is grown at a temperature of 30°C.
3. The phage titer is determined at 30°C (**Section** 3.3). These titer plates will serve as the source of *ts* mutants.
4. Two LB Petri plates are seeded with BV7001 by adding one drop of the bacteria to 3 ml of TA and pouring over the surface of the LB plate. One plate will be incubated at the permissive temperature (30°C) while the other will be incubated at the restrictive temperature (39°C). Each phage mutant will be picked and patched onto two plates, one labeled 30°C and the other labeled 39°C.

5. Once the TA of the seeded plate labeled 39°C has solidified, it will be placed on a slide warmer whose temperature is set at ~ 39°C or so that the plate remains warm. This plate, which will define the restrictive temperature for our mutants, will remain on the plate-warmer throughout the picking and patching it onto both plates. If a plate-warmer is not available, this step should be done quickly so that the phages do not begin to grow on the plate at ambient temperatures (**Note 11**).
6. From one of the phage titer plates, gently poke a plaque through its center with a sterile toothpick and touch the surface of the two plates both seeded with BV7001 (su-) and one labeled 30°C and the other labeled 39°C in an analogous position. This step is repeated for as many of the plaques that can be picked and patched on the surfaces of the two seeded plates.
7. The sets of plates are incubated at the respective temperatures for at least 16 h.
8. Phage stocks of the putative *ts* mutants are prepared (**Section 3.3**).
9. Once the ts phage stocks are prepared, reversion analysis will be done to determine if each mutant contains just one *ts* mutation. It will be done by a similar analysis as carried out for the putative amber mutants. One phage mutation reverts in bacterial systems at a frequency of about 10^{-5}–10^{-7}. Simultaneous reversion of two *ts* mutations would not be detected.
10. Reversion frequency is determined as the titer of the phage under non permissive conditions (such as different temperatures) divided by the titer under permissive conditions. Table 20.1 displays the reversion frequencies obtained from an amber phage mutant isolation from a hydroxylamine treatment of ε^{34}c16 phage but this type of table can be made for the *ts* mutants.

3.6 Plate Complementation Assay: Defining Genes from Mutations in the Phage ε^{34}

In order to determine the minimum number of essential genes present in a pool of mutants, a complementation assay is done. A complementation study simply places two mutations in the same bacterial cell generally under restrictive conditions. Such conditions would be high temperature for a *ts* phage mutant or the wild-type host for the amber phage mutants. This assay is a shortcut whose results are reliable but should be confirmed by an infection of two phages into the same bacterial strain *(15)*. To explain how these assays are performed two sets of studies are described. Set 1 is a set in which hydroxylamine mutagenesis was used to isolate a number of mutations in the lysogeny functions of the ε^{34} phage particle (**Fig. 20.2** with two panels; 9). These clear plaque phage mutants were isolated from WT ε^{34} phage and resulted in a change in the appearance of the plaque from turbid or cloudy (WT) to clear. Complementation in the clear mutants resulted in turbid plaques, thus instead of "restrictive" conditions, these conditions are more phenotypic. Set 2 contains amber mutations isolated after hydroxylamine mutagenesis (as described above). For both sets of mutants the protocol is virtually identical.

1. Using a sterile transfer pipette, 1 drop of a BV7001 overnight culture was added to a small tube containing 3 ml of TA. This was quickly mixed and poured uniformly over the surface of a room temperature LB plate. This was allowed to solidify for approximately 20 min.
2. Five clear plaque ε^{34} phage mutants, ε^{34}c20, ε^{34}c26, ε^{34}c51, ε^{34}c61, and ε^{34}c82, were diluted to $\sim 10^6$ pfu/ml.
3. To the top of a seeded plate, held at an angle of about 45°, a drop of a clear plaque mutant was added as close to the left edge of the plate as possible and allowed to drip almost to the bottom of the plate. This procedure was repeated until all of the phage mutants to be tested on this side of the plate have been dripped (**Note 12**). The plate was allowed to sit on the bench for up to 20 min to allow the streaks to dry.
4. The plate was then rotated 90° and the process is repeated. The result is that each clear plaque mutant intersects other clear plaque mutants. The point of intersection is equivalent to a multiply infected cell (in this case with two types of clear plaque mutants). In **Fig.** 20.2 (left panel), from the vertical streaks from left to right it is evident that the clear plaque mutant complements both phage mutants ε^{34}c26 and ε^{34}c51 since the intersection of both of these phages produces a turbid

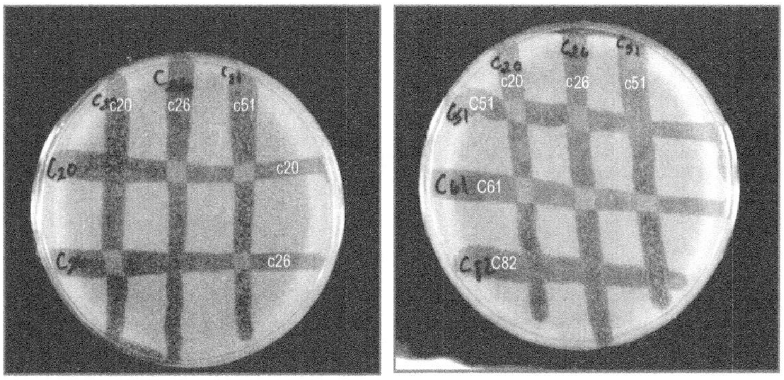

Fig. 20.2. Defining four clear plaque ε^{34} genes by plate complementation analysis. Left Panel: Vertical phage streaks from left to right are c20, c26, and c51 and the horizontal streaks from top to bottom are c20 and c26. The turbid patches indicate that complementation has taken place. This left panel indicates that complementation has occurred between c20, c26, and c51. For the Right Panel, the vertical phage streaks from left to right are c20, c26, and c51 and the horizontal streaks from top to bottom are c51 and c82. The turbid patches, where two perpendicular streaks intersect, indicate that complementation has taken place. The Right Panel indicates that complementation has occurred between c20, c51, and c61.

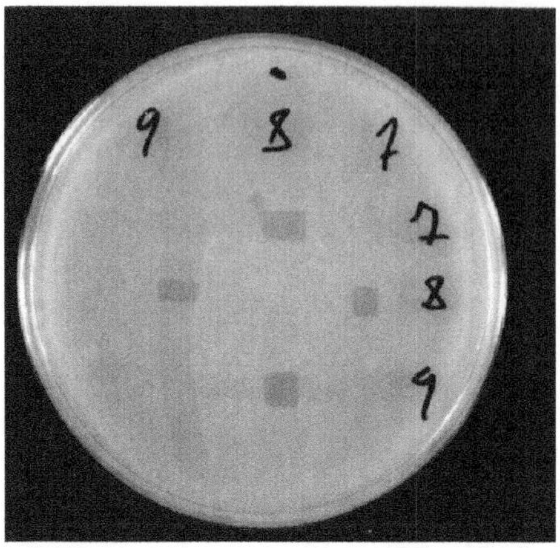

Fig. 20.3. Complementation between ε^{34} amber mutations. Amber alleles am7, am8, and am9 are complemented. The data indicate that the three amber alleles represent two genes.

square patch at the point of the intersection of the doubly infected cells. As indicated on this panel, the intersection of this phage with itself produces a clear square patch, as expected (**Note 13**). The clear plaque mutants, ε^{34}c20, ε^{34}c26, and ε^{34}c51 represent clear plaque alleles in three different phage genes. In the rightmost panel, these three clear plaque mutants are used to complement two other clear plaque mutants, ε^{34}c61 and ε^{34}c82. The results indicate that the ε^{34}c61 phage represents another gene-defining allele. This Petri plate drop complementation assay has shown that ε^{34} phage contains at least four genes involved in the formation of lysogens. We have termed these cI, cII, cIII, and cIV *(9)*. The rightmost panel gives us a bonus. The phage mutant ε^{34}c82 does not complement any of the other clear plaque phage mutants. This is strongly suggestive that the site of repressor binding has been altered in this mutant. This simple Petri plate drop complementation assay has indicated the existence of four clear plaque genes and the repressor binding site.

5. In **Fig. 20.3**, it is shown that amber mutants designated in this figure as 7, 8, and 9, define two genes. Phages 7 and 9 are alleles in the same gene. The cells used in this study were WT su− cells which are restrictive conditions for these conditional lethal mutants. Clear square patches indicate that complementation has occurred between two phage streaks where they intersect.

3.7 HA General Mutagenesis of Plasmid-Borne Phage Genes

1. A HA general mutagenic protocol was developed to mutate plasmid-borne genes *(4)*. This plasmid HA mutagenic procedure has been extensively used.
2. Set water bath at 65°C and do not start experiment until it reaches this temperature.

3. For each mutagenesis reaction to be carried out on a separate plasmid DNA, three sterile Eppendorf microtubes (1.5 ml) were needed. One sample would be a control with no mutagenic mixture, while the second tube contained samples that would be incubated for 30 min while the last tube would contain plasmid samples incubated with HA for 60 min.
4. One volume samples of pure plasmid DNA is added to five volumes of Phosphate–EDTA buffer pH6 and to four volumes of hydroxylamine hydrochloride (1 M adjusted to pH6 with NaOH) containing 1 mM EDTA. The plasmid amount normally used is 3 μg pJS28 DNA (DNA is at 1 μg/10 μl). One volume sample thus becomes 30 μl and the total volume of the mixtures was usually 200 μl.
5. To each reaction tube is added:
   30 μl DNA (~ 3 μg pJS28 DNA; N.B. DNA is at 1 μg/10 μl);
   150 μl Phosphate–EDTA buffer pH 6
   120 μl 1M hydroxylamine pH 6/1mM EDTA (adjust with NaOH)
6. Keep mixture on ice for 45 min;
7. The mutagenic samples are incubated at 65°C for two time periods: 30 and 60 min.
8. The samples that are withdrawn from the 65°C bath are immediately placed in labeled dialysis cassettes (SLIDE-A-LYZER – Pierce, Cat No. 66415, 66430) dialyzed against two changes of TE (2 l of TE, each time) at 4°C.
9. The mutated plasmid samples are now ready for study (**Note 14**).

4 Notes

1. Bacterial cells are stored long-term at −80°C in 50% glycerol. Petri plate bacterial streaked culture was used for only one month before replacing with a fresh culture. Storage for more than one month, in our hands, has sometimes resulted in losing strain viability and in Petri plate desiccation.
2. The density should be checked by refractive index using a refractometer. Please note that other phage may require different CsCl concentrations.
3. Before an infection of this type, the viable bacterial cell count was determined by dilution and titration on Petri plates containing LB agar.
4. After 4h infected cells containing phages were lysed with the added chloroform. Enough chloroform has to be added so as to settle to the bottom of the flash.
5. Residual PEG on the walls of the Oak Ridge tubes will hinder efforts at resuspending the PEG–NaCl–phage pellet by

gentle vortexing (Section 3.2.9). Often the area around the pellet is swabbed with a Q-tip to remove the PEG after the tubes have been stored upside down.

6. The term "relative" is used because one can measure only those phage particles that are able to form plaques; however, it is known that a substantial fraction of phage are produced that are not infectious, especially when one lyses bacterial cells as part of their phage stock preparation protocol.
7. The temperature of around 55°C is the lowest temperature that still permits the TA to remain in liquid form. The TA should not remain at this temperature for over 3 h as some of it may precipitate.
8. These dilutions required for determining phage concentrations can be done on a much smaller scale by using Eppie tubes where dilutions can be done using 500 μl of Phage Dilution Buffer in these tubes. Dilutions of 1:100 are achieved by adding 5 μl of the phage solution, etc. However, if it done in this fashion one has to consider the phage that can be adhered to the barrel of the pipetteman used in these dilutions and so the possibility of cross-contamination also increases.
9. When dialysis bags are used, they are boiled several times, in a 1 mm EDTA solution and finally stored in that solution at 4°C. Gloves have to be used at all times due to the presence of nucleases and proteases on the hand.
10. If the mutagenized phage is inadvertently grown in BV7001 su- then the selection will be against the isolation of an amber mutant. In this strain, amber mutations within genes will cause the production of protein fragments which will normally be degraded by host proteases.
11. Temperature sensitive mutants can be obtained in various grades of temperature defectiveness. Depending on the point in the phage lifecycle in which the ts defective gene is expressed and the nature of its defectiveness, if enough time is allowed at the non-restrictive temperature, the ts phenotype will not be seen. A slide warmer is recommended for isolation of variously defective ts alleles (Fisher and VWR are among the many suppliers that have an assortment).
12. Care must be taken not to have the streaks to mix at the bottom since it might cause mixing in all of the streaks.
13. It is possible for one to obtain slightly turbid cross-sections with a single allele because of intragenic complementation which is also taken as evidence of the multimeric nature of the defective gene *(15)*.
14. This plasmid mutagenesis protocol is most efficient when there is some sequence information about the cloned fragment and / or facile phenotypic assay for function or lack of function.

References

1. Freese, E., Bautz, E. and Freese, E. B. (1961) The chemical and mutagenic specificity of hydroxylamine. *Proc. Natl. Acad. Sci.* **47**, 845–855.
2. Tessman, I. (1968) Mutagenic treatment of double- and single-stranded DNA phages T4 and S13 with hydroxylamine. *Virology.* **34**, 330–333.
3. Venza Colon, C. J., Vasquez Leon, A. Y. and Villafane, R. J. (2004) Initial interaction of the P22 phage with the Salmonella typhimurium surface. *P. R. Health Sci. J.* **23**, 95–101.
4. Humphreys, G. O., Willshaw, G. A., Smith, H. A. and Anderson, E. S. (1976) Mutagenesis of plasmid DNA with hydroxylamine: Isolation of mutants of multi-copy plasmids. *Molec. Gen. Genet.* **145**, 101–108.
5. McClain, W. H., Gabriel, K., Lee, D. and Otten, S.(2004) Structure-function analysis of tRNA-gln in an Escherichia coli knockout strain. *RNA* **10**, 795–804.
6. Sapunaric, F. and Levy, S. B. (2003) Second-site suppressor mutations for the serine-202 to phenylalanine substitution wihin the interdomain loop of the tetracycline efflux protein Tet(C). *J. Biol. Chem.* **278**, 28588–28592.
7. Schwarz, J. J. and Berget, P. B. (1989) The isolation and sequence of missense and nonsense mutations in the cloned bacteriophage P22 tailspike protein gene. *Genetics* **121**, 635–649.
8. Salgado, C. J., Zayas, M. and Villafane, R. (2004) Homology between two different *Salmonella* phages: Salmonella enterica serovar Typhimurium phage P22 and Salmonella enterica serovar Anatum var, 15+ phage ε^{34}. *Virus Genes* **29**, 87–98.
9. Villafane, R. and Black, J. (1994) Identification of four genes involved in the lysogenic pathway of the *Salmonella newington* bacterial virus ε^{34}. *Arch. Virol.* **135**, 179–183.
10. Maloy, S. (1990) *Experimental techniques in bacterial genetics.* Jones and Barlett, Boston, MA.
11. Greenberg, M., Dunlap, J. and Villafane, R. (1995) Identification of the tailspike protein from the *Salmonella newington* phage ε^{34} and partial characterization of its phage-associated properties. *J. Struct. Biol.* **115**, 283–289.
12. Villafane, R. and King, J. (1988). The nature and distribution of temperature sensitive folding mutations in the tailspike gene of bacteriophage P22. *J. Mol. Biol.* **204**, 607–619.
13. Yamamoto, K., Alberts, B. M., Benzinger, R., Lawhorne, L. and Treiber, G. (1970) Rapid bacteriophage sedimentation in the presence of polyethylene glycol and its application to large scale virus purification. *Virology.* **40**, 734–744.
14. Davis, R. W., Botstein, D. and Roth, J. (1982) In *Advances in Bacterial Genetics.* Cold Spring Harbor, New York, pp. 94–97.
15. Botstein, D., Chan, R. K. and Waddell, C. H. (1972) Genetics of bacteriophage P22. II. Gene order and gene function. *Virology* 49, 268–282.

Chapter 21

Modifying Bacteriophage λ with Recombineering

Lynn C. Thomason, Amos B. Oppenheim, and Donald L. Court†

Abstract

Recombineering is a recently developed method of in vivo genetic engineering used in *Escherichia coli* and other Gram-negative bacteria. Recombineering can be used to create single-base changes, small and large deletions, and small insertions in phage λ as well as in bacterial chromosomes, plasmids, and bacterial artificial chromosomes (BACS). This technique uses the bacteriophage λ generalized recombination system, Red, to catalyze homologous recombination between linear DNA and a replicon using short homologies of 50 base pairs. With recombineering, single-stranded oligonucleotides or double-stranded PCR products can be used to directly modify the phage λ genome in vivo. It may also be possible to modify the genomes of other bacteriophages with recombineering.

Key words: Recombination, genetic engineering, bacteriophage λ, Red system, recombineering, mutation.

1 Introduction

Genome manipulation is an invaluable tool for the microbial geneticist. The ability to isolate mutations in specific genes has permitted analysis of microbial development and function. For example, in the study of bacteriophage λ, mutations in each morphogenetic gene enabled elucidation of the pathways for capsid *(1)* and tail assembly *(2)*. Classically, mutations in bacteriophages were often created by some form of global mutagenesis, such as UV irradiation or chemical treatment with a DNA-damaging agent (*see* **Chapter 17** for chemical mutagenesis protocols for bacteriophages). Genetic engineering allowed the creation of localized mutations by first cloning the region of interest into a plasmid and subjecting only that region to mutagenesis,

Martha R. J. Clokie, Andrew M. Kropinski (eds.), *Bacteriophages: Methods and Protocols, Volume 1: Isolation, Characterization, and Interactions, vol. 501*, © 2009 Humana Press, a part of Springer Science+Business Media
DOI 10.1007/978-1-60327-164-6_21 Springerprotocols.com

and then re-introducing the altered DNA back into the host bacterium and ultimately into the phage genome.

Recently, a new in vivo technology for introducing genetic changes, namely recombineering *(3–6)*, has been developed. In this method a specifically designed single- or double-stranded oligonucleotide is electroporated into a bacterial cell in such a way that it will recombine with an infecting phage. Recombineering has been facilitated by recent advances in genomic analysis, as it is only possible to recombineer an organism for which the DNA sequence is known. When used in combination with classical methods of screening and selection for the mutant genotypes, recombineering provides a powerful new tool to manipulate microbial genomes and episomes. The protocol described here is designed to enable modification of the bacteriophage λ genome, but we expect it to also work for at least some other bacteriophages.

Recombineering utilizes the bacteriophage λ-encoded generalized recombination functions, collectively termed Red to allow phage DNA to recombine with target DNA. As the 'target' is DNA that is electroporated into the cells before infection, this technique allows phage mutants to be created without having to make specific constructs for each desired mutation. The Red system can catalyze in vivo recombination using linear DNA as a substrate *(3–6)*. Red consists of three λ proteins: Gam which inhibits the *E. coli* RecBCD enzyme, thus preventing degradation of double-strand DNA (dsDNA); Exo which is a 5'–3' dsDNA exonuclease; and Beta which is a single-strand DNA (ssDNA) annealing protein *(7)*. Together, Exo and Beta process and recombine the substrate DNA with the target DNA *(8)*.

Only one protein, Beta, is necessary for single-strand oligonucleotide (oligo) recombination *(4)*. Using single-strand oligos, point mutations can be made or corrected and deletions can be created. Heterologous insertions of double-stranded DNA (dsDNA) can also be made with recombineering; this reaction requires all three λ Red proteins *(3)*. Usually this dsDNA is provided in the form of a PCR product.

Whether the substrate for recombineering is an oligo or a dsDNA, the Red functions act at short regions of homology to recombine the substrate DNA with the target *(3, 4)*. We routinely use oligos which are 70 bases in length; these contain the alteration to be introduced at a central base(s). The dsDNA is usually in the form of a PCR product with 50 base pairs (bp) of homology to a target in the cell or to the phage chromosome at each end of the dsDNA. This homology is introduced during the PCR amplification and often flanks a drug cassette although other DNA elements can also be inserted (**Fig. 21.1**).

In vivo expression of the bacteriophage recombination system, Red, is required for recombineering. A λ lysogen with a

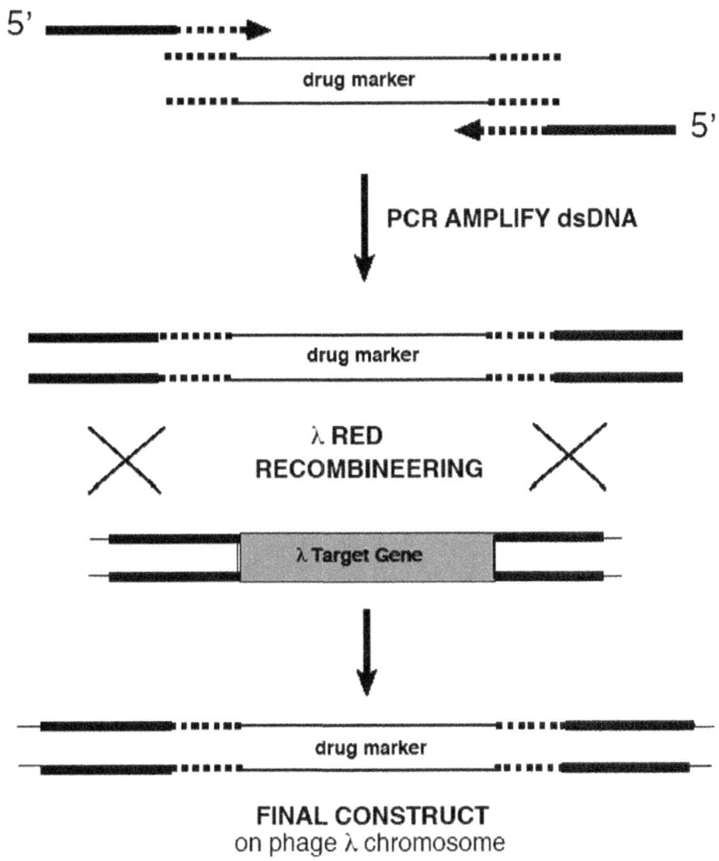

Fig. 21.1. Recombineering with dsDNA. Hybrid primers are used for PCR to amplify the DNA to be inserted, in this case a drug marker, which is targeted to the phage chromosome by recombineering. The heavy lines indicate the homology used for targeting the DNA; the dotted lines indicate the portion of primer used to amplify the DNA to be inserted.

defective prophage can be used to express Red from the strong λ p_L promoter *(3, 4)*, this mode of expression preserves the normal *red* gene regulation. The essential components of this prophage have been recently transferred to a number of different plasmids *(6, 9)*. Zhang et al., *(10)* and Datsenko and Wanner *(11)* have created plasmid constructs expressing the Red genes from the arabinose promoter, p_{BAD}. Some of the plasmids have temperature-sensitive (*ts*) origins of DNA replication *(9, 11)*, which permit them to be inactivated in the cell when they are no longer needed. Plasmids allow the Red functions to be readily transferred from one bacterial strain to another by transformation. This is especially convenient for introduction into *recA* mutant strains. The plasmid-borne prophage constructs *(9)* contain the *ts* *c*I*857* allele of the phage λ repressor, allowing Red expression to be controlled by temperature, whether it is located on the *E. coli* chromosome or on a plasmid. At low temperatures (30–34°C), the Red recombination functions are strongly repressed, but when the temperature of the bacterial culture is shifted to 42°C, they are highly expressed. Very short times of expression (15 min) are required for fully active recombination.

Any phages to be modified can be supplied by the infection of a defective lysogen. Alternatively an intact λ *c*I857 prophage can be used for recombineering. In the latter situation the phage itself both supplies the Red functions and is targeted by the recombineering *(12)*. We have recently found that in many cases it is unnecessary to use a prophage-containing host when targeting the phage λ, since the infecting phage can itself supply the Red functions (unpublished data from this laboratory).

The desired DNA sequence of the final recombinant construct should be decided before attempting to modify the phage chromosome with recombineering. A computer program for DNA analysis is very helpful in this regard. Both the original phage genome and the modified genome should be maintained as electronic files; this facilitates the design of oligonucleotides for use as PCR primers or as ssDNA recombination substrates. Gene-regulation issues should be considered at the design stage. For example, when inserting a drug resistance gene, bringing in its promoter will help establish drug resistance, however, care must be taken to prevent transcription from this promoter from extending beyond the drug marker and thus affecting downstream genes. Yu et al., *(3)* describe drug cassettes containing a promoter, open reading frame, and a transcription terminator; primers for amplification of these cassettes are listed in **Table 21.1**. The computer file of the recombinant construct is also used to design primers needed to verify the recombinant candidates.

The desired change will determine the DNA substrate used for recombineering. For large heterologous insertions, such as a drug cassette, a pair of synthetic chimeric primers is used to amplify the desired PCR product. Built into the 5' end of each primer are 50 bases of homology to the phage chromosome which are required for targeting; these homologous bases are followed by around 20 bases used to prime and PCR amplify the drug cassette or other DNA region to be inserted. When modifying phage DNA to make deletions, small substitutions, or base changes a synthetic single-strand oligo of ~ 70–100 nt is used and 35–40 bases of complete homology to the phage should flank the alteration. Recombinant frequencies of several percent have been obtained for oligo recombination onto phage λ *(13)*. Frequencies for dsDNA are lower *(6)* and so a selection or screen should be applied.

Table 21.1
PCR primers and possible source of template for drug cassette amplification *(3)*

Gene	Source	Primer sequence
ampicillin	pBluescript SK(+) (Stratagene)	5' CATTCAAATATGTATCCGCTC
		5' AGAGTTGGTAGCTCTTGATC
chloramphenicol	pPCR-Script Cam (Stratagene)	5' TGTGACGGAAGATCACTTCG
		5' ACCAGCAATAGACATAAGCG
kanamycin	*Tn*5	5' TATGGACAGCAAGCGAACCG
		5' TCAGAAGAACTCGTCAAGAAG
spectinomycin	DH5αPRO(Clontech)	5'ACCGTGGAAACGGATGAAGGC
		5' AGGGCTTATTATGCACGCTTAA
tetracycline	*Tn10*	5' CAAGAGGGTCATTATATTTCG
		5' ACTCGACATCTTGGTTACCG

2 Materials

1. A bacterial strain expressing the λ Red recombination system (**Table 21.2**; request strains from Dr. Donald Court; *court@ncifcrf.gov*; **Note 1**).
2. Purified PCR product with 50 bases of flanking homology on both sides of the desired change, or an oligonucleotide primer ~70 nucleotides (nt) in length with the desired alteration centrally located.
3. High-titer lysate of the bacteriophage to be engineered *(14)*.
4. LB medium, tryptone plates and tryptone top agar *(14)*.
5. Electroporator (we use the Bio-Rad *E. coli* Pulser) and chilled 0.1-cm electroporation cuvettes (Bio-Rad Laboratories; Hercules, CA; http://www.bio-rad.com/).
6. TM (10 mM Tris base, 10 mM $MgSO_4$, adjust pH to 7.4 with HCl) or TMG (10 mM Tris base, 10 mM $MgSO_4$, 0.01% gelatin, adjust pH to 7.4 with HCl).
7. Bacterial strains suitable for plating phage λ *(14)*.
8. Appropriate selective LB plates containing the concentrations of antibiotic required for drug resistance supplied by a single copy gene: 30 μg/ml ampicillin, 30 μg/ml kanamycin, 10 μg/ml chloramphenicol, 12.5 μg/ml tetracycline, and 50 μg/ml spectinomycin.
9. Sterile 82-mm nitrocellulose filters if needed.

Table 21.2
Bacterial strains available for recombineering

Strain	Genotype
DY329	W3110 Δ*lacU169 nadA* :: *Tn*10 *gal490 pgl*Δ8 λ*c*I857 Δ(*cro bioA*) (TetR)
DY330	W3110 Δ*lacU169gal490 pgl*Δ8 λ*c*I857 Δ(*cro-bioA*)
DY331	W3110 Δ*lacU169* Δ(*srlA-recA*)*301* :: *Tn*10 *gal490 pgl*Δ8 λ*c*I857 Δ(*cro-bioA*) (Tet^R)
DY378	W3110 λ*c*I857 Δ(*cro-bioA*)
DY380[1]	*mcrA* Δ(*mrr-hsdRMS-mcrBC*) *ϕ80dlacZ*ΔM15 *lacX74 deoR recA*1 *endA*1 *araD*139 Δ(*ara, leu*) 7697 *galU gal490 pgl*Δ8 *rpsL nupG* λ(*c*I857*ind*1) Δ{(*cro-bioA*) <> *tetRA*}(Tet^R)
DY441	DY329 with *cat – sacB* inserted between *c*I857 and *rexA* (Tet^R, Cm^R)
HME5	W3110 Δ*lacU169* λ*c*I857 Δ(*cro-bioA*)
HME45[2]	W3110 *gal490 pgl*Δ8 λ*c*I857 Δ(*cro-bioA*)
HME63	W3110 Δ*lacU169* λ*c*I857 Δ(*cro-bioA*) *galKam mutS* <> *amp*
HME64	W3110 Δ*lacU169* λ*c*I857 Δ(*cro-bioA*) *galKam uvrD* <> *kan*

[1] DH10B derivative (Invitrogen Corp.; Carlsbad, CA; http://www.invitrogen.com/)
[2] Gives less background on low concentrations of chloramphenicol than DY378.

3 Methods

The first method described is for modifying the DNA of an infecting phage. In this method, Red is expressed from a defective prophage, located either on the *E. coli* chromosome or on a plasmid, and the cells are infected with the phage to be altered. Variations of the procedure that are necessary when using an intact prophage are indicated in the Notes section. In the alternative method (described second), an intact λ prophage serves as both the source of Red and the target for recombineering. The general procedure is similar for both recombineering variations: cells are grown to exponential phase, the Red system is induced, then the cells are thoroughly washed and the appropriate substrate DNA carrying the desired genetic change is introduced by electroporation.

It is useful to try to design the construction so that the plating properties or plaque morphology of the modified phage differ from that of the parent phage and thus allowing visual screening. For example, sometimes a PCR product can introduce the desired mutation while at the same time correcting a known mutation on the phage. Amber or *ts* alleles are commonly available throughout the λ phage genome. Select for correction of the pre-existing mutation and screen among these recombinants for the mutation of interest. If the alteration introduces or removes a restriction site, the region can be amplified with PCR and the product digested with the appropriate restriction enzyme to verify the change. When inserts or deletions are made, often plaque hybridization can also be used to identify recombinant phages if no selection or screen can be devised.

3.1 Preparation of DNA for Transformation

1. Design and procure the oligos to use for PCR-mediated generation of a dsDNA product, or for use in single-stranded oligo engineering (**Note 2**). Primer sequences used to PCR amplify commonly used drug cassettes are listed in **Table 21.2**. Remember to add the targeting homology to the 5' ends of the oligos used to amplify PCR products.
2. Amplify the PCR product, examine it by gel electrophoresis and gel purify by isolating the desired band if unwanted products are obtained. Do not expose the PCR product to UV light during gel purification, since damaged DNA affects recombination frequency. Either use a non-UV based dye or if only one PCR product is obtained then just run 5 μl of product on a gel to check it is the correct size and then purify the remainder of the PCR product. Purify the PCR product by ethanol precipitation or using a commercially available kit to remove salt. It is better to avoid using a plasmid template to create a PCR-amplified drug cassette, since residual intact circular plasmid transforms the cells very efficiently and gives unwanted background that can obscure recombinant detection. If a plasmid template must be used, first linearize the plasmid with a restriction enzyme and digest the completed PCR with the restriction enzyme *Dpn*I before using it for electroporation. *Dpn*I will not cut the PCR product but will cleave DNA isolated from a Dam $^+$ host (**Note 3**).

3.2 Preparation of Bacterial Cultures

1. Inoculate a suitable bacterial strain (**Table 21.2** and **Note 4**) from a frozen glycerol stock or a single colony into 3–5 ml LB medium. Grow with aeration at 30 – 32°C overnight.
2. Equilibrate two shaking water baths to 30 –32 and 42°C, respectively. Add ~0. 25 ml of the overnight culture to 20 ml of LB medium in a 125 ml (preferably baffled) Erlenmeyer flask (**Note 4**).
3. Place the flask in the 32°C shaking water bath and grow cells with shaking until the A600 is between 0.4 and 0.6 (**Note 5**).

3.3 Adsorption of Bacteriophage to be Modified (see Note 6 for a variation)

1. Harvest the cells by centrifuging for 7 min at 4600 × *g*, at 4°C.
2. Resuspend cell pellet in 1 ml TM buffer.
3. Infect the cells with the phage to be engineered at a multiplicity of infection of 1–3 phages per cell. The cells will be

at $\sim 1-2 \times 10^9$/ml. Let the phage adsorb to the cells for 15 min at room temperature (*see* **Note 7** for variation).

3.4 Induction of Recombination Functions

1. Transfer the infected cells to 5 ml of 42°C LB medium in a 50 ml Erlenmeyer flask and shake at 220 rpm for 15 min (*see* **Note 8** for variation) to induce the Red functions and allow the adsorbed phage to inject its DNA into the cells. While the cells are inducing at 42°C, fill an ice bucket with an ice-water slurry.
2. Immediately after the 42°C induction, rapidly cool the flask in the ice-water slurry with gentle swirling. Leave on ice for 5–10 min. While the cells are on ice, chill the necessary number of labeled 35- to 50-ml plastic centrifuge tubes in preparation for spinning in a pre-chilled (4°C) centrifuge.

3.5 Preparation of Electrocompetent Cells

1. Transfer the cultures to the appropriately labeled chilled 35- to 50-ml centrifuge tubes. Centrifuge for 7 min at 4600 × *g* at 4°C. Aspirate or decant supernatant.
2. Add 1 ml ice-cold distilled water to the cell pellet in the bottom of each tube and gently resuspend cells with a large bore pipette tip or by flicking the tube (do not vortex). Add 30 ml of ice-cold distilled water to each tube, seal, and gently invert to mix, again without vortexing (**Note 9**). Centrifuge tubes again as in step 1.
3. Decant the 30 ml supernatant very carefully so as not to lose or disturb the soft pellet in each tube (**Note 10**) and suspend each cell pellet gently in 1 ml ice-cold distilled water.
4. Transfer the suspended cells to cold microcentrifuge tubes. Centrifuge the cells in a microfuge 30–60 s at maximum speed at 4°C. Very carefully remove the supernatant again taking care not to lose the cells. In each of the tubes, suspend the cell pellet in 200 μl cold distilled water. There should be enough cells for four or five electroporations. If more electroporations are needed increase the number of starting cultures rather than changing the initial culture volume.

3.6 Introduction of DNA by Electroporation

1. Chill the required number of 0.1-cm electroporation cuvettes on ice. Turn on the electroporator and set to 1.80 kV (**Note 11**).
2. Mix 100–150 ng of salt-free PCR fragment or 10–100 ng of oligo with 50 – 100 μ l of the suspension of induced or uninduced cells in microcentrifuge tubes on ice. The mixing and subsequent electroporation steps should be done rapidly; do not leave the DNA-cell mixes on ice for extended periods. Be sure to include the following electroporation reactions and controls:

   a. Phage-infected cells plus DNA. This reaction should yield the phage recombinants that you designed.
   b. Phage-infected cells without DNA. This control will help to identify contamination in the phage stock, determine the reversion frequency of alleles being scored, and obtain some idea of the efficiency of the reaction.
3. Introduce the DNA into the cells by electroporation. The time constant should be greater than 5 ms for optimal results. Low time constants can indicate problems such as inadequate washing of the cells, impurities in the DNA, or technical problems with the electroporator.
4. Immediately after electroporation, add 1 ml LB medium to the cuvette using a micropipette with a 1000 μl pipette tip (*see* **Note 12** for variation).

3.7 Preparation of the Recombinant Phage Lysate

1. Add the 1ml LB-electroporation mix to 5 ml LB medium in a 50 ml baffled Erlenmeyer flask and aerate by shaking in a 39°C water bath for 90 min.
2. Add a few drops of chloroform and continue aeration a few minutes more to complete cell lysis and release the phage particles, then pellet the debris in a 4°C centrifuge at 8000 rpm for 10 min. Transfer the recombinant lysate into a storage tube.

3.8 Plating the Recombinant Lysate for Plaque Selection or Screening

1. Make 10-fold serial dilutions of the lysates in TM or TMG through 10^{-6} and spot 10 μl of each dilution on freshly poured lawns of a bacterial indicator permissive for phage growth using TB plates. Incubate plates at the appropriate temperature. Also use a selective indicator strain here if possible (**Note 13**). Store the diluted lysate at 4°C.
2. Analyze the spot plates, comparing plating properties of the recombinant lysate with those of the control lysate (no DNA added). Using the spot plates as a guide, plate from the appropriate dilution tube to obtain single plaques of the recombinant phage.
3. If you cannot select or screen visually for the recombinant plaques wanted, you may be able to use plaque hybridization *(15)* to detect them. Identification by plaque hybridization will require the presence of a short (≥15 bp) unique DNA sequence present in the recombinant and absent in the parent to be used as a target for the probe. Thus point mutations cannot be detected with plaque hybridization.

3.9 Recombinant Phage Analysis

1. Resuspend a purified candidate plaque in 50 μl of TM and use 20 μl of this suspension as template for a PCR (**Note 14**). Reserve the remainder of the plaque suspension to grow a stock of the correct isolate.
5. Prepare a lysate of the confirmed recombinant phage using standard methodologies *(14)* (**Note 15**).

4 Notes

1. Most of the strains containing the defective lambdoid prophage are W3110 derivatives; however the prophage can be moved into other backgrounds by P1 transduction *(3)*. Any strain of choice can be transformed with plasmids expressing the recombination functions *(9, 11)*. If the same complete prophage that provides Red is also to be engineered, it must have a functional integration and excision system, and the *exo*, *beta*, and *gam* genes with their regulatory circuitry must be intact. To recombineer using an intact phage, first construct and confirm the bacterial lysogen *(14)* with the *c*I857 phage of choice. We have recently found that *red* expression from an infecting phage is sufficient to catalyze recombineering, providing the Red functions are intact, and in some cases recombinant yields are better in the absence of the prophage. We recommend executing the entire procedure in parallel with two bacterial strains, one containing the defective prophage, and one that lacks it, such as C600, which is a good host for phage λ.
2. We routinely used salt-free oligos that are otherwise unpurified for many applications. However, the oligos are the greatest source of adventitious mutations that occur during recombineering *(13)* and for some applications Polyacrylamide Gel Electrophoresis (PAGE) purification may be helpful.
3. If the recombinant from a dsDNA (e.g., PCR product) recombination is selected as a lysogen, and the PCR product was amplified from a plasmid template, it is important to include a control reaction of un-induced cells transformed with the PCR product; drug resistant colonies appearing in the control reaction demonstrate that intact plasmid was present in the transformed DNA.
4. When subculturing the cells, the dilution should be at least 70-fold. If you are expressing the Red system from a plasmid, add the appropriate antibiotic to maintain selection during growth. If arabinose or some other inducer of the recombination functions is used, be sure to add it to the medium. 10 mM arabinose (for Ara^+strains) induces expression of the λ Red functions from the plasmids of Datsenko and Wanner *(11)*. Also include an additional flask of an un-induced culture as a negative control. The temperature shift is unnecessary when arabinose or another chemical is used for Red induction.
5. Usually the cells will grow to the correct optical density in about 2 h. Don't let them enter early stationary phase ($>0.7 - 0.8$ OD_{600}), since this will result in poor expression of the Red functions.

6. If a lysogen is with an intact prophage that will both supply Red and be targeted by recombineering, skip this step and proceed directly to step 3.4.
7. Adsorption conditions may vary for phages other than λ. For example, a nonlambdoid phage such as T4 may require adsorption on ice.
8. The 15 min induction time will result in expression of the phage replication functions and resultant cell killing. Thus, for an intact prophage, the full 15 min induction time should only be used only when phages are being induced and a recombinant lysate prepared. If you desire to retain the targeted phage as a prophage and plate the recombinant bacterial colonies (selecting, for example, insertion of a drug resistance cassette onto the prophage), reduce the induction time to 4–5 min. Because the level of Red expression will be lower with the shorter heat pulse, recombinant frequencies will be correspondingly reduced.
9. Throughout the procedure, resuspend the cells gently and without vortexing. The washing steps remove any chemical added to induce the Red system.
10. Remove tubes from the centrifuge promptly after the distilled water wash. The pellet is very soft and care should be taken not to dislodge it, especially when processing multiple tubes.
11. Electroporation conditions are crucial and although other brands of cuvettes and electroporator may work, we have not confirmed this. 0.2 cm cuvettes will require different electroporation conditions and standardization to obtain optimal recombination frequencies by empirical methods (see the instruction manual for your electroporator).
12. When recombinant lysogens of a complete prophage are selected (such as by drug resistance), the yield will be very low, thus the entire contents of a single electroporation mix are spread on one Petri plate as described below. Lining the plate with an 82-mm diameter sterile nitrocellulose filter allows both the nonselective outgrowth and the selection to be done on Petri plates. Carefully lay a sterile 82-mm diameter nitrocellulose filter atop a rich nonselective (i.e., LB) plate with sterile forceps. Immediately after electroporation, add $\sim 0.3\,\text{ml}$ LB to the cells in the cuvette and spread them on the filter. Incubate the plates for 3 h or more at $30 - 32°\text{C}$; then transfer the filter to the appropriate drug plate, again using sterile forceps. The number of recombinants is generally less than 500 per electroporation mix. Usually approximately half of the surviving cells will have spontaneously lost the prophage. It is possible to screen for non-lysogens by testing candidate colonies for their ability to plate λ, since the prophage renders the cells immune to phage infection; the cured cells will also be viable at 42°C while those containing the prophage will not.

13. If recombinant phages can be selected from the lysates by growth under some condition, spot the same dilutions on the selective indicator strain. Several bacterial strains and other genetic tricks are very useful for selecting certain phage λ genotypes *(14)*. For example, strains with the appropriate suppressor tRNA will allow a phage with an amber mutation in an essential gene to grow, while the isogenic non-suppressing strain will prevent growth. If the recombineering repairs such an amber mutation, the recombinant can be selected for directly by plating on the non-suppressing strain. If an amber mutation is introduced into an essential gene, the double-layer technique *(16)* can be used to identify phages with the amber by plaque morphology. Another example of a useful selective strain is a bacterial host mutant for *recA* function, since λ phages doubly mutant for both the *red* and *gam* genes it will not form plaques on this host. Conversely, a P2 lysogen is restrictive for wild- type λ and will plate only *red gam* mutant phages. A *ts* mutation can be selected against by plating at the appropriate temperature.
14. The genetic alteration created by recombineering determines the PCR primers needed to confirm recombinants. If a heterologous DNA such as a drug marker is inserted, design four primers, two flanking the insert and two pointing outwards from within the cassette, all with compatible annealing temperatures. Flanking primers are paired with the internal cassette primers to amplify the two junctions at the insertion site. The outside primer pair, which hybridizes to the external flanking sequences rather than to the insert itself, can also be used to demonstrate loss of the target sequences and presence of the insertion, as long as the relative sizes of the two possible PCR products differ. If a single base change alters a restriction site, recombinants bearing the change can be identified by generation of a PCR product encompassing the region and digestion of that product with the relevant restriction enzyme, followed by agarose gel electrophoresis.
15. Unwanted mutations can be introduced during the chemical synthesis of the oligo or primer population *(13)*; therefore, it is important to confirm the final construct by sequence analysis, especially the regions derived from the original oligos or primers.

Acknowledgements

Comments from Jim Sawitzke improved the manuscript.

References

1. Georgeopolous, C., Tilly, K. and Casjens, S. 1983. Lambdoid phage head assembly. *In* Lambda II (R. Hendrix, J. Roberts, F. Stahl, and R. Weisberg, eds.) pp. 279–304. Cold Spring Harbor Laboratory Press, Cold Spring Harbor, New York.
2. Katsura, I. 1983. Tail assembly and injection. *In* Lambda II (R. Hendrix, J. Roberts, F. Stahl, and R. Weisberg, eds.) pp. 331–346. Cold Spring Harbor Laboratory Press, Cold Spring Harbor, New York.
3. Yu, D., Ellis, H. M., Lee, E. C., Jenkins, N. A., Copeland, N. G., and D. L. Court. 2000. An efficient recombination system for chromosome engineering in *Escherichia coli. Proc. Natl. Acad. Sci. U.S.A.* **97,** 5978–5983.
4. Ellis, H.M., Yu, D., DiTizio, T., and Court, D.L. 2001. High efficiency mutagenesis, repair, and engineering of chromosomal DNA using single-stranded oligonucleotides. *Proc. Natl. Acad. Sci. U.S.A.* **98,** 6742–6746.
5. Court, D.L., Sawitzke, J.A., and Thomason L.C. 2002. Genetic engineering using homologous recombination. *Annu. Rev. Genet.* **36,** 361–388.
6. Thomason L.C., Myers, R.S., Oppenheim, A., Costantino, N., Sawitzke, J.A. Datta, S., Bubenenko, M. and Court D.L. (2005). Recombineering in prokaryotes. *In* Phages: Their Role in Bacterial Pathogenesis and Biotechnology. pp. 383–399. ASM Press, Herndon, Va.
7. Smith, G.R. General recombination. 1983. *In* Lambda II (R. Hendrix, J. Roberts, F. Stahl, and R. Weisberg, eds.) pp. 175–209. Cold Spring Harbor Laboratory Press, Cold Spring Harbor, New York.
8. Murphy, K.C., Campellone, K.G., and Poteete, A.R. 2000. PCR-mediated gene replacement in *Escherichia coli. Gene.* **246,** 321–330.
9. Datta, S, Costantino, N., and Court, D.L. 2006. A set of recombineering plasmids for gram-negative bacteria. *Gene.* **379**, 109–115.
10. Zhang, Y., Buchholz, F., Muyrers, J. P. P., and Stewart, F. 1998. A new logic for DNA engineering using recombination in *Escherichia coli. Nature Genetics.* **20,** 123–128.
11. Datsenko, K.A., and Wanner, B.L. 2000. One-step inactivation of chromosomal genes in *Escherichia coli* K-12 using PCR products. *Proc. Natl. Acad. Sci. U.S.A.* **97,** 6640–6645.
12. Court, D.L., Swaminathan, S., Yu, D., Wilson, H., Baker, T., Bubunenko, M., Sawitzke, J., and Sharan, S.K. 2003. Mini-lambda: a tractable system for chromosome and BAC engineering. Gene. **315,** 63–69.
13. Oppenheim, A.B., Rattray, A.J., Bubunenko, M., Thomason, L.C., and Court, D.L. 2004. In vivo recombineering of bacteriophage λ by PCR fragments and single-strand oligonucleotides. *Virology.* **319,** 185–189.
14. Arber, W., Enquist, L., Hohn, B., Murray, N.E., and Murray, K. 1983. Experimental methods for use with lambda. *In* Lambda II (R. Hendrix, J. Roberts, F. Stahl, and R. Weisberg, eds.) pp. 433–471. Cold Spring Harbor Laboratory Press, Cold Spring Harbor, New York.
15. Sambrook, J. and Russell, D. W. 2001. Bacteriophage λ and its vectors. *In* Molecular Cloning: A Laboratory Manual *(Third Edition)* pp. 2.25–2.110. Cold Spring Harbor Laboratory Press, Cold Spring Harbor, New York.
16. Campbell, A. 1971. Genetic structure. *In:* The Bacteriophage Lambda (A.D. Hershey, ed.) pp. 13–44. Cold Spring Harbor Laboratory Press, Cold Spring Harbor, New York.

Chapter 22

Identification and Isolation of Lysogens with Induced Prophage

Jonathan Livny, Christopher N. LaRock, and David I. Friedman

Abstract

The fate of lysogens following prophage induction has assumed added significance with the finding that in many pathogens virulence genes are carried on prophages and, in some, the production and/or release of the virulence factor is under control of the phage lytic regulatory program. We outline a method for identifying and characterizing from a total lysogen population, the subpopulation in which the prophage is induced. The prophage is genetically altered so that on induction it does not go through the lytic pathway, but does express a resolvase that acts at a reporter cassette located at another site on the bacterial chromosome to irreversibly change the resistance of the bacterium from tetracycline to chloramphenicol. Thus, induced derivatives survive and are easily identified even if they make up a small fraction of the population.

Key words: SIVET, prophage, induction, tandem prophages, resolvase, *res* sites.

1 Introduction

Interest in phage biology has been heightened by findings that many pathogenic bacteria harbor prophages that carry gene encoding factors contributing to virulence *(1)*. Moreover, in some Shiga toxin producing *E. coli* (STEC), production and release of Shiga toxin (Stx) results from induction of the prophage carrying the *stx* genes *(2)*. These observations strongly suggest that it is the subpopulation of lysogens that undergoes induction in the intestinal environment that is the major contributor of toxin production and release during infection. Studying the subpopulation of induced lysogens both in culture and during infection presents significant challenges as it requires a method to isolate cells that are normally killed by the induced prophage

Martha R. J. Clokie, Andrew M. Kropinski (eds.), *Bacteriophages: Methods and Protocols, Volume 1: Isolation, Characterization, and Interactions, vol. 501*, © 2009 Humana Press, a part of Springer Science+Business Media
DOI 10.1007/978-1-60327-164-6_22 Springerprotocols.com

that may only be transiently exposed to an inducing stimulus, and that, in many cases, may represent a small proportion of the total lysogen population. To address these challenges, our laboratory developed an approach known as SIVET (selective in vivo expression technology). This method, adapted from RIVET (recombination-based in vivo expression technology *(3)*), allows selection for the rare bacteria in which expression from a promoter of interest occurs, in our case the early P_R promoter of phage H-19B *(4)*. While the SIVET approach can be applied to study any bacterial promoter of interest, this section will focus on employing SIVET to study lambdoid phage induction.

1.1 General Approach for Construction of SIVET Lysogen Strains

It is essential that a SIVET lysogen survive prophage induction. Therefore, if the prophage encodes functions that cause host cell death following induction, the first step in constructing a SIVET lysogen is to inactivate genes encoding lethal functions. This can be achieved by deleting most of the prophage, leaving intact the operator regions, the gene encoding the repressor with its promoter, and the promoter being studied. Alternatively, prophage-encoded genes whose products lead to host cell lethality can be eliminated. For example, the λ and H-19B prophages (and presumably other lambdoid prophages) can be disarmed by deleting the *O* or *P* replication genes and the *N* transcription regulation gene *(5)*. Any available method of allelic replacement, such as those employing positive-selection suicide vectors *(6)*, can be used to make these genetic alterations. For construction of *E. coli* SIVET lysogen strains, we have found recombineering to be the most effective method for allelic replacement. Recombineering is a method developed in *E. coli* for obtaining efficient recombination with either single or double stranded linear oligonucleotides. The oligonucleotides include in addition to the sequences to be inserted, flanking sequences homologous to the target sequences (**Fig. 22.4**). For SIVET construction, the target sequences are chromosomal. We have included an abbreviated protocol for recombineering below; the reader is referred to chapter 18. Strain construction will be outlined using the antibiotic resistance elements used in our work; however, different antibiotic markers may be used depending on the resistance characteristics of the bacterium being studied.

If a bacterium has prophages in addition to the one being studied that on induction are lethal to the host, elimination of lethal genes in the prophage under study will obviously not render the lysogen resistant to phage-mediated killing following induction. Therefore, it is essential to determine if the bacterium has other prophages that cause host cell death upon induction. This can be accomplished by inactivating the prophage under study and then determining if the host bacterium is killed under inducing conditions. This test of inducing lethality can be done using sublethal doses of an inducing agent such as mitomycin C. If other lethal prophages are identified, their lethal functions must also be inactivated. Alternatively the prophage to be studied can be moved to another bacterial host, if one is available, that is free of lethal prophages. Regardless of the origin of the starting strain, it is important that this strain has a single copy of the prophage to be studied and does not have multiple tandem copies of the prophage. Therefore, we begin the methods section of this chapter with a general method to distinguish between strains carrying a single copy and those carrying tandem copies of the prophage. This approach is adapted from one developed in the Court lab *(7)* that is based on PCR and can easily be applied to phages that have not been extensively characterized. The one addition from the reported method that we found to be essential for reproducible results is the use of a recombination deficient bacterium (*recA*$^-$). In our study, the *recA*$^-$ allele was moved into the strain being studied by P1 transduction. However, the change can be made using any method available for allelic replacement.

We next discuss the genetic elements responsible for the essential feature of a SIVET strain, its ability to undergo a heritable and selectable phenotypic change upon induction. These genetic elements are located at different sites on the bacterial genome (**Fig. 22.1**). The first is a gene encoding the TnpR resolvase of the γδ transposon *(8)* inserted under the control of the promoter of interest. In our study, *tnpR* was inserted downstream of the *cro* gene of H-19B under the control of the early phage promoter P_R. The phage repressor directly controls transcription from P_R; thus TnpR is expressed only following induction of the prophage. The second genetic element is a reporter construct that includes a *cat* cassette (encoding chloramphenicol acetyltransferase) interrupted by a *tetR* cassette flanked by *resC* sites (**Fig. 22.1**). In the absence of TnpR, this construct confers resistance to tetracycline but not to chloramphenicol. TnpR acts at the two *resC* sites to catalyze a recombination reaction that removes the *tetR* cassette and one *resC* site. Although one copy of *resC* (~ 150 bp) is retained in the *cat* gene, the *resC* sites have a nucleotide change that removes a block in Cat expression that would be imposed by the wild-type *res* site. Thus, excision of *tetR*-*resC* re-establishes a *cat* gene that encodes an active product. Re-establishment of the *cat* gene following induction results in an irreversible genetic change of the SIVET bacterium making it sensitive to tetracycline and resistant to chloramphenicol (**Fig. 22.1**). The proportion of SIVET lysogens that have undergone induction can then be determined simply by comparing the number of chloramphenicol resistant bacteria to the total number of bacteria isolated.

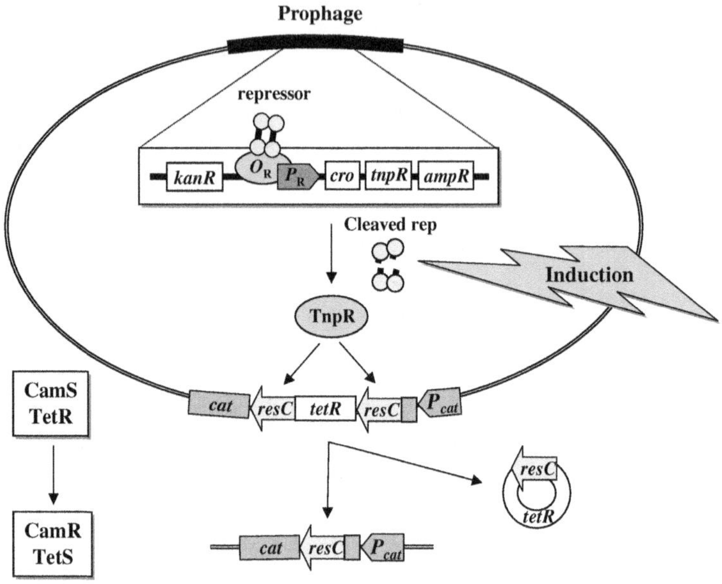

Fig. 22.1. The SIVET construct designed to identify lysogens with induced prophage. Top represents altered prophage with *kanR* and *ampR* replacements and *tnpR* gene. When prophage is induced, repressor is cleaved and transcription from P_R ensues. TnpR is synthesized and acts at *resC* sites in the *cat::resC-tetR-resC::cat* cassette to excise the *resC-tetR* insert releasing the *tetR-resC* circle. The bacterium goes through an irreversible change from TetR CamS to CamR TetS.

2 Materials

2.1 Buffers

1. TBE: Prepare 5X stock with 450 mM Tris base, 450 mM boric acid, and 10 mM EDTA pH 8.0, store at room temperature. Prepare 1X working solution by dilution of one part with four parts water.
2. PCR buffer (10X): 500 mM KCl, 200 mM Tris–HCl, pH8.3, 15 mM$MgCl_2$, 0.1% gelatin, 2 mM each deoxynucleotides (dATP, dCTP, dGTP, dTTP). Aliquot and store at −70°C.

2.2 Media

1. LB: 10 g tryptone, 5 g yeast extract, 5 g NaCl.
2. M9 sucrose plates (for testing whether bacterium carries a functional *sacB* gene); 10 g agar, 100 g sucrose, 15 ml 40% glycerol, 1 ml 2% CAA, 2 mg/ml biotin, 100 ml 10 X M9 salts (per liter: 60 g Na_2HPO_4, 30 gKH_2PO_4, 5 g NaCl, 10 gNH_4Cl).

2.3 Reagents, Equipment, and Bacteria

1. Taq DNA polymerase (Invitrogen, Carlsbad, CA).
2. Prestained molecular weight markers: 1 kb Plus DNA ladder (Invitrogen, Carlsbad, CA).
3. Agarose gel: 1% agarose (w/v) in TBE.

4. Mitomycin C made up to 2 mg/ml in EtOH. Dilute 1:1000 to 2 μg/ml for induction of prophages of the λ family.
5. Electroporation cuvettes, 0.1 cm gap (BioRad, Hercules, CA).
6. Qiagen PCR purification column.
7. Bacterium- DY378 and DY330 *(9)*.

3 Methods

The method for construction of SIVET strains described below was designed to study induction of lambdoid phages in *E. coli*. Hence, certain aspects of these methods will likely have to be modified to apply the technology for construction of SIVET strains to study prophage induction in other bacterial species. In constructing new SIVET lysogen strains, the individual SIVET components such as the *tnpR* gene and the *cat::resC-tetR-resC* cassette can be amplified by PCR using as a template our H-19B SIVET strain. Please send requests for this strain to D.I. Friedman at davidfri@umich.edu. Sequences of primers to be used in PCR amplification of these components are listed in **Table 22.1**.

3.1 Determining Whether Prophage is Present in Single Copy

1. The sequence of the *attP* site and one of the *attB* sites must be determined to design appropriate primers to produce the required PCR products (**Fig. 22.2**). The two PCR products should be of different sizes so that they can be distinguished following gel electrophoresis. For example, our primers for the 933 W prophage were designed to yield an *attP* fragment of 365 bp and an *attL* fragment of 643 bp.
2. Using a disposable pipette tip, pick a small amount of a colony formed from the lysogen to be examined and distribute into 500 μl of sterile H_2O in a microfuge tube.

Table 22.1
Primer Sequences

Product	Forward primer sequence	Backward primer sequence
cat::resC-tetR-resC	5′-ggcgctggcgatgagacgttg	5′-ttcagcgacagcttgctgtacg
cat-sacB	5′-atgagacgttgatcggcacg	5′-actgtccatatgcacagatg
tnpR	5′-ttgatttaggatacatttttatgcgac	5′-ttagttgctttcatttattactttatatactgttg
tnpr168	5′-ttgatttaagatacatttttatgcgac	5′-ttagttgctttcatttattactttatatactgttg
tnpr184	5′-ttgatttagcatacatttttatgcgac	5′-ttagttgctttcatttattactttatatactgttg
ampR	5′-ttcggggaaatgtgcgcgga	5′-acggggtctgacgctcagtg
kanR	5′-agcgtaatgctctgccagtg	5′-aggaattccccggatccgtc

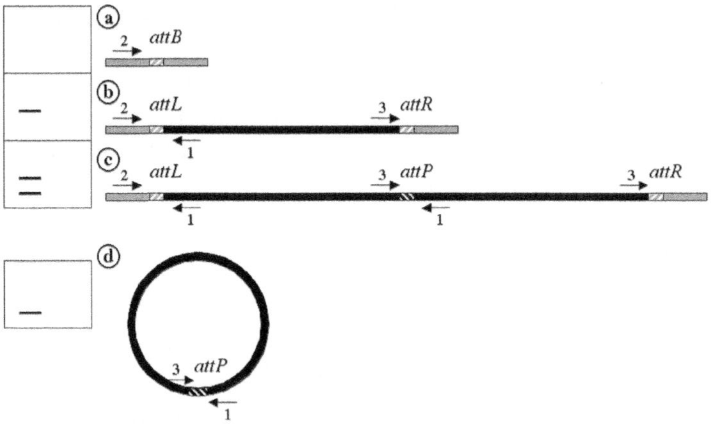

Fig. 22.2. Production of amplification products designed to distinguish between single and tandem multiple copies of a unique prophage. Lines are representative of bacterial DNA (*grey*) and prophage DNA (*black*), with *att* sites (*hatched*). The *attL* and *attR* sites are generated by insertion of the phage into the bacterial *att* site and the *attP* site is generated by the recombination of two phage genomes (for details on the molecular events in phage integrative recombination, see Weisberg and Landy *(12)*). Primer annealing sites and orientations are indicated on their respective strand by arrows pointing in the 3′ direction. On the left is a representation of the expected results when PCR products are observed following gel electrophoresis analysis. The number and position of bands provides information on whether there is more than one copy of the unique prophage: (**a**) no copies, (**b**) single copy, (**c**) 2 or more copies, (**d**) free phage.

3. Suspend cells by vortexing at maximum setting for 10 s.
4. Pellet cells for 1 min at 8000 × *g* and carefully decant the supernatant without disturbing the cell pellet.
5. Wash cells three times, each time resuspending the pellet in 500 μl of ddH_2O. After the final wash resuspend in 100 μl sterile H_2O (**Note 1**). Add 20 μl of the washed cell mixture to a PCR mix of final volume 50 μl containing 50 pmols of each of the three primers (1, 2, and 3), 5 μl 10X PCR buffer solution (with deoxynucleotides), and 0. 5 μl Taq polymerase or as specified by polymerase manufacturer (**Note 2**).
6. Amplify using a thermal cycling program appropriate for the designed fragment sizes (**Note 3**).
7. Examine 5 μl of amplified DNA products by electrophoresis using a 1% agarose gel in 1X TBE running buffer (**Note 4**). As shown in **Fig. 22.3**, no bands should be observed with a nonlysogen. A lysogen containing a single integrated prophage should yield a single band. In our example, the *attL* fragment was amplified (**Note 5**). A lysogen carrying multiple prophage integrated in tandem should produce two bands. In our example, the *attL* and *attP* products were amplified.

3.2 Building a SIVET Strain

The methodology described below applies to the construction of SIVET lysogen in *E. coli* strains carrying lambdoid prophages.

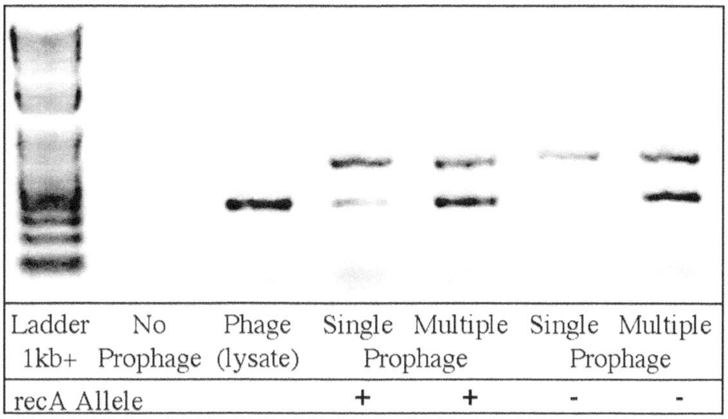

Fig. 22.3. Distinguishing single and multiple tandem copies of prophages (here phage 933 W). PCR products were run on a 1% agarose. Indicated below is the phage status of sample being studied and *recA* allele of lysogens. Size ladder (*lane 1*). A strain without a prophage forms no band (*lane 2*). A phage lysate produces only the *attP* fragment (*lane 3*). A strain with a single integrated prophage in a *recA*$^+$ strain shows an *attB* band, with a background *attP* band (*lane 4*) that is not present in a *recA*– strain (*lane 6*). A strain with two or more integrated prophage amplifies both the *attB* and the *attP* fragments (*lanes 5 and 7*), unaffected by the nature of the *recA* allele.

Hence, the reader is reminded that some of the methods and specific genetic alterations described may have to be modified in applying this approach to the study of other families of temperate phages and/or other species of lysogenic bacteria.

1. To induce recombination function required for recombineering, 30 ml of L broth is inoculated with 100 μl of an overnight culture of a recombineering strain (e.g., DY378) lysogenized with a single copy of the prophage to be altered (**Note 6**). The culture is grown in L broth at 32°C in a 500 ml flask to early exponential phase (OD_{600} =. 300).
2. Half of the culture is transferred to 42°C for 10 min to induce expression of λ recombination proteins while the other half remains at 32°C serving as a negative control.
3. The two cultures are cooled by rotating in an ice slurry and sedimented.
4. Cells are washed 3 times with 1 ml of ice-cold ddH_2O, and resuspended in 100 μl ice-cold ddH_2O.
5. Genes affecting host survival are replaced by antibiotic resistance cassettes (e.g., kanamycin resistance). Antibiotic cassettes are PCR amplified using as the template bacteria carrying the cassette either chromosomally or on a plasmid (**Note 7**). The primer oligonucleotides include 50 terminal 5′ nucleotides that are homologous to sequences flanking the gene(s) to be eliminated and 20 terminal 3′ nucleotides homologous to the ends of the amplified cassette (**Fig. 22.4**). The amplicon is purified using a Qiagen

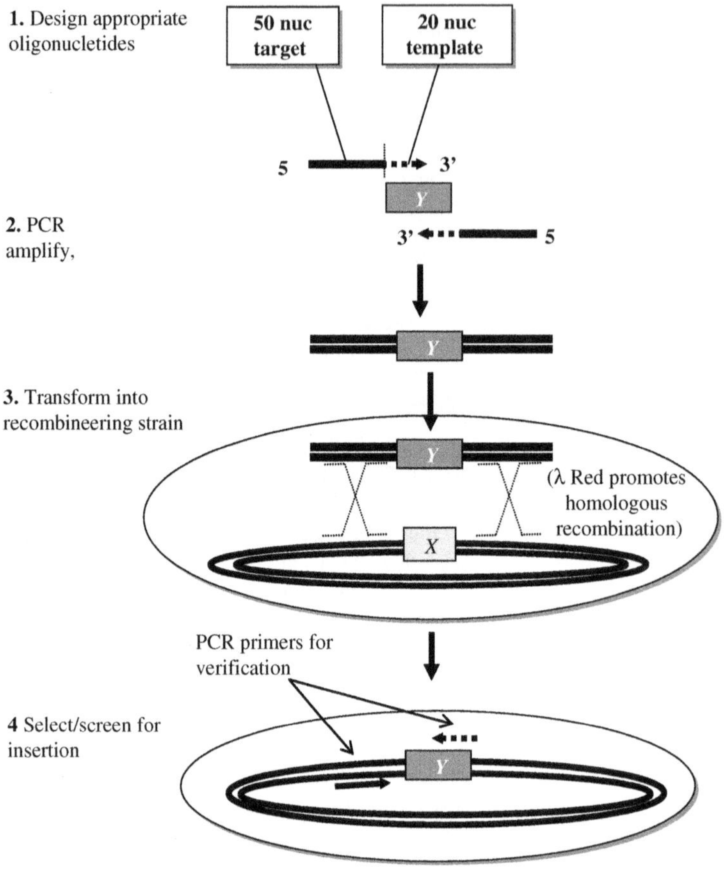

Fig. 22.4. Using recombineering for allelic replacement, in step 4, if *Y* encodes antibiotic resistance, select for resistance. If *X* encodes *cat-sacB*, select for resistance to sucrose and screen for sensitivity to chloramphenicol. Verify insertion is in the correct location by PCR.

PCR purification column and eluted in 50 μl ddH_2O. The amplicon is then recombined into the prophage genome as described above.

6. The prepared cells are electroporated with 2 μl of purified PCR amplicon. For best results, electroporation should be conducted at 1.8 kV using electroporation cuvettes with a 0.1 cm gap. Following electroporation, cells are immediately resuspended in 1 ml L broth and grown at 32°C for 1–2 h. Cultures may also be incubated overnight to increase the likelihood of obtaining desired recombinants.
7. Following growth, 20–300 μl of culture are spread on LB agar plates containing the appropriate antibiotic and incubated at 32°C overnight.
8. Gene replacement can be verified by PCR amplification using one primer complimentary to a site inside the antibiotic resistance gene and a primer complimentary to a site inside the prophage near the site of insertion (**Fig. 22.4**).

9. When prophage genes encoding functions lethal to the host are deleted or rendered inactive, the *tnpR* gene is inserted under control of the targeted promoter. The first step in introducing *tnpR* into the prophage genome is to introduce a *cat-sacB* cassette *(10)*, which confers resistance to chloramphenicol and sensitivity to sucrose, downstream of the targeted promoter (**Note 8**). Again, recombineering, as described above, was used in our construction.
10. The *tnpR* gene with its ribosomal binding site (RBS) is synthesized by PCR using a bacterium carrying that gene as the template. The primers include the same 50 nucleotides of homology to the prophage that were on the 5′ ends of the *cat-sacB* cassette with the 3′ ends having 20 nucleotides of homology to the ends of the *tnpR* gene. The *cat-sacB* cassette is replaced by this amplicon, again in our study by recombineerring. Lysogens with this replacement will be resistant to sucrose and sensitive to chloramphenicol.
11. The *cat::resC-tetR-resC* cassette is crossed into the bacterial chromosome at a position unlinked to the promoter element. In our study, recombineering was used to cross this construct into the *galK* gene. The *cat::resC-tetR-resC* cassette inserted in the *galK* gene can be moved into another strain by P1 transduction. To move it into a different locus, the *cat::resC-tetR-resC* cassette can be amplified from the SIVET strain by PCR and inserted into the desired chromosomal location by recombineering or by other methods of allelic replacement.
12. If tetracycline resistance cannot be used as a selection, the *tetR* cassette in *cat::resC-tetR-resC* can be replaced by a gene encoding another antibiotic resistance. To make this replacement using recombineering, the *cat::resC-tetR-resC* cassette must first be moved into a recombineering strain (e.g., DY378 or DY330) by P1 transduction. Alternatively, a DY330 derivative with *cat::resC-tetR-resC* can be obtained from the Friedman lab. The *tetR* gene should be replaced from its start to stop codons by the gene encoding the desired antibiotic resistance. For example, we have successfully substituted a *kanR* gene for the *tetR* gene.
13. The individual elements of the SIVET lysogen are combined in a single strain. In our studies both elements were moved into a new background by P1 transduction.
14. For certain applications, a SIVET strain carrying both the SIVET and wt versions of the same prophage is required (**Section 3.3 Step 5**). To construct such a strain, the SIVET prophage is moved (by P1 transduction) into a lysogen carrying multiple copies of the wt version of the prophage integrated in tandem (**Note 6**). It is important to be aware that even if the recipient strain carries multiple copies of the wt prophage, transduction may result in the entire array of wt

prophages replaced by the SIVET prophage. Either of two methods can be used to verify that the SIVET strain maintains a wt prophage following introduction of the SIVET prophage. First, PCR amplification spanning the site of the *tnpR* insertion in the SIVET prophage should yield two bands, a larger one corresponding to the SIVET prophage and a smaller one corresponding to the wt prophage (without *tnpR*). Second, exposure of the strain to an inducing agent, such as mitomycin C, should result in host cell lysis and phage production if there is a wild-type copy.

3.3 Using SIVET Lysogen Strains to Measure the Frequency of Prophage Induction

1. Cultures of the SIVET strain are grown under optimal conditions corresponding to overnight growth of *E. coli* at 37°C in L broth. The broth should be supplemented with the antibiotic required to maintain selection of bacteria with the complete *cat::resC-tetR-resC* cassette (for our construct 5 μg/ml tetracycline was required).
2. Overnight cultures are diluted 1:100 in L broth (without antibiotic) and grown overnight at 37°C .
3. Following overnight growth, 10 μl and 100 μl of 10^{-1}, 10^{-3}, 10^{-5}, and 10^{-7} dilutions of these cultures are transferred to LB agar plates supplemented with appropriate concentrations of antibiotics to select for all SIVET lysogens (in our experiment, kanamycin, 30 μg/ml, and ampicillin, 25 μg/ml) or to select for induced SIVET lysogens (in our experiments kanamycin, ampicillin, and chloramphenicol, 8 μg/ml). The low amount of chloramphenicol was necessary because the Cat protein with the insertion resulting from the remaining *resC* site confers less effective resistance than the wild-type Cat protein.
4. Plates are incubated at 37°C for 48 h. The frequency of induction is determined by dividing the number of CamR lysogens per ml of culture by the number of KanR, AmpR lysogens per ml of culture.
5. Calibrating the SIVET: Studies with RIVET have shown that extremely low levels of TnpR expression can result in excision of *res-tet-res (11)*, suggesting that acquisition of CamR by SIVET strains may occur at levels of derepression insufficient to induce the wt prophage. To verify that the level of derepression required to excise *resC-tet-resC* is similar to that required to activate the wt prophage, the frequency of CamR lysogens in a SIVET strain carrying only a SIVET prophage can be compared to that in a SIVET strain carrying both a SIVET and a wt prophage of the same type. If the proportion of the lysogen population that is CamR is similar in the two strains, it suggests that in the majority of cells in which *resC-tetR-resC* is excised the wt prophage was not induced and thus acquisition of CamR occurs at a level of derepression

significantly lower than that required to induce phage lethal functions. Alternatively, if the proportion of lysogens that are CamR is significantly lower in the SIVET strain carrying the wt prophage, it suggests that the level of derepression resulting in excision of *resC-tetR-resC* is similar to or greater than the level of derepression required for expression of lytic functions. If the acquisition of CamR by the SIVET lysogen is found to be sensitive to low levels of derepression, a SIVET prophage encoding a *tnpR* gene with a mutant ribosomal binding site can be used (**Note 9**).

4 Notes

1. If present in relatively large amounts, free phage DNA in the reaction may result in synthesis of significant amounts of the *attP* oligonucleotide yielding an observable band in the gel (**Fig. 22.2**), an observation that would lead to a false conclusion of multiple tandem prophages. The formation of this band results from an *attP* site in genomes of phage particles. Hence, this activity cannot be reduced by additional washes.
2. A lysate of the phage can serve as a control template for PCR production of the *attP* band. Because the *att* site is most likely internal to the ends (e.g., λ), this is possible even if the genome in the phage particle is linear. One or two μl of phage is sufficient in a PCR reaction with the *attP* primers.
3. The thermal cycling program used for amplification (the oligonucleotides in our experiments were each composed of $\sim$ 350 nucleotides) included the following steps: 95°C for 5′, followed by 30 rounds of 95°C for 30 s., 54°C for 45 s, 72°C for 45 s and 72°C for 10′ before holding at room temperature. This program serves as an example and the program chosen for all amplifications should be adjusted as appropriate for the template, primer annealing temperature (tm), and fragment size.
4. The percent agarose should be adjusted according to fragment size; likewise, any staining method for visualization of bands is acceptable. Here, bands were visualized using ethidium bromide staining of a minigel and viewing with UV, after electrophoresis for 30 min at 120 V.
5. The use of a *recA*$^-$ derivative significantly reduces the background of circular phage DNA, a major source of unwanted *attP* site DNAs by eliminating spontaneous prophage induction.
6. Verify that the recombineering strain carries only a single copy of the prophage using the method described in **Section 3.1**.

7. For best results, chromosomally inserted antibiotic genes, such as those encoded by the H-19B SIVET strain, should be used as template for PCR amplification. If plasmid DNA is used as the template for PCR, treat the plasmid with a restriction enzyme that does not cleave in the genetic element serving as the template for the PCR amplicon. Because circular supercoiled plasmid DNA transforms more efficiently than linear DNA and does not need to integrate into the chromosome to confer antibiotic resistance, even a very small amount of plasmid contamination of the PCR reaction will lead to a high background of resistant colonies.
8. *sacB* is a large gene susceptible to inactivation by PCR caused mutation. Thus, CamR colonies encoding *cat-sacB* should be screened for sucrose sensitivity. Resuspend the colony in a drop of buffer and streak on a minimal plate supplemented with sucrose (**Section 2.2. Step 2**). If the bacteria are sucrose sensitive individual colonies will not be formed.
9. Three variations of the tnpR genes have been used in RIVET and SIVET, *tnpRWT*, *tnpR168*, and *tnpR184 (11)*. The latter two encode point mutations in their RBS that lower the translational efficiency of *tnpR* and thus increase the threshold of promoter activation required to elicit *res-tetR-res* excision. Our studies have shown that t*npR168* is the optimal *tnpR* derivative for measuring H-19B prophage induction *(4)*. However, for other applications, other *tnpR* derivatives may more effectively allow excision of *res-tetR-res* to match induction of lytic functions.

Acknowledgements

Research from this laboratory was supported by Public Health Grant AI11459-10. J.L. was supported in part by NIH Training Grants GM007315 and AI07528.

References

1. Waldor, M. K., Friedman, D. I. & Adhya, S. L. (2005) *Phages; Their role in bacterial pathogenesis and biotechnology* (ASM Press, Washington, D.C.).
2. Tyler, J. S., Livny, J. & Friedman, D. I. (2005) in *Phage: Role in Pathogenesis and Biotechnology*, eds. K., W. M., Friedman, D. I. & Adhya, S. (ASM Press, Washington, D.C.).
3. Camilli, A. & Mekalanos, J. J. (1995) Use of recombinase gene fusions to identify *Vibrio cholerae* genes induced during infection. *Mol. Microbiol.* **18,** 671–683.
4. Livny, J. & Friedman, D. I. (2004) Characterizing spontaneous induction of Stx encoding phages using a selectable reporter system. *Mol. Microbiol.* **51,** 1691–1704.
5. Eisen, H. A., Fuerst, C. R., Siminovitch, L., Thomas, R., Lambert, L., Pereira da Silva, L. & Jacob, F. (1966) Genetics and physiology of defective lysogeny in K12 (lambda): studies of early mutants. *Virology* **30,** 224–241.
6. Quandt, J. & Hynes, M. F. (1993) Versatile suicide vectors which allow direct selection for

gene replacement in gram-negative bacteria. *Gene* **127,** 15–21.
7. Powell, B. S., Rivas, M. P., Court, D. L., Nakamura, Y. & Turnbough, C. L., Jr. (1994) Rapid confirmation of single copy lambda prophage integration by PCR. *Nucleic Acids Res* **22,** 5765–5766.
8. Grindley, N. D. (1983) Transposition of Tn3 and related transposons. *Cell* **32,** 3–5.
9. Yu, D., Ellis, H. M., Lee, E. C., Jenkins, N. A., Copeland, N. G. & Court, D. L. (2000) An efficient recombination system for chromosome engineering in *Escherichia coli. Proc. Natl. Acad. Sci. U S A* **97,** 5978–5983.
10. Ellis, H. M., Yu, D., DiTizio, T. & Court, D. L. (2001) High efficiency mutagenesis, repair, and engineering of chromosomal DNA using single-stranded oligonucleotides. *Proc. Natl. Acad. Sci. U S A* **98,** 6742–6746.
11. Lee, S. H., Hava, D. L., Waldor, M. K. & Camilli, A. (1999) Regulation and temporal expression patterns of *Vibrio cholerae* virulence genes during infection. *Cell* **99,** 625–634.
12. Weisberg, R. A. & Landy, A. (1983) in *Lambda II*, eds. Hendrix, R. W., Roberts, J. W., Stahl, F. W. & Weisberg, R. A. (Cold Spring Harbor Laboratory, Cold Spring Harbor, NY), pp. 211–250.

Chapter 23

Generalized Transduction

Anne Thierauf, Gerardo Perez, and Stanley Maloy

Abstract

Transduction is the process in which bacterial DNA is transferred from one bacterial cell to another by means of a phage particle. There are two types of transduction, generalized transduction and specialized transduction. In this chapter two of the best-studied systems – *Escherichia coli*-phage P1, and *Salmonella enterica*-phage P22 – are discussed from theoretical and practical perspectives.

Key words: Generalized transduction, specialized transduction, transducing phage, P22, P22 HT, pac, P1, P1vir, Salmonella, E.coli, homologous recombination, site-specific recombination, prophage plasmid, temperate phage, phage lysate, complete transductant, abortive transductant, cotransduction, two-factor cross, three-factor cross, pseudolysogens.

1 Introduction

Transduction is the process in which bacterial DNA is transferred from one bacterial cell to another by means of a phage particle. There are two types of transduction, generalized transduction and specialized transduction.

Generalized transduction occurs when bacterial DNA is packaged into phage heads instead of phage DNA. Phage particles that have encapsidated bacterial DNA are called transducing particles (TPs). Like any typical phage particle, a TP can adsorb to the surface of its host and eject its DNA into the host cytoplasm. Once inside the host, stable inheritance of the donor DNA can be attained by either integrating into the host chromosome through homologous recombination or by faithful replication and transfer to daughter cells (in cases wherein the donor DNA is a plasmid). The process is called generalized transduction because the bacterial DNA fragment that is packaged in the phage head can be

Martha R. J. Clokie, Andrew M. Kropinski (eds.), *Bacteriophages: Methods and Protocols, Volume 1: Isolation, Characterization, and Interactions, vol. 501*, © 2009 Humana Press, a part of Springer Science+Business Media
DOI 10.1007/978-1-60327-164-6_23 Springerprotocols.com

derived from essentially any part of the host genome. Generalized transducing particles can be produced during lytic growth of either virulent or temperate phage *(1–3)*. Generalized transduction is one of the most important techniques for genetic analysis of bacteria facilitating mapping, strain construction, mutagenesis, backcrosses, and many other genetic uses.

In contrast to generalized transduction, specialized transduction results from the aberrant excision of a prophage from a specific integration site, packaging both phage DNA and adjacent DNA from the bacterial genome into a single phage particle. Specialized transducing phage can be isolated from many different regions of the bacterial genome by genetically manipulating the phage to integrate at a variety of sites in the genome. Stable inheritance of donor DNA by specialized transduction may occur either by site-specific recombination mediated by the phage integrase or by homologous recombination through bacterial recombinases.

1.1 Generalized Transduction by Phage P22

Phage P22 is the most widely used transducing phage in *Salmonella*. P22 is a temperate phage that binds to the O-antigen residues comprising the lipopolysaccharide on the outer membrane of *Salmonella*. After phage DNA translocation into the host cytoplasm, the linear double-stranded P22 genome circularizes by recombination of its terminally redundant ends. During conditions conducive for lytic growth, the circular genome of P22 undergoes several rounds of theta-replication to provide a template for mRNA and protein synthesis. P22 then switches to rolling circle replication to produce long concatemers of double-stranded P22 DNA that are packaged into phage heads by a "headful" mechanism. A specific 8–10 bp sequence of DNA called *pac* serves as the site for initiation of packaging. A phage-encoded endonuclease initially cuts the DNA at the *pac* site and initiates encapsidation of the DNA into a phage head until the phage head is filled with 44 Kb of DNA. The concatameric DNA protruding from the base of the phage head is then cleaved by the endonuclease, and transferred to an empty phage head. These packaging reactions are processively repeated until 3–5 phage heads are filled with DNA *(4)*. The complete tail apparatus which is composed of the portal protein, six trimeric tail spikes and a coiled-coil tail fiber, plugs the phage head to prevent the release of the encapsidated DNA *(5)*. For detailed reviews on the biology of phage P22, see Susskind and Botstein *(6)* and Poteete *(7)*.

Fifty to a hundred new phage particles are released when a host cell lyses. New phage particles will infect other host cells in the culture until the broth contains a high concentration of phage ($\sim 10^{10}$–10^{11} pfu/ml). Chloroform is added to burst host cells that survived phage lysis and cell debris is removed by centrifugation, yielding a supernatant called a "phage lysate" or "phage stock." If cells or cell debris are allowed to remain in a lysate, the phage will adsorb to them, causing a drop in the effective titer of phage. When stored in an appropriate medium, P22 lysates remain stable for many years at 4°C. They may also be stored frozen in 2.5 M glycerol at –70°C.

Occasionally host DNA is packaged into P22 heads due to the erroneous recognition of sequences in the *Salmonella* genome that are homologous to the P22 *pac* site. When P22 infects a cell, occasionally the P22 nuclease cuts one of these chromosomal pseudo-*pac* sites and initiates processive packaging of 44 Kb chromosomal DNA fragments into P22 phage heads *(1, 2)*. Because pseudo-*pac* sites are distributed unevenly around the *Salmonella* chromosome, the frequency of transduction by wild-type P22 varies markedly for different regions of the chromosome. The P22 particles carrying chromosomal DNA (transducing particles) can inject this DNA into a new host. The DNA can then recombine into the chromosome of the new host by homologous recombination. Although different regions of the genome are transferred at distinct frequencies, wild-type P22 can transfer DNA fragments from nearly all regions of the chromosome *(1, 2)*.

A P22 derivative that is very useful for generalized transduction is called P22 HT105/1 *int-201 (8)*. This phage has a high transducing (HT) frequency due to a mutant nuclease that has less specificity for the *pac* sequence *(9)*. About 50% of the P22 HT phage heads carry random transducing fragments of chromosomal DNA *(10, 11)*. The *int* mutation prevents formation of stable lysogens that would interfere with subsequent transductions. This P22 derivative packages all regions of the bacterial genome at similar frequencies. P22 HT105/1 *int-201* also efficiently transduces multicopy plasmids, packaged as concatemers produced during rolling circle replication (*see* **Section 1.6** below).

The use of P22 generalized transduction has until recently been limited to sensitive serovars of *Salmonella enterica*, largely because of the unique structure of the O-antigen component of *Salmonella* lipopolysaccharide (LPS), which provides the cell-surface receptor for P22 adsorption. O-antigen is a polysaccharide, and its synthesis is directed by the *rfb* gene cluster (which encodes enzymes for synthesis of the oligosaccharide units) and the *rfc* gene (which encodes the enzyme responsible for polymerization of the oligosaccharide units) *(12)*.

The potential for extending P22 host range to other genera comes from the work by Neal et al. *(13)*, who constructed a cosmid that carries the *Salmonella enterica* serovar Typhimurium *rfb* cluster and the *rfc* gene. *Escherichia coli* strains carrying this cosmid are sensitive to P22 infection, and thus chromosomal genes can be transduced by P22. This cosmid extends the host range of P22 to other species that allow replication of the cosmid, expression of the Typhimurum LPS, and morphogenesis of P22. In cases wherein the organism's indigenous LPS might interfere in the recognition of the *Salmonella* LPS by P22, rough mutants lacking LPS may have to be used.

P22 has not been useful for transductional analysis of the human pathogen *Salmonella enterica* serovar Typhi. Plasmids can be transduced into Typhi, indicating that P22 can recognize Typhi LPS and inject its DNA. Typhi strains that are deficient in methyl-directed mismatch repair (*mutL* or *mutS* mutants) act as efficient recipients for interspecies transduction from other *Salmonella* serovars *(14)*. However, P22 does not replicate in Typhi, so additional techniques are needed for intra- and interspecies genetic exchange from this serovar.

1.2 Generalized Transduction by Phage P1

Phage P1 was isolated from a lysogenic strain of *E. coli (15)*, and its use as a generalized transducing phage followed soon after the discovery of P22-mediated transduction *(16, 17)*. Since then, P1 has been the workhorse of *E. coli* genetics. P1 is a temperate phage that infects a variety of Gram-negative bacteria by binding to the terminal glucose residues of the lipopolysaccharide core on the outer membrane. Adsorption of P1 tail fibers to the receptor is Ca^{+2}-dependent. Upon translocation into the host cytoplasm, the linear double-stranded P1 DNA circularizes by recombination. During lytic growth, the circular genome of P1 initiates several rounds of bidirectional replication before switching to rolling circle replication that produces long concatemers of double-stranded P1 DNA. A phage-encoded endonuclease recognizes a specific 162 bp *pac* sequence on the phage DNA and cuts the DNA at this site to initiate headful packaging into a phage head. After the phage head is filled with about 110–115 Kb of DNA, the excess DNA is cleaved by a sequence independent mechanism. Subsequent rounds of packaging are then initiated from the cleaved DNA and these packaging reactions continue processively until 3–5 phage heads are filled with DNA. When a cell lyses, it releases 25–150 new phage particles. A typical P1 lysate contains about 10^9 pfu/ml of phage. The biology of phage P1 is reviewed in Yarmolinsky and Sternberg *(18)*.

P1 transducing particles carry DNA from different regions of the *E. coli* chromosome at about equal frequencies. This implies that P1 transducing phages are not generated by packaging of the host chromosome at pseudo-*pac* sites. Packaging of the host DNA probably initiates at random breaks that arise during the infection *(19)*.

Unlike P22, phage P1 exists as an extrachromosomal plasmid during lysogeny. P1 lysogens express a Type III restriction system that can interfere with subsequent transductions. Lysogens can be avoided by using a derivative called P1 vir^S that constitutively produces a phage antirepressor that interferes with the c1 repressor function of phage P1. Other derivatives of P1 are available that use alternative ploys to avoid stable lysogeny.

A most useful property of phage P1 is its potentially wide host range. Some strains, such as *Klebsiella oxytoca* M5a1, are naturally sensitive to P1 infection *(20)*. Mutant strains of *Salmonella* lacking galactose epimerase (*galE*) produce a truncated LPS that is effectively recognized by the P1 tail fiber. These strains are resistant to P22. However, growth of a *galE* mutant in the presence of 1% glucose and 1% galactose restores synthesis of the normal LPS. Thus, *Salmonella galE* mutants can be infected by P1, or by P22, depending on the composition of the medium in which they are grown *(21, 22)*.

1.3 Generalized Transduction in Other Bacteria

In addition to P22 and P1, other generalized transducing phages have been isolated in *Salmonella* and *E. coli*. Specific generalized transducing phages have also been isolated for a variety of other bacteria (**Table 23.1**). In addition, mutations that alter the

Table 23.1
Examples of generalized transducing phage

Bacteria	Phage	Reference
Acetobacter methanolicus	Acm1	*(40)*
Acinetobacter	P78	*(41)*
Actinobacillus actinomycetemcomitans	Aa phi 23	*(42)*
Bacillus anthracis, B. cereus, B. thuringiensis	CP-51, CP-54	*(43)*
Bacillus cereus 13472 or 4415, *B. thuringiensis subsp. aizawai*	TP-17	*(44)*
Bacillus licheniformis	SP-15	*(45)*
Bacillus megaterium	MP13	*(46)*
Bacillus pumilis	AR9, PBP1	*(47, 48)*
Bacillus stearothermophilus	TP-42, TP-56 (temperate) TP-68 (virulent)	*(49)*
Bacillus subtilis	AR9, PBS1, SP-15	*(45, 47, 50)*
Bacillus thuringiensis var. berliner 1715	CP-54Ber	*(51)*
Bacterium anitratum	BP1	*(52)*
Bordetella avium	Ba1	*(53)*
Bosea thiooxidans	TPC-1	*(54)*
Caulobacter crescentus	phi Cr30	*(55)*
Corynebacterium renale	RP28	*(56)*
Erwinia carotovora	phage 59	*(57)*

(continued)

Table 23.1 (continued)

Bacteria	Phage	Reference
Erwinia chrysanthemi	phi EC2	*(58, 59)*
Escherichia coli	P1, phi w39, T1, T4, Mu, D108	*(3, 60–62)*
Klebsiella pneumoniae	P1	*(20)*
Listeria monocytogenes	P35 (phiLMUP35), U153(phiCU-SI153/95)	*(63)*
Methanobacterium thermoautotrophicum	psi M1	*(64)*
Mycobacterium smegmatis, M. vaccae	Bxz1	*(65)*
Myxococcus xanthus	Mx4, Mx8	*(66–69)*
Proteus mirabilis	phim and pi1, 5006M	*(70, 71)*
Pseudomonas aeruginosa	UT1, B86, F116, DMS3, M6a, pf20	*(72–79)*
Pseudomonas cepacia	CP75	*(80)*
Pseudomonas maltophilia	M6a	*(78)*
Pseudomonas putida	Pf16h2, pf20	*(79, 81)*
Rhizobium meliloti	phage 11	*(82)*
Salmonella enterica	P22, SE1, KB1, L, ES18	*(3, 32, 83–85)*
Salmonella enterica sv. Typhi	j2	*(86, 87)*
Serpulina hyodysenteriae	VSH-1	*(88)*
Serratia marcescens	3M	*(89, 90)*
Shigella dysenteriae	D108	*(62)*
Shigella flexneri	PE5	*(91)*
Staphylococcus aureus	phage 80 alpha, phi 11	*(92–94)*
Streptococci GroupA	A25, CA1, C1	*(95, 96)*
Streptococci GroupC	C1	*(96)*
Streptococci GroupG	A25	*(97)*
Streptococcus thermophilus	phi 17α, phi 56α	*(98)*
Streptomyces avermitilis, S. coelicolor, S. verticillus	DAH2, DAH4, DAH5, DAH6	*(99)*
Streptomyces hygroscopicus	SH10	*(100)*
Streptomyces venezuelae	SV1	*(101, 102)*
Vibrio alginolyticus, V. parahaemolyticus	phi VP253, phi VP143	*(103)*
Vibrio cholerae	CP-T1	*(104)*
Vibrio sp. strain 60	As3	*(105)*
Xanthomonas campestris	XTP1	*(106)*

host-range of phage by allowing binding to a different host receptor can expand the host-range for generalized transducing phages *(3)*. Although some phage only yield low frequencies of generalized transduction, an approach described by Schmieger can be used to isolate higher transducing phage derivatives *(10)*.

1.4 Transduction Without Plaque Formation in Bacteria

Certain phages adsorb to a bacterial host and inject their DNA but fail to replicate or establish lysogeny. Although these phages cannot replicate, they can still transfer plasmid DNA or DNA fragments into these recipient bacteria. For example, phage P1 will adsorb to *Myxococcus xanthus* and inject its DNA, but it fails to replicate or establish lysogeny. However, P1 serves as a convenient way of introducing plasmids or transposons from *E. coli* to *M. xanthus (23)*. A list of other examples is shown in **Table 23.2**.

Table 23.2
Examples of transducing but non-plaque-forming phages

Phage	Donor bacteria	Recipient bacteria	Reference
P22	*Salmonella enterica* sv. Typhimurium	*Salmonella enterica* sv. Typhi	*(14)*
P1	*Escherichia coli*	*Agrobacterium tumefaciens*	*(18, 107)*
P1	*Escherichia coli*	*Alcaligenes faecalis*	*(18, 107)*
P1	*Escherichia coli*	*Flavobacterium* sp. M64	*(18, 107)*
P1	*Escherichia coli*	*Myxococcus xanthus*	*(18, 23, 108, 109)*
P1	*Escherichia coli*	*Pseudomonas aeruginosa*	*(18, 107, 110)*
P1	*Escherichia coli*	*Serratia marcescens*	*(18, 107, 110)*
P1	*Escherichia coli*	*Yersinia pestis*	*(18, 111)*
P1	*Escherichia coli*	*Yersinia pseudotuberculosis*	*(18, 111)*
FP43	*Streptomyces griseofuscus*	*Chainia ochracea* ATCC 15814	*(112)*
FP43	*Streptomyces griseofuscus*	*Chainia olivacea* ATCC 15722	*(112)*
FP43	*Streptomyces griseofuscus*	*Saccharopolyspora erythraea* ATCC 11635	*(112)*
FP43	*Streptomyces griseofuscus*	*Saccharopolyspora hirsuta* ATCC 27875	*(112)*
FP43	*Streptomyces griseofuscus*	*Streptomyces cirratus* ATCC 21731	*(112)*
FP43	*Streptomyces griseofuscus*	*Streptomyces coelicolor* A3(2)	*(112)*

(continued)

Table 23.2 (continued)

Phage	Donor bacteria	Recipient bacteria	Reference
FP43	*Streptomyces griseofuscus*	*Streptomyces felleus* ATCC 19752	*(112)*
FP43	*Streptomyces griseofuscus*	*Streptomyces lividans* 1326	*(112)*
FP43	*Streptomyces griseofuscus*	*Streptomyces phaeochromogenes* NRRL B35	*(112)*
FP43	*Streptomyces griseofuscus*	*Streptomyces thermotolerans* ATCC 2550	*(112)*
FP43	*Streptomyces griseofuscus*	*Streptomyces venezuelae* ATCC 10712	*(112)*
FP43	*Streptomyces griseofuscus*	*Streptoverticillium kentuckense* NRRL B-1831	*(112)*
Λ	*Escherichia coli*	*Vibrio cholerae*	*(113)*
Λ	*Escherichia coli*	*Vibrio harveyi*	*(114)*
SP02	*Bacillus subtilis* 168	*Bacillus amyloliquefaciens* H	*(115)*
SP02	*Bacillus subtilis* 168	*Bacillus amyloliquefaciens* N	*(115)*
SP02	*Bacillus subtilis* 168	*Bacillus amyloliquefaciens* DSM7	*(115)*
SP02	*Bacillus subtilis* 168	*Bacillus subtilis* DSM704	*(115)*
5006MHFTk, 5006MHFTak	*Proteus mirabilis*	*Proteus vulgaris* PV127	*(71)*

1.5 Inheritance of Chromosomal DNA via Homologous Recombination

Although there are several different pathways that mediate homologous recombination in bacteria, inheritance of double-stranded, linear DNA through generalized transduction primarily occurs via the RecBCD pathway. The exonuclease and helicase activities of RecBCD initiate the processing of a DNA fragment from a free double-stranded end, and facilitate loading of RecA protein onto the invading strand. The invading strand is aligned to the homologous DNA molecule by RecA to promote genetic exchange *(24)*.

Homologous recombination between the chromosome of the recipient cell and the linear, double-stranded DNA that came from a transducing particle occurs as soon as the DNA fragment enters the recipient bacterium. In approximately 10% of these infections, the injected DNA is incorporated into the recipient chromosome by homologous recombination *(25)*. The resulting transductants are termed complete transductants. The other 90% are termed abortive transductants *(26,27)*. In these cells, a phage-encoded protein binds to the ends of the DNA fragment *(28)*, and because this DNA lacks a free end, it is not an efficient substrate for homologous recombination. Such abortive transductants are segregated from the population during subsequent cell division.

Generalized transduction is an efficient way of moving mutations between strains to construct derivatives with different genotypes. The genetic or biochemical effects of a particular mutation can only be determined if the mutant strain is compared to a strain that only differs from the mutant strain by a single mutation (an "isogenic strain"). The most common way to ensure that two strains are isogenic is to perform a "back-cross." Back-crossing entails transferring of a small region of DNA containing the mutation into the parental strain by recombination. If a mutation confers a selectable phenotype, the transfer from the donor to an appropriate recipient strain can easily be detected. If the mutation cannot be selected directly, linkage to an adjacent, selectable marker can be used to move the mutation into a recipient strain.

Transduction can also be useful in determining the relative map location of genes. The frequency with which two markers are coinherited in a transductional cross depends on the distance between them, so measurements of cotransduction frequency represent the simplest means of expressing genetic distances. The cotransduction frequency of two markers provides a measure of their linkage. The Wu formula *(29)* provides an approximate correlation between cotransduction frequency and physical distance (**Fig. 23.1**). Determining the linkage of two genetic markers is called a two-factor cross. It is also possible to determine the relative location of genetic mutations by using three genetic markers (three-factor crosses) or by genetic crosses versus a set of defined deletion mutations (deletion mapping). Compared to DNA hybridization or DNA sequencing, genetic mapping often provides a simpler and less expensive way to rapidly determine the location of mutations in bacteria.

Transduction can also be used to isolate new mutations produced by localized mutagenesis *(12)*. The donor bacteria can be treated with a mutagen before preparing a generalized transducing lysate, or a phage lysate can be directly treated with a mutagen. The mutagenized lysate is then used to infect an appropriate recipient strain followed by selecting for a dominant genetic marker and screening for coinheritance of mutations in a linked gene.

1.6 Transduction of Plasmids

In addition to chromosomal genes, plasmid DNA can also be transduced from one bacterial cell to another. Although a variety of phage can transfer plasmid DNA between *Salmonella* strains, P22 has been used the most. One mechanism for transfer of high copy number plasmid relies upon phage gene products that inhibit the host RecBCD exonuclease functions. This allows plasmids to replicate via rolling circle replication, generating long

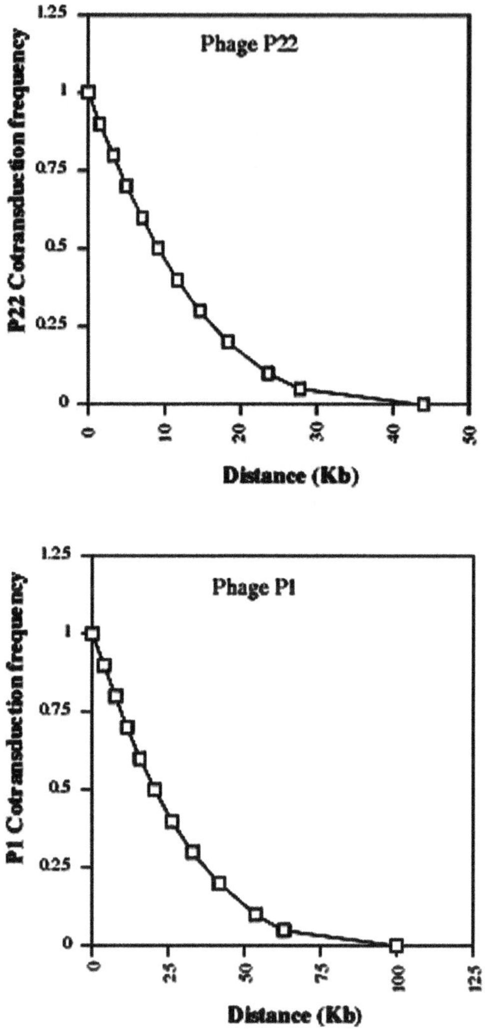

Fig. 23.1. Cotransduction frequency vs distance calculated by the Wu formula. The Wu formula is based upon the relationship $d = L - -(L \times \sqrt[3]{C})$, where d = distance between markers, L = length of phage DNA, and C = cotransduction frequency.

plasmid concatemers that are substrates for packaging by P22 HT into phage heads *(30)*. A second mechanism requires that the plasmid possess some homology with the phage DNA *(31)*. Homologous recombination within the shared sequence results in integration of the plasmid into the phage genome and subsequent transfer of the plasmid into a recipient cell. A third mechanism for transfer of low copy number plasmids relies upon integration of the plasmid into the host chromosome, packaging of the fragment of host chromosome by P22 HT, and subsequent transfer of the fragment into a recipient cell *(32)*. Upon transduction into a recipient host via any of these three mechanisms, recombination between the tandemly repeated sequences results in re-circularization of the plasmid DNA. The circularized DNA can subsequently replicate as an autonomous plasmid.

1.7 Isolation of Transducing Phage

Two general approaches have been taken to find generalized transducing phage from a number of bacterial species. One method is to identify lysogenic strains of a given species, recover the resident temperate phage, and then test it for transduction. This is the method used to identify phages P22 and P1 *(15, 33)*. Useful transducing phage for *P. aeruginosa* was also initially recovered from lysogens *(34)*.

The other approach is to screen the environmental sample where the bacterial species is normally found. The *M. xanthus* transducing phage Mx4 and Mx8 were identified in collections of myxophages that have been isolated from soil and dung, two characteristic habitats for myxobacteria *(35, 36)*.

2 Materials

2.1 Bacterial Strains and Phage

Salmonella strains and P22 HT *int* phage can be obtained from the *Salmonella* Genetic Stock Centre at the University of Calgary (http://www.ucalgary.ca/~kesander/). P22 HT *int* has a turbid plaque phenotype. Rare clear plaque mutants that reproduce more rapidly than the turbid plaque phage can accumulate in a P22 HT *int* lysate after it is repeatedly propagated. Such clear plaque mutants rapidly lyse infected cells, decreasing the number of transductants obtained. Therefore, P22 HT *int* phage stocks should be occasionally checked for clear plaque mutants, and if this is a problem a new lysate should be prepared from a single isolated turbid plaque.

E. coli strains and P1 phage can be obtained from the *E. coli* Genetic Stock Center at Yale University (http://cgsc.biology.yale.edu/).

2.2 Reagents, Media, and Solutions

LB medium (aka Lysogeny Broth or Luria-Bertani broth), EBU medium, TS top agar, and P22 broth can be prepared as described in Maloy et al. *(12)*. Salt solutions should be prepared in deionized H_2O.

3 Methods

Transduction requires three steps: growth of a phage lysate on the donor strain, infection of a recipient strain, and selection of transductants. It is often necessary to avoid coinfection with viable phage. The methods used depend upon the properties of the specific transducing phage.

3.1 P22 Transduction (modified from Maloy, (37)

3.1.1 Preparation of P22 HT Int phage Lysates

1. Pick a single colony and start an overnight culture of the donor strain in 1 ml LB at 37°C.
2. Add 1 ml P22 broth to 200 μl (kill second 200) of the overnight culture. The final multiplicity of infection (MOI) should be about 0.01–0.1 pfu/cell.
3. Incubate 8–16 h in a 37°C shaker. (Temperature sensitive mutants can be grown at 30°C.)
4. Add several drops of chloroform and vortex. Transfer the supernatant to a microfuge tube and centrifuge for 1 min at 12,000 rpm to pellet the cell debris.
5. Transfer the supernatant to a new microfuge tube. Add several drops of chloroform and vortex. Store at 4°C. (A good lysate should contain $10^{10} - 10^{11}$ pfu/ml).

3.1.2 P22 Transduction Without Phenotypic Expression

1. Grow the recipient strain overnight in 1 ml LB.
2. Dilute the phage by adding 50 μl (kill second 50) of the P22 HT *int* lysate to 450 μl (kill second 450) 0.85% NaCl. Mix the cells and freshly diluted phage lysate on selection plates as follows:

Plate	μl cells	μl P22	
A	200 μl	–	No phage control
B	–	50 μl	No cell control
C	200 μl	10 μl	
D	200 μl	50 μl	
E	200 μl	200 μl	

3. Sterilize a glass spreader by dipping it into alcohol then briefly passing it through a flame to burn off the residual alcohol. Thoroughly spread the cells and phage on the plates, re-sterilizing the glass spreader between each plate.
4. Place the plates in an incubator upside-down. Incubate 1–2 days at 37°C. (Temperature sensitive mutants can be grown at 30°C.)
5. Count the colonies on each plate. There should be no growth on the cell or phage control plates. Any transductants that will be saved should be immediately purified on EBU plates and cross-streaked against phage P22-H5 to purify phage sensitive colonies (**Section 3.1.5**).

3.1.3 P22 Transduction with Phenotypic Expression

For certain transductions (for example, when selecting Kan^R or Str^R) phenotypic expression is required before plating on the selective medium. Phenotypic expression can be done in two ways:

- Spread the cells and phage on nonselective medium (e.g., LB plate), incubate 4–8 h, then replica plate onto the selective medium. This replica plating approach usually gives more colonies and ensures that different colonies are not due to siblings.

- Mix cells and phage in a microfuge tube and incubate about 1 h at 37°C before plating on the selective medium. Although this broth approach may yield somewhat fewer colonies and may produce some siblings, it is easier and faster than the replica plating approach. The broth method is detailed below.

1. Perform steps 1–2 as described in **Section 3.1.2** above.
2. Leave at room temperature for 15–30 min to allow phage adsorption.
3. Add 1 ml of LB to each tube. Incubate at 37°C for 1 h to allow phenotypic expression.
4. Centrifuge each tube for 1 min in a microfuge.
5. Pour off the supernatant. Add 100 μl LB, and vortex to resuspend the pellet.
6. Continue with steps 3–5 described in **Section 3.1.2** above.

3.1.4 Spot Titering P22 Lysates

When P22 HT *int* is used, phage stocks do not usually need to be titered. However, it is a good idea to check the titer of your phage stock if a transduction does not work.

1. Mark 4 quadrants on the bottom of an EBU plate. The surface of the plate should not be noticeably wet. If the plate is wet, place it in an incubator or oven with the lid ajar until dry.
2. Melt TS top agar in a microwave. For each phage to be titered, add 3.0 ml of the melted top agar to a test tube and place in a 50°C heating block. After the top agar cools to about 50°C, add 0.1 ml of an overnight culture of a P22-sensitive *Salmonella* strain to each tube.
3. Immediately swirl and pour onto the EBU plate.
4. Allow the top agar to solidify for 15–30 min.
5. Gently spot 10 μl of appropriate phage dilutions onto quadrants on the plate (usually $10^{-6}, 10^{-7}, 10^{-8}, 10^{-9}$ dilutions in sterile 0.85% NaCl).
6. Leave the plate on the bench for about 30 min or until the drops of phage dry.
7. Incubate the plates upside-down at 30–37°C overnight.
8. Confluent growth of the bacteria in the top agar will result in a "lawn" and plaques will appear where phage has been spotted. Count the number of plaques in each spot. Each viable phage (plaque forming unit or pfu) will produce one plaque. Calculate the phage titer as follows:

$$\frac{\text{pfu}}{\text{ml}} = \frac{\text{number of plaques} \times \text{dilution factor}}{10\,\mu\text{l}} \times \frac{1000\,\mu\text{l}}{\text{ml}}$$

3.1.5 Checking Salmonella for P22 Sensitivity

In addition to receiving a transducing fragment, some of the transductants may have also been infected with P22 phage. Although P22 HT *int* is unable to integrate into the chromosome due to the lack of integrase, the phage can form unstable, episomal pseudolysogens. Pseudolysogens can be differentiated from nonlysogens and true lysogens on EBU plates *(38)*. EBU plates contain pH indicators, which are green at neutral pH but turn dark blue at low pH. When streaked on green plates, nonlysogens and true lysogens form light-colored colonies. However, in a colony containing pseudolysogens many cells undergo lysis, which lowers the pH of the medium resulting in dark blue colonies. The phage DNA in pseudolysogens is replicated as a low copy number plasmid, so it is possible to obtain "phage-free" segregants by simply streaking for isolated colonies on EBU plates. The "phage-free" segregants will form light-colored colonies while the pseudolysogens will remain blue.

P22 HT *int* is used for transductions because the *int* mutation prevents formation of stable lysogens. However, when cells are left on plates with lytic phage, there is a strong selection for revertants that form stable lysogens. Since stable lysogens cannot be reinfected with P22, such transductants are not very useful for genetic studies. Therefore, it is important to isolate "phage free" transductants as soon as possible after colonies arise. EGTA can be included in plates to chelate the Ca^{+2} required for phage adsorption, thereby preventing reinfection of transductants *(19)*. However, EGTA cannot be added during the initial phage infection or it will prevent adsorption of the transducing phage. Even colonies from medium containing EGTA must be cleaned up on EBU plates. Once light-colored colonies have been purified from EBU plates, they should be checked to make sure they are not true lysogens. This is done by cross-streaking against a P22 *c2* mutant called H5. The P22 *c2* gene encodes a repressor equivalent to cI of phage lambda. Phage free cells are infected by P22-H5 and lysed, but P22 lysogens are resistant to P22-H5.

Avoid "digging into" the agar when streaking EBU plates because when growing anaerobically the bacteria ferment more glucose and all of the colonies will appear dark blue. Also, the EBU phenotype should be observed promptly after growth appears because when left on EBU plates for many days, all of the colonies will turn dark colored.

3.2 Transduction with P1*vir* (modified from Silhavy et al. *(39)*)

3.2.1 Preparation of P1 Phage Lysates

1. Pick a single colony and start an overnight culture of the donor strain in 1 ml LB. Grow with aeration at 37°C.
2. Subculture 10 µl of the overnight culture into 1 ml LB containing 0.2% glucose and 5 mM $CaCl_2$.
3. Incubate for 30 min with aeration at 37°C. An early exponential phase culture is optimal for preparation of a high titer P1 lysate.
4. Add 10 µl of a P1 lysate with approximately 5×10^8 P1 *vir* per ml.

5. Incubate with aeration at 37°C until the culture lyses or for 2–3 h. Typically the cells will lyse during this time, but a usable phage titer may be obtained even if the culture does not clear.
6. Add 0.1 ml chloroform and vortex.
7. Transfer the supernatant to a sterile microfuge tube and centrifuge at 12,000 rpm for 1 min to pellet the debris.
8. Transfer the supernatant to a sterile tube. Add 0.1 ml chloroform, vortex, and store at 4°C.

3.2.2 P1 transduction Without Phenotypic Expression

1. Pick a single colony and start an overnight culture of the recipient strain in 1 ml LB. Grow with aeration at 37°C.
2. Centrifuge the overnight culture for 30 s at 12,000 rpm in a microfuge.
3. Resuspend the cell pellet in 0.5 ml of CaMg (5 mM $CaCl_2$ + 10 mM $MgSO_4$).
4. Mix the cells and phage in sterile microfuge tubes as follows:

Plate	μl cells	μl P1	
A	100 μl	–	No phage control
B	–	100 μl	No cell control
C	100 μl	10 μl	
D	100 μl	50 μl	
E	100 μl	100 μl	

5. Leave at room temperature for about 30 min to allow phage adsorption.
6. Add 100 μl of 1 M sodium citrate to each tube and vortex briefly. (The citrate chelates the divalent cations, preventing subsequent reinfection.)
7. Pipette the contents of each tube to a separate plate of agar medium formulated to select for the desired transductants.
8. Dip a glass hockey stick into alcohol then briefly pass it through a flame to burn off the residual alcohol. Thoroughly spread the plates with the alcohol-flamed glass spreader.
9. Place the plates in an incubator upside-down. Incubate overnight at 30–37°C.
10. Count the colonies on each plate. There should be no growth on the cell or phage control plates.

3.2.3 P1 Transduction with Phenotypic Expression

1. Perform steps 1–5 as described in **Section 3.2.2** above.
2. Add 1 ml of LB to each tube. Incubate at 37°C for 1 h to allow phenotypic expression.
3. Centrifuge for 30 s at 12,000 rpm in a microfuge to pellet the cells.

4. Pour off the supernatant. Resuspend the cells in 0.1 ml LB containing 20 mM sodium citrate.
5. Continue with steps 7–9 from **Section 3.2.2**.

3.2.4 Spot titering P1 Lysates

Because using several different multiplicities of infection (MOI) with a P1 sensitive *E. coli* recipient typically yields transductants, titering phage lysates is usually unnecessary. However, it may be useful to titer a phage stock if the transduction fails. Although LB plates will work for this procedure, P1 plaques are small and it is easier to count plaques on EBU plates because the plaques produce dark blue spots on a green background.

1. Prepare an overnight culture of a sensitive *E. coli* strain in 1 ml LB at 37°C.
2. Centrifuge the cells for 30 s at 12,000 rpm in a microfuge.
3. Resuspend the cell pellet in 0.5 ml of 10 mM $MgCl_2$. Store on ice until use.
4. Mark 4 quadrants on the bottom of an EBU plate. The surface of the plate should not be noticeably wet. If the plate is wet, place it in an incubator or oven with the lid ajar until dry.
5. Melt TS top agar in a microwave. For each phage to be titered, add 3.0 ml of the melted top agar to a test tube and place in a 50°C heating block. After the top agar cools to about 50°C, add 0.1 ml of the *E. coli* suspension in 10 mM $MgCl_2$ to each tube.
6. Immediately swirl and pour onto the EBU plate. Allow the top agar to solidify for 15–30 min.
7. Gently spot 10 μl of appropriate phage dilutions onto quadrants of the plate (usually $10^{-6}, 10^{-7}, 10^{-8}, 10^{-9}$ dilutions in sterile 0.85% NaCl).
8. Leave the plate on the bench for about 30 min or until the drops of phage dry.
9. Incubate the plates upside-down at 30 – 37°C overnight.
10. Confluent growth of the bacteria in the top agar will result in a "lawn" and plaques will appear where phage has been spotted. Count the number of plaques in each spot. Each viable phage (plaque forming unit or pfu) will produce one plaque. Calculate the phage titer as follows:

$$\frac{\text{pfu}}{\text{ml}} = \frac{\text{number of plaques} \times \text{dilution factor}}{10\,\mu\text{l}} \times \frac{1000\,\mu\text{l}}{\text{ml}}$$

Acknowledgements

Gerardo Perez was supported by an NIGMS MBRS Scholarship.

References

1. Margolin P (1987) Generalized transduction. In: Neidhardt F, Ingraham J, Low K, et al. (eds): *Escherichia coli and Salmonella Typhimurium: Cellular and Molecular Biology*. American Society for Microbiology, Washington, D.C.
2. Masters M (1985) Generalized transduction. In: Scaife J, Leach D, Galizzi A (eds): *Genetics of Bacteria*. Academic Press, New York, 197–205
3. Masters M (1996) Generalized transduction. In: Neidhardt FC (ed): *Escherichia coli and Salmonella: Cellular and Molecular Biology*. American Society for Microbiology, Washington, D.C., 2421–441
4. Casjens S, Hayden M (1988) Analysis in vivo of the bacteriophage P22 headful nuclease. *J Mol Biol* 199: 467–74
5. Tang L, Marion WR, Cingolani G, et al. (2005) Three-dimensional structure of the bacteriophage P22 tail machine. *EMBO* 24:2087–95
6. Susskind MM, Botstein D (1978) Molecular genetics of bacteriophage P22. *Microbiol Rev* 42: 385–413
7. Poteete AR (1988) Bacteriophage P22. In: Calender R (ed): *The Bacteriophages*. Plenum Press, New York, 647–82
8. Davis R, Botstein D, Roth RJ (1980) *Advanced Bacterial Genetics*. Cold Spring Harbor Laboratory, Cold Spring Harbor, New York
9. Casjens S, Sampson L, Randall S, et al. (1992) Molecular genetic analysis of bacteriophage P22 gene 3 product, a protein involved in the initiation of headful DNA packaging. *J Mol Biol* 227: 1086–99
10. Schmieger H (1972) Phage P22-mutants with increased or decreased transduction abilities. *Mol Gen Genet* 119: 75–88
11. Schmieger H, Backhaus H (1976) Altered cotransduction frequencies exhibited by HT-mutants of *Salmonella*-phage P22. *Mol Gen Genet* 143: 307–9
12. Maloy S, Stewart V, Taylor R (1996) *Genetic analysis of pathogenic bacteria*. Cold Spring Harbor Laboratory, Cold Spring Harbor, NY
13. Neal BL, Brown PK, Reeves PR (1993) Use of *Salmonella* phage P22 for transduction in *Escherichia coli*. *J Bacteriol* 175: 7115–18
14. Zahrt T, Mora G, Maloy S (1994) Inactivation of mismatch repair overcomes the barrier of transduction between *Salmonella typhimurium* and *Salmonella typhi*. *J Bacteriol* 176:1527–29
15. Bertani G (1951) Studies on lysogenesis. I. The mode of phage liberation by lysogenic *Escherichia coli*. *J Bacteriol* 62:293–300
16. Lennox ES (1955) Transduction of linked genetic characters of the host by bacteriophage P1. *Virology* 1:190–206
17. Yanofsky C, Lennox ES (1959) Linkage relationship of the genes controlling tryptophan synthesis in *Escherichia coli*. *Virology* 8: 425–47
18. Yarmonlinsky MB, Sternberg N (1988) Bacteriophage P1. In: Calendar R (ed): *The Bacteriophages*. Plenum Press, New York, 291–438
19. Sternberg NL, Maurer R (1991) Bacteriophage-mediated generalized transduction in *Escherichia coli* and *Salmonella typhimurium*. *Methods Enzymol* 204: 18–43
20. Streicher S, Gurney E, Valentine RC (1971) Transduction of the nitrogen-fixation genes in *Klebsiella pneumoniae*. *Proc Natl Acad Sci U S A* 68:1174–7
21. Enomoto M, Stocker BAD (1974) Transduction by phage P1kc in *Salmonella typhimurium*. *Virology* 60:503–14
22. Ornella EP, Stocker BAD (1974) Relation of lipopolysaccharide character to P1 sensitivity in *Salmonella typhimurium*. *Virology* 60: 491–502
23. Shimkets LJ, Gill RE, Kaiser D (1983) Developmental cell interactions in *Myxococcus xanthus* and the *spoC* locus. *Proc Natl Acad Sci U S A* 80:1406–10
24. Maloy SR, Cronan Jr. JE, Freifelder D (1994) *Microbial Genetics,* 2nd edn. Jones and Bartlett, Boston, MA
25. Ebel-Tsipis J, Fox MS, Botstein D (1972) Generalized transduction by phage P22 in *Salmonella typhimurium*. II. Mechanism of integration of transducing DNA. *J Mol Biol* 71:449–69
26. Demerec M, Ozeki H (1959) Tests for alleleism among auxotrophs of *Salmonella typhimurium*. *Genetics* 44:269–78
27. Ozeki H (1959) Chromosome fragments participating in transduction in *Salmonella typhimurium*. *Genetics* 44:457–70
28. Benson NR, Roth J (1997) A *Salmonella* phage-P22 mutant defective in abortive transduction. *Genetics* 145:17–27
29. Wu TT (1966) A model for three-point analysis of random general transduction. *Genetics* 54:405–10
30. Garzon A, Cano DA, Casadesus J (1995) Role of Erf recombinase in P22-mediated plasmid transduction. *Genetics* 140:427–34

31. Orbach MJ, Jackson EN (1982) Transfer of chimeric plasmids among *Salmonella typhimurium* strains by P22 transduction. *J Bacteriol* 149: 985–94
32. Mann BA, Slauch JM (1997) Transduction of low-copy number plasmids by bacteriophage P22. *Genetics* 146: 447–56
33. Zinder ND, Lederberg J (1952) Genetic exchange in *Salmonella. J Bacteriol* 64: 679–99
34. Holloway BW (1969) Genetics of *Pseudomonas. Bacteriol Rev* 33:419–43
35. Campos JM, Geisselsoder J, Zusman DR (1978) Isolation of bacteriophage Mx4, a generalized transducing phage for *Myxococcus xanthus. J Mol Biol* 119:167–78
36. Martin S, Sodergren E, Matsuda T, Kaiser D (1978) Systematic isolation of transducing phages for *Myxococcus xanthus. Virology* 88:44–53
37. Maloy S (1989) *Experimental Techniques in Bacterial Genetics.* Jones and Bartlett, Boston, MA
38. Bochner B (1984) Curing bacterial cells of lysogenic viruses by using UCB indicator plates. *BioTechniques* 2:234–40
39. Silhavy TJ, Berman ML, Enquist LW, et al. (1984) *Experiments with gene fusions.* Cold Spring Harbor Laboratory, Cold Spring Harbor, NY
40. Kiesel B, Wunsche L (1993) Phage Acm1-mediated transduction in the facultatively methanol- utilizing *Acetobacter methanolicus* MB 58/4. *J Gen Virol* 74:1741–5
41. Herman NJ, Juni E (1974) Isolation and characterization of a generalized transducing bacteriophage for *Acinetobacter. J Virol* 13: 46–52
42. Willi K, Sandmeier H, Kulik EM, et al. (1997) Transduction of antibiotic resistance markers among *Actinobacillus actinomycetemcomitans* strains by temperate bacteriophages Aa phi 23. *Cell Mol Life Sci* 53:904–10
43. Thorne CB (1978) Transduction in *Bacillus thuringiensis. Appl Environ Microbiol* 35:1109–15
44. Reynolds RB, Reddy A, Thorne CB (1988) Five unique temperate phages from a polylysogenic strain of *Bacillus thuringiensis subsp. aizawai. J Gen Microbiol* 134:1577–85
45. Taylor MJ, Goldberg ID (1971) Growth and cultivation of the unusual generalized transducing *Bacillus* bacteriophage SP-15. *Appl Microbiol* 22:113–9
46. Vary PS, Garbe JC, Franzen M, et al. (1982) MP13, a generalized transducing bacteriophage for *Bacillus megaterium. J Bacteriol* 149:1112–9
47. Sargent MG, Bennett MF (1985) Amplification of a major membrane-bound DNA sequence of *Bacillus subtilis. J Bacteriol* 161:589–95
48. Lovett PS (1972) PBPI: a flagella specific bacteriophage mediating transduction in *Bacillus pumilus. Virology* 47:743–52
49. Welker NE (1988) Transduction in *Bacillus stearothermophilus. J Bacteriol* 170:3761–4
50. Lepesant-Kejzlarova J, Lepesant JA, Walle J, et al. (1975) Revision of the linkage map of *Bacillus subtilis* 168: indications for circularity of the chromosome. *J Bacteriol* 121: 823–34
51. Lecadet MM, Blondel MO, Ribier J (1980) Generalized transduction in *Bacillus thuringiensis* var. berliner 1715 using bacteriophage CP-54Ber. *J Gen Microbiol* 121:203–12
52. Twarog R, Blouse LE (1968) Isolation and characterization of transducing bacteriophage BP1 for *Bacterium anitratum* (*Achromobacter* sp.). *J Virol* 2:716–22
53. Shelton CB, Crosslin DR, Casey JL, et al. (2000) Discovery, purification, and characterization of a temperate transducing bacteriophage for*Bordetella avium. J Bacteriol* 182:6130–6
54. Deb C, Chakraborty R, Ghosh AN, et al. (2003) A generalized transducing thiophage (TPC-1) of a facultative sulfur chemolithotrophic bacterium, *Bosea thiooxidans* CT5, of alpha-Proteobacteria, isolated from Indian soil. *FEMS Microbiol Lett* 227:87–92
55. Bender RA (1981) Improved generalized transducing bacteriophage for *Caulobacter crescentus. J Bacteriol* 148:734–5
56. Hirai K, Yanagawa R (1970) Generalized transduction in *Corynebacterium renale. J Bacteriol* 101:1086–7
57. Romaniuk LV, Mukvich NS, Kishko, IaG (1985) Characteristics of transduction in *Erwinia* by phage 59. *Mol Gen Mikrobiol Virusol* 10:34–9
58. Franza T, Enard C, van Gijsegem F, et al. (1991) Genetic analysis of the *Erwinia chrysanthemi* 3937 chrysobactin iron-transport system: characterization of a gene cluster involved in uptake and biosynthetic pathways. *Mol Microbiol* 5:1319–29
59. Hugouvieux-Cotte-Pattat N, Reverchon S, Robert-Baudouy J (1989) Expanded linkage map of *Erwinia chrysanthemi* strain 3937. *Mol Microbiol* 3:573–81
60. Yoshida Y, Mise K (1984) Characterization of generalized transducing phage phi w39 heteroimmune to phage P1 in *Escherichia coli* W39. *Microbiol Immunol* 28:415–26

61. Young KK, Edlin G (1983) Physical and genetical analysis of bacteriophage T4 generalized transduction. *Mol Gen Genet* 192: 241–6
62. Mise K, Suzuki K (1970) New generalized transducing bacteriophage in *Echerichia coli. J Virol.* 6:253–5
63. Hodgson DA (2000) Generalized transduction of serotype 1/2 and serotype 4b strains of *Listeria monocytogenes. Mol Microbiol* 35: 312–23
64. Meile L, Abendschein P, Leisinger T (1990) Transduction in the archaebacterium *Methano- bacterium thermoautotrophicum* Marburg. *J Bacteriol* 172:3507–8
65. Lee S, Kriakov J, Vilcheze C, et al. (2004) Bxz1, a new generalized transducing phage for *mycobacteria. FEMS Microbiol Lett* 241:271–6
66. Campos JM, Geisselsoder J, Zusman DR (1978) Isolation of bacteriophage Mx4, a generalized transducing phage for *Myxococcus xanthus. J Mol Biol* 119:167–78
67. Geisselsoder J, Campos JM, Zusman DR (1978) Physical characterization of bacteriophage Mx4, a generalized transducing phage for *Myxococcus xanthus. J Mol Biol* 119: 179–89
68. Tojo N, Sanmiya K, Sugawara H, et al. (1996) Integration of bacteriophage Mx8 into the *Myxococcus xanthus* chromosome causes a structural alteration at the C-terminal region of the IntP protein. *J Bacteriol* 178: 4004–11
69. Avery L, Kaiser D (1983) In situ transposon replacement and isolation of a spontaneous tandem genetic duplication. *Mol Gen Genet* 191:99–109
70. Nakamura M, Horiuchi S, Nakaya R (1975) Comparative studies on generalized transducing bacteriophages of *Proteus mirabilis*, phim and pi1. *Jpn J Microbiol* 19:123–31
71. Coetzee JN (1975) Transduction of a *Proteus vulgaris* strain by a *Proteus mirabilis* bacteriophage. *J Gen Microbiol* 89:299–309
72. Kilbane JJ, Miller RV (1988) Molecular characterization of *Pseudomonas aeruginosa* bacteriophages: identification and characterization of the novel virus B86. *Virology* 164:193–200
73. Ripp S, Ogunseitan OA, Miller RV (1994) Transduction of a freshwater microbial community by a new *Pseudomonas aeruginosa* generalized transducing phage, UT1. *Mol Ecol* 3: 121–6
74. Byrne M, Kropinski AM (2005) The genome of the *Pseudomonas aeruginosa* generalized transducing bacteriophage F116. *Gene* 346:187–94
75. Caruso M, Shapiro JA (1982) Interactions of Tn7 and temperate phage F116L of *Pseudomonas aeruginosa. Mol Gen Genet* 188: 292–8
76. Morrison WD, Miller RV, Sayler GS (1978) Frequency of F116-mediated transduction of *Pseudomonas aeruginosa* in a freshwater environment. *Appl Environ Microbiol* 36: 724–30
77. Budzik JM, Rosche WA, Rietsch A, O'Toole GA (2004) Isolation and characterization of a generalized transducing phage for *Pseudomonas aeruginosa* strains PAO1 and PA14. *J Bacteriol* 186:3270–3
78. Holloway BW, Krishnapillai V, Morgan AF (1979) Chromosomal genetics of *Pseudomonas. Microbiol Rev* 43:73–102
79. Chakrabarty AM, Gunsalus IC (1970) Transduction and genetic homology between *Pseudomonas* species*putida and aeruginosa. J Bacteriol* 103:830–2
80. Matsumoto H, Itoh Y, Ohta S, et al. (1986) A generalized transducing phage of *Pseudomonas cepacia. J Gen Microbiol* 132: 2583–6
81. Morgan AF, Dean HF (1985) Chromosomal map of *Pseudomonas putida* PPN, and a comparison of gene order with the *Pseudomonas aeruginosa* PAO chromosomal map. *J Gen Microbiol* 131:885–96
82. Sik T, Horvath J, Chatterjee S (1980) Generalized transduction in *Rhizobium meliloti. Mol Gen Genet* 178:511–6
83. Sander M, Schmieger H (2001) Method for host-independent detection of generalized transducing bacteriophages in natural habitats. *Appl Environ Microbiol* 67:1490–3
84. Schicklmaier P, Schmieger H (1995) Frequency of generalized transducing phages in natural isolates of the *Salmonella typhimurium* complex. *Appl Environ Microbiol* 61: 1637–40
85. Llagostera M, Barbe J, Guerrero R (1986) Characterization of SE1, a new general transducing phage of *Salmonella typhimurium. J Gen Microbiol* 132:1035–41
86. Mise K, Kawai M, Yoshida Y, et al. (1981) Characterization of bacteriophage j2 of *Salmonella typhi* as a generalized transducing phage closely related to coliphage P1. *J Gen Microbiol* 126:321–6
87. Mise K, Yoshida Y, Kawai M (1983) Generalized transduction between *Salmonella typhi* and *Salmonella typhimurium* by phage j2 and characterization of the j2 plasmid in *Escherichia coli. J Gen Microbiol* 129: 3395–400
88. Humphrey SB, Stanton TB, Jensen NS, et al. (1997) Purification and

characterization of VSH-1, a generalized transducing bacteriophage of *Serpulina hyodysenteriae*. *J Bacteriol* 179:323–9

89. Regue M, Fabregat C, Vinas M (1991) A generalized transducing bacteriophage for *Serratia marcescens*. *Res Microbiol* 142: 23–7
90. Matsumoto H, Tazaki T, Hosogaya S (1973) A generalized transducing phage of *Serratia marcescens*. *Jpn J Microbiol* 17:473–9
91. Financsek I, Ketyi I (1976) Generalized transduction of *Shigella flexneri* by converting phage PE5. *Acta Microbiol Acad Sci Hung* 23:317–24
92. Schroeder CJ, Pattee PA (1984) Transduction analysis of transposon Tn551 insertions in the trp-thy region of the *Staphylococcus aureus* chromosome. *J Bacteriol* 157: 533–7
93. Novick RP, Edelman I, Lofdahl S (1986) Small *Staphylococcus aureus* plasmids are transduced as linear multimers that are formed and resolved by replicative processes. *J Mol Biol* 192:209–20
94. Bachi B (1980) Physical mapping of the BglI, BglII, PstI and EcoRI restriction fragments of staphylococcal phage phi 11 DNA. *Mol Gen Genet* 180:391–8
95. Totolian AA, Boitsov AS, Kol' K, et al. (1981) Comparative characters of the transducing virulent streptococcal phages A25 and CA1 *Mol Biol (Mosk)* 15:894–900
96. Wannamaker LW, Almquist S, Skjold S (1973) Intergroup phage reactions and transduction between group C and group A streptococci. *J Exp Med* 137:1338–53
97. Colon AE, Cole RM, Leonard CG (1972) Intergroup lysis and transduction by streptococcal bacteriophages. *J Virol* 9:551–3
98. Mercenier A, Slos P, Faelen M, Lecocq JP (1988) Plasmid transduction in *Streptococcus thermophilus*. *Mol Gen Genet* 212: 386–9
99. Burke J, Schneider D, Westpheling J (2001) Generalized transduction in *Streptomyces coelicolor*. *Proc Natl Acad Sci U S A* 98: 6289–94
100. Suss F, Klaus S (1981) Transduction in *Streptomyces hygroscopicus* mediated by the temperate bacteriophage SH10. *Mol Gen Genet* 181:552–5
101. Stuttard C (1983) Localized hydroxylamine mutagenesis, and cotransduction of threonine and lysine genes, in *Streptomyces venezuelae*. *J Bacteriol* 155:1219–23
102. Stuttard C, Atkinson L, Vats S (1987) Genome structure in *Streptomyces* spp.: adjacent genes on the *S. coelicolor* A3(2) linkage map have cotransducible analogs in *S. venezuelae*. *J Bacteriol* 169:3814–6
103. Muramatsu K, Matsumoto H (1991) Two generalized transducing phages in *Vibrio parahaemolyticus* and *Vibrio alginolyticus*. *Microbiol Immunol* 35:1073–84
104. Hava DL, Camilli A (2001) Isolation and characterization of a temperature-sensitive generalized transducing bacteriophage for *Vibrio cholerae*. *J Microbiol Methods* 46: 217–25
105. Ichige A, Matsutani S, Oishi K, et al. (1989) Establishment of gene transfer systems for and construction of the genetic map of a marine *Vibrio* strain. *J Bacteriol* 171: 1825–34
106. Weiss BD, Capage MA, Kessel M, et al. (1994) Isolation and characterization of a generalized transducing phage for *Xanthomonas campestris* pv. campestris. *J Bacteriol* 176:3354–9
107. Murooka Y, Harada T (1979) Expansion of the host range of coliphage P1 and gene transfer from enteric bacteria to other Gram-negative bacteria. *Appl Environ Microbiol* 38:754–7
108. Kaiser D, Dworkin M (1975) Gene transfer to myxobacterium by *Escherichia coli* phage P1. *Science* 187:653–4
109. O'Connor KA, Zusman DR (1983) Coliphage P1-mediated transduction of cloned DNA from *Escherichia coli* to *Myxococcus xanthus*: use for complementation and recombinational analyses. *J Bacteriol* 155:317–29
110. Amati P (1962) Abortive infection of *Pseudomonas aeruginosa* and *Serratia marcescens* with coliphage P1. *J Bacteriol* 83:433–4
111. Lawton WD, Molnar DM (1972) Lysogenic conversion of *Pasteurella* by *Escherichia coli* bacteriophage P1Cm. *J Virol* 9:708–9
112. McHenney MA, Baltz RH (1988) Transduction of plasmid DNA in *Streptomyces* spp. and related genera by bacteriophage FP43. *J Bacteriol* 170:2276–82
113. Harkki A, Hirst TR, Holmgrn J, Palva ET (1986) Expression of the *Escherichia coli lamB* gene in *Vibrio cholerae*. *Microb Pathog* 1:283–8
114. Jasiecki J, Czy A, Gabig M, Wegrzyn G (2001) Construction and use of a broad-host range plasmid expressing the *lamB* gene for utilization of bacteriophage lambda vectors in the marine bacterium *Vibrio harveyi*. *Mar Biotechnol* (NY) 3:336–45
115. Marrero R, Young FE, Yasbin RE (1984) Characterization of interspecific plasmid transfer mediated by *Bacillus subtilis* temperate bacteriophage SP02.*J Bacteriol* 160: 458–61.

Chapter 24

Preparation and Characterization of Anti-phage Serum

Thomas E. Waddell, Kristyn Franklin, Amanda Mazzocco, and Roger P. Johnson

Abstract

This chapter describes a method for the generation of polyclonal antibodies against bacteriophages and how these may be assayed immunochemically and biologically.

Key words: Bacteriophage, phage, Freund's incomplete adjuvant, ELISA, transduction, lytic transduction, *Escherichia coli* O157:H7 antiserum, phage neutralization.

1 Introduction

Antiserum has been used to serologically group bacteriophages *(1–8)*. For example *Pseudomonas aeruginosa* phages D3 and F116 belong to serologically cross-reacting groups D and F, respectively. They have also been used to neutralize nonadsorbed phages in one-step growth experiments *(1)*; and, to inactivate lytic phage in transductional analyses *(9, 10)*.

2 Materials

1. Cesium chloride-purified test phage (*Escherichia coli* O157:H7 lytic phage rV5) and control phage (*E. coli* O157:H7 temperate phage V10).
2. Freund's Incomplete Adjuvant (Invitrogen, Grand Island, NY; http://www.invitrogen.com/) or equivalent.

Martha R. J. Clokie, Andrew M. Kropinski (eds.), *Bacteriophages: Methods and Protocols, Volume 1: Isolation, Characterization, and Interactions, vol. 501*, © 2009 Humana Press, a part of Springer Science+Business Media
DOI 10.1007/978-1-60327-164-6_24 Springerprotocols.com

3. Flat-bottomed 96-well ELISA plates (Nunc Immunlon II, Nalgene Nunc International, Rochester, NY; http://www.nalgenunc.com/)
4. ELISA plate reader
5. Goat anti-rabbit IgG (H + L) Horseradish Peroxidase Conjugate (GAR-HRP; Jackson ImmunoResearch Laboratories Inc., West Grove, PA; http://www.jacksonimmuno.com/home.asp)
6. 50 mM bicarbonate buffer (pH 9.0) (Sigma-Aldrich St. Louis, MO; http://www.sigmaaldrich.com)
7. 1× Phosphate Buffered Saline [PBS: 2 g/l KH_2HPO_4, 2 g/l KCl, 8 g/l NaCl, 21.6 g/l $Na_2HPO_4 \cdot H_20$, supplemented with 0.05% (vol/vol) Tween 20
8. 1× PBS supplemented with 0.05% (vol/vol) Tween 20 and 0.1% (wt/vol) Bovine Serum Albumin (MP Biomedicals, Solon, OH; http://www.mpbio.com/)
9. Tetramethylbenzidene Horseradish Peroxidase (HRP) Substrate Solution (ALERCHEK, Portland, ME; http://www.alerchek.com/) or equivalent.
10. 0.2 M H_2SO_4
11. *Escherichia coli* O157:H7 or other suitable host bacterium.
12. 1 M $MgSO_4 \cdot 7H_2O$
13. Luria-Bertani (LB) broth containing various amounts of antiserum, prepared as described below.
14. MNA agar plates containing 10 mM $MgSO_4$ and various amounts of antiserum, prepared as described below.

MNA:Per liter:	**Nutrient Broth**	**180.0 g**
	NaCl	76.5 g
	Agar No.1 (Oxoid)	90.0 g
	$CaCl_2$	74.7 mg
	$FeCl_3$	9.9 mg
	$MgSO_4 \cdot 7H_2O$	4.5 g
	Distilled water to:	1000 ml
	30% sterile glucose	10 ml, after autoclaving

Dispense 20–30 ml into Petri plates and allow them to solidify at room temperature. Store in plastic sleeves at 4°em C for up to 4 months.

3 Methods

3.1 Prepare heat-inactivated anti-phage serum by repeated intramuscular injection of New Zealand white rabbits with cesium chloride-purified test phage in Freund's Incomplete Adjuvant using conventional methods *(11)*. Collect pre- and post-immunization blood samples (2–3 ml) for testing at each dose and prepare, heat-inactivate and store sera at 4°C. Typically, two or three monthly injections will stimulate strong serum IgG antibody responses.

3.2 Assess the titer of the serum against the cesium chloride-purified test phage by ELISA. Methods for performing and optimizing ELISA may be found in Harlow and Lane *(11)*, or similar reference work.

1. Coat ELISA plates test wells with 100 μl bicarbonate buffer containing test phage ($\sim 10^7$ pfu/ml) for 1 h at 37°C. Coat control wells with 100 μl of bicarbonate buffer.
2. Wash the wells five times with 300 μl of 1× PBS.
3. Block wells with 300 μl of 1× PBS containing 0.1% BSA.
4. Dilute sera 2-fold serially in 1× PBS containing 0.05% Tween 20 and 0.1% BSA. Appropriate dilutions to test range are from ~1 : 1000 to 1:200,000.
5. Add 100 μl of diluted serum to the test and control wells coated with or without phage. Incubate for 1 h at 37°C.
6. Wash the wells five times with 300 μl of 1× PBS containing 0.05% Tween 20.
7. Dilute the secondary antibody, GAR-HRP, in 1× PBS containing 0.05% Tween 20 and 0.1% BSA. The appropriate dilution of secondary antibody can be predetermined by titration of the conjugate against rabbit antibody immobilized in ELISA plate wells.
8. Add 100 μl of diluted secondary antibody to the wells. Incubate for 30 min at 37°C.
9. Wash the wells seven times with ~300 μl of 1× PBS containing 0.05% Tween 20.
10. Add 100 μl of soluble HRP substrate solution. Incubate for 10 min or less at 37°C.
11. Stop the reaction by adding 100 μl of 0.2 M H_2SO_4.
12. Measure the absorbance (OD) of the wells at 450/630 nm in an ELISA plate reader.
13. Determine the titer of the serum as the highest dilution that results in an OD greater than the mean +3 standard deviations of the mean of the negative control wells.

3.3 After the titer of the antiserum becomes elevated, collect serum from the rabbits at regular intervals. Store the serum at 4°C for later pooling into a single lot that will be used in the transduction experiments. Since some experiments such as lytic transductions (**Chapters 23** and **25**) may consume a large volume of serum, it is best to qualify and store a large pool of antiserum.

3.4 Assess the performance of the antiserum under the conditions that will be encountered in the assay system. This assessment will require determining what dilution (if any) will neutralize the test phage in broth, and in agar plates (see below). Judiciously pool the high titer antiserum before performing these tests.

3.1 Assaying Activity of Anti-Phage Serum in Broth

1. Prepare an overnight culture of *E. coli* O157:H7 EC990298 in approximately 10 ml of LB broth. Assume that the culture contains ~10^9 cfu/ml.
2. Supplement the culture with 10 mM $MgSO_4$ by adding 100 μl of 1 M $MgSO_4$.
3. Add ~10^9 pfu of the test phage. Mix well and incubate for 20 min at 37°C to allow adsorption of the phage to the cells. Note that this is the latent period, which may vary among phages.
4. Prepare 250 μl aliquots of LB broth containing none, 1:25, 1:50, 1:100, 1:200, and 1:400 dilutions of antiserum (v/v).
5. Split the bacterial culture into 5 × 1 ml samples in centrifuge tubes labeled with "none, 1:25, 1:50, 1:100, 1:200, and 1:400." Retain the remaining culture for titration later.
6. Pellet the bacterial cells by centrifugation at 6000 × *g* for 10 min. Remove the supernatant and filter a small amount for later titration of free phage.
7. Resuspend each pellet in an aliquot containing none, 1:25, 1:50, 1:100, 1:200, and 1:400 antiserum (v/v). Incubate the tubes for 40 min at 37°C for neutralization to occur.
8. Titrate the number of viable cells (C), free phage (F) and total phage (T) in each tube.
9. For viable cells (C), dilute 10 μl of each culture 10-fold and spread 100 μl aliquots of the dilutions onto duplicate LB agar plates.
10. For free phages (F) dilute 50 μl of each culture in 450 μl of LB broth, filter the diluted sample through 0. 22 μm syringe filters, dilute the filtrates 10-fold and spread 100 μl of the dilutions onto duplicate MNA plates with *E. coli* O157:H7 EC990298 as the host.
11. For total phages (T), dilute 10 μl of each culture 10-fold and spread 100 μl of the dilutions onto duplicate MNA plates with *E. coli* O157:H7 EC990298 as the host.
12. Incubate the titration plates overnight at 37°C.

13. Examine the plates and record the results.
14. Determine the highest dilution of antiserum at which the bacteria have formed a complete lawn without any plaques, indicating complete neutralization of any free phages liberated into the broth during the 40 min incubation period. This is the dilution that will be used during the next step in qualifying the antiserum and in the transduction assay, if required.

3.2 Assaying Activity of Anti-Phage Serum in Agar

1. Prepare duplicate LB agar plates containing 10 mM $MgSO_4$ and none, 1:25, 1:50, 1:100, 1:200, and 1:400 dilutions of antiserum (v/v). These plates are made by first preparing and autoclaving LB agar, cooling the agar to 55–60°C in a water bath, adding 1 M $MgSO_4$ to 10 mM, and the required volume of antiserum to obtain the above dilutions. Air-dry them for 15 min in a biohood before use. These plates must be prepared freshly on the day of use to ensure no loss of activity of the antiserum.
2. Prepare an overnight culture of *E. coli* O157:H7 EC990298 in approximately 10 ml of LB broth. Assume that the culture contains $\sim 10^9$ cfu/ml, and hold a portion for later titration.
3. Supplement the culture with 10 mM MgSO4 by adding 1 M $MgSO_4$.
4. Add $\sim 10^9$ pfu of the test phage. Mix well and incubate for 20 min at 37°C to allow adsorption of the phage to the cells.
5. Prepare 2.5 ml of LB broth containing the amount of antiserum required to neutralize the phage in broth, as determined above.
6. Pellet the bacterial cells by centrifugation at 6000 × *g* for 10 min. Remove the supernatant and filter a small amount for later titration of free phage.
7. Resuspend the pellet in 2.5 ml of LB broth containing antiserum. Incubate the tube for 40 min at 37°C.
8. Titrate the number of viable cells (C), free phage (F), and total phage (T) in the culture.
9. For viable cells (C), dilute 10 μl of the culture 10-fold and spread 100 μl aliquots of the dilutions onto duplicate LB agar plates.
10. For free phages (F) dilute 50 μl of the culture in 450 μl of LB broth, filter the diluted sample through 0. 22 μm syringe filters, dilute the filtrates 10-fold and spread 100 μl of the dilutions onto duplicate MNA plates with *E. coli* O157:H7 EC990298 as the host.
11. For total phages (T), dilute 10 μl of the culture 10-fold and spread 100 μl of the dilutions onto duplicate MNA plates with *E. coli* O157:H7 EC990298 as the host.
12. Also, similarly titrate the number of viable cells in the original culture from Step II, and free phages from Step V1.

13. Incubate the plates overnight at 37°C and record results.
14. After the 40 min incubation period (Step VII), spread 250 μl aliquots of the phage-infected cells onto duplicate plates containing none, 1:25, 1:50, 1:100, 1:200, and 1:400 dilutions of antiserum (v/v). Incubate the plates overnight at 37°C.
15. Examine the plates and record the results.
16. Determine the highest dilution of antiserum at which the bacteria have formed a complete lawn without any plaques, indicating complete neutralization of any phages liberated during overnight incubation on the agar plates. This is the dilution of antiserum that will be used in the transduction assay.

References

1. Adams, M.D. 1959. Bacteriophages. Interscience Publishers, Inc., New York.
2. Nutter, R.L., L.R. Bullas, and R.L. Schultz. 1970. Some properties of five new*Salmonella* bacteriophages. Journal of Virology *5*: 754–764.
3. Betz, J.V. and K.E. Anderson. 1964. Isolation and characterization of bacteriophages active on *Clostridium sporogenes*. Journal of Bacteriology *87*:408–415.
4. Feary, T.W., E. Fisher Jr, and T.N. Fisher. 1964. Isolation and preliminary characteristics of three bacteriophages associated with a lysogenic strain of *Pseudomonas aeruginosa*. Journal of Bacteriology *87*:196–208.
5. Przondo-Hessek, A., S. Slopek, and J. Miodonska. 1967. Serologic characterization of the *Klebsiella* bacteriophages of the Przondo-Hessek collection. Archivum immunologiae et therapiae experimentalis (Warsaw) *15*:557–562.
6. Guice, M.B. and F.S. Newman. 1969. Comparative characterization of choleraphages by serologic, pH stability and thermal stability methods. Journal of Infectious Diseases *119*:2–10.
7. Atkins, G.J. 1973. Some bacteriophages active against *Rhizobium trifolii* strain W19. Journal of Virology *12*:149–156.
8. Spanier, J.G. and J.F. Timoney. 1977. Bacteriophages of *Streptococcus equi*. Journal of General Virology *35*:369–375.
9. Canosi, U., G. Luder, and T.A. Trautner. 1982. SPP1-mediated plasmid transduction. Journal of Virology *44*:431–436.
10. Kiesel, B. and L. Wunsche. 1993. Phage Acm1-mediated transduction in the facultatively methanol-utilizing *Acetobacter methanolicus* MB 58/4. Journal of General Virology *74*:1741–1745.
11. Harlow, E. and D. Lane. 1988. Antibodies: A Laboratory Manual. Cold Spring Harbor Press, Cold Spring Harbor, NY.

Chapter 25

Generalized Transduction by Lytic Bacteriophages

Thomas E. Waddell, Kristyn Franklin, Amanda Mazzocco, Andrew M. Kropinski and Roger P. Johnson

Abstract

As interest in lytic phages as antimicrobial therapies or as treatments to reduce environmental contamination with pathogenic bacteria has increased, so has the need to determine if the use of lytic phages may lead to dissemination of virulence factors through generalized transduction, as occurs with temperate phages. Here we describe simple methods we have developed to determine if a lytic phage, rV5, can mediate generalized transduction in *Escherichia coli* O157:H7. These sensitive methods can be easily adapted to study generalized transduction between virulent and avirulent strains of bacteria.

Key words: Bacteriophage, phage, transduction, lytic transduction, virulent, *Escherichia coli* O157:H7, phage therapy; safety.

1 Introduction

Phage packaging in infected bacteria is an imprecise process that can result in the encapsidation of bacterial DNA due to recognition of pseudo-*pac* sites on the host chromosome. In theory, these host DNA-containing viral particles could act to transfer (transduce) genes from one bacterium to another. This phenomenon is unlikely with virulent phages because the host genome is usually extensively degraded during infection by virulent phages and cells coinfected by transducing particles (TP) and plaque-forming particles (PFP) will be lysed. Under unusual test conditions however, numerous virulent phages including *Serratia* phage φIF3 *(1)*, coliphages T1 *(2–5)*, T4 *(6, 7)*, RB43 *(8)*, and RB49 *(9)*; *Salmonella typhi* phage Vi I *(10)*, *Pseudomonas* phages φKZ *(11)*, E79 *(12)*, and pf16 *(13–15)*; *Xanthomonas* phage XTP1 *(16)*, *Methanobacterium* archaeophage Ψ M1 *(17)*, *Caulobacter* phage φCr30T

Martha R. J. Clokie, Andrew M. Kropinski (eds.), *Bacteriophages: Methods and Protocols, Volume 1: Isolation, Characterization, and Interactions, vol. 501*, © 2009 Humana Press, a part of Springer Science+Business Media
DOI 10.1007/978-1-60327-164-6_25 Springerprotocols.com

(18), *Citrobacter rodentium* phage φCR1 *(19)*, and *Bacillus* phage SPP1 *(20, 21)* have been demonstrated to mediate generalized transduction.

In order to demonstrate transduction by lytic phages, a low multiplicity of infection (MOI) has to be employed to reduce the incidence of coinfection by TP and PFP. In addition to MOI, a variety of other techniques have been used to reduce the phage's virulence, including UV irradiation *(22)*, antiserum *(14)*, the presence of certain plasmids in the recipient *(11, 12)*, as well as conditional-lethal (frequently *ts* or *am*) *(9, 14, 18)* or other *(7)* phage mutants.

The method involves generating potential TP by growing the lytic phage and a control temperate phage on a strain containing a chromosomal antibiotic resistance marker; infecting large numbers of an isogenic strain lacking the selectable marker at a low MOI; and isolating transductants from large numbers of cells infected with PFP in the presence of phage-neutralizing antiserum. The frequencies of generalized transduction of the marker gene by the virulent and temperate control phages are compared. The methods are illustrated with the lytic phage rV5, which has been used successfully to eliminate *E. coli* O157:H7 infections in calves *(23)* (**Note 1**), a control temperate phage v10, and an isogenic pair of *E. coli* O157:H7 differing in the presence of kanamycin resistance in the chromosomal gene, *malM*.

The following sections are organized in sequence with materials and methods for generation of TP from temperate V10 and virulent rV5, and assays for generalized transduction of these TP into a host *E. coli* O157:H7 strain.

2 Materials

2.1 Temperate (V10) and Lytic (rV5) Phages

1. Difco Luria Bertani (LB) broth agar.
2. LB Top Agarose (9.5 g/l NaCl, 15.5 g/l Difco LB Miller Broth Base, 5 g/l Difco Agar supplemented with 10 mM $MgSO_4$) (**Note 2**)
3. Lambda Diluent (10 mM Tris–HCl, pH 7.5 containing 2 g/l $MgSO_4 \cdot 7H_2O$).
4. High titer stock of temperate phage – *E. coli* O157 temperate phage V10.
5. *E. coli* O157:H7 strain containing a selectable chromosomal marker, for example *E. coli* O157:H7 298Kan-1; *malM*::*kan* (referred to as *E. coli* O157:H7 298Kan-1)
6. Isogenic *E. coli* O157:H7 strain for titrating V10, for example *E. coli* O157:H7 EC990298.
7. Difco LB Broth

8. Kanamycin (Sigma-Aldrich) (25 mg/ml in H_20; filter sterilized)
9. LB Agar plates with and without kanamycin and antiserum (**Note 3**).
10. 1 M $MgSO_4$, filter sterilized.
11. Chloroform

2.2 Generalized Transduction Assay (Phages V10 and rV5 and Negative Control)

1. Difco LB Agar
2. Difco LB Broth
3. Modified Nutrient Agar (MNA) – *see* **Chapter 7** Titration for recipe.
4. Lambda Diluent (10 mM Tris–HCl, pH 7.5 containing 2 g/l $MgSO_4 \cdot 7H_2O$)
5. High titer stock of purified temperate phage prepared from *E. coli* O157:H7.
6. High titer stock of purified lytic phage rV5 prepared from rV5 on *E. coli* O157:H7 298Kan-1.
7. *E. coli* O157:H7 strain for titrating V10 and rV5, for example, *E. coli* O157:H7 EC990298.
8. Kanamycin (Sigma-Aldrich) (25 mg/ml in H_20; filter sterilized)
9. LB agar plates without supplements, with kanamycin (100 μg/ml), with the appropriate dilution of anti-phase serum (prepared and tested as in **Chapter 24**), and with both kanamycin (100 μg/ml) and antiserum.
10. 1 M $MgSO_4$

3 Methods

3.1 Propagation of Temperate Phage V10

1. Grow *E. coli* O157:H7 298Kan-1 in LB broth (without kanamycin) at 37°C with shaking until the culture is visibly turbid. This will typically take 2–3 h if a single colony is subcultured in ~ 4 ml of medium.
2. Make serial 1:10 (v/v) dilutions of temperate phage V10 in Lambda Diluent. Dispense 100 μl of the 10^{-3}–10^{-6} dilution into duplicate tubes. This dilution range may differ for other phages.
3. Add 100 μl of bacterial culture and incubate the mixture for 20 min at 37°C.
4. Transfer the tubes to a 48°C water bath.
5. Add 4 ml of 0.5% LB top agarose containing 10 mM $MgSO_4$ to a single tube and mix the contents well by gentle inversion. Immediately pour the mixture onto the surface of the LB agar plate and quickly but gently tilt the plate so that the melted agarose covers the surface completely. Incubate the plate on the bench in the upright position until the agarose solidifies.

6. Incubate the plates overnight at 37°C in the upright position to allow growth of the phage.
7. Examine the plates for the presence of turbid bacterial plaques due to growth of the temperate phage. Select the plates in which there is near confluence of the turbid areas of growth.
8. Add ~5 ml of Lambda Diluent to each of the selected plates. Break up the top agar layer well with a sterile plastic spreader. Cover the plates and incubate them for ~30 min at 37°C.
9. Transfer the liquid from the plate to a sterile chloroform-resistant centrifuge tube.
10. Add chloroform to the tube to ~1% (v/v). Cap the tube and incubate with gentle agitation for ~10 min at room temperature.
11. Centrifuge the material at 6000 × *g* for 15 min.
12. Remove the supernatants from the tubes, being careful not to disturb the chloroform. Pool the supernatants.
13. Filter the supernatant through sterile 0.45 and 0.2 μm low-protein binding syringe filters and check the filtrate for sterility by plating onto LB agar.
14. Label the filtered material with the date, phage name and the host strain.
15. Titrate the phage in the filtered supernatant on *E. coli* O157:H7 EC990298 as described in **Chapter 7**, and store at 4°C (**Note 4**).

3.2 Propagation of Lytic Phage rV5

1. Propagate the test phage once only in *E. coli* O157:H7 298Kan-1 at a MOI of 0.01–0.1, as described in **Chapter 7** (**Note 5**).
2. Purify the phage on a cesium chloride gradient as described in **Chapters 22** and Volume 2 **Chapter 13**, then titrate and store at 4°C and at −70°C

3.3 Generalized Transduction Assay with Temperate Phage

See **Fig. 25.1** for a flow diagram of this procedure.

1. Prepare an overnight culture of *E. coli* O157:H7 EC990298 in ~30 ml of LB broth. Assume that the culture contains ~10^9 colony forming units (cfu)/ml, and determine the titer by standard plate counts. Subculture the broth at 1/100–1/1000 in 30 ml fresh LB broth and incubate with shaking at 37°C for 2–4 h to produce a fresh culture in exponential growth for titration of free and cell-bound phages (i.e., for calculation of the number of infected bacterial cells). It is preferable to conduct transduction experiments with lytic and temperate phage simultaneously.
2. Transfer 0.5 ml of the culture to each of two sterile centrifuge tubes. Add ~2.5×10^{10} pfu of phage V10 (~ 2.5 ml) to one tube labeled "V10." Add an equal volume of LB broth alone to the other tube labeled "negative." Supplement the cultures with 10 mM $MgSO_4$ by adding 30 μl of 1 M $MgSO_4$. Mix the

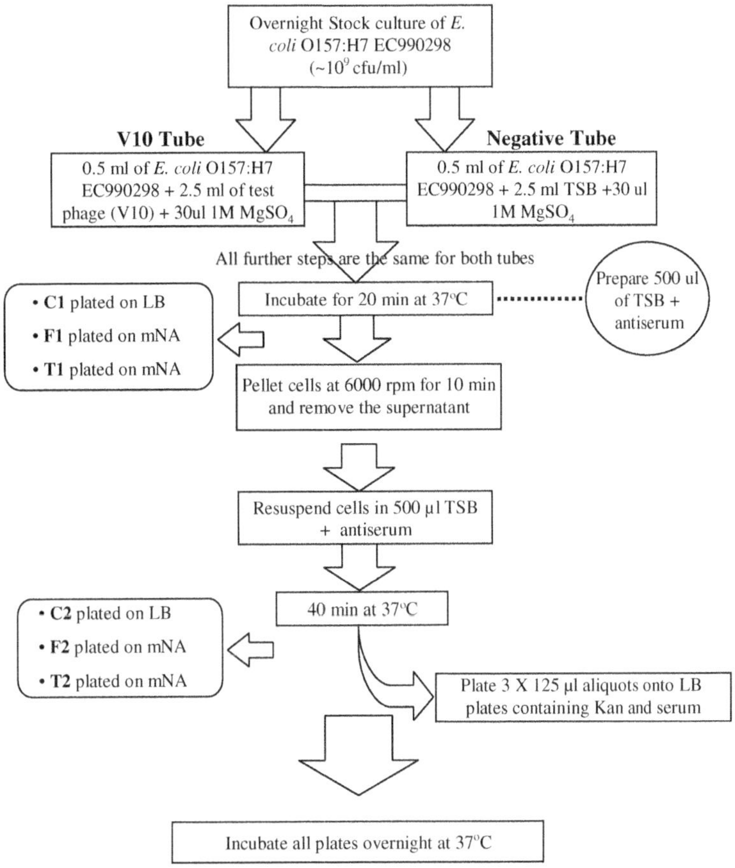

Fig. 25.1. Flow chart for testing temperate control phage V10 for generalized transduction of kanamycin resistance. The letters C, T, and F refer to the concentrations of viable bacteria (C), total phages (T), and free phages (F) as explained in the text.

contents of the tubes well and incubate for 20 min at 37°C with shaking at ~200 rpm to allow adsorption of the phage to the cells.

3. During this incubation period, prepare 2 × 0.5 ml of broth containing the amount of antiserum required to neutralize the test phage in broth, as determined and prepared in **Chapter 24**.
4. After the 20-min incubation period, remove the following samples from the cultures for titration of total (cell-associated and cell-free) phages (T1), cell-free phages (F1), and viable bacterial cells (C1). Hold the samples on ice and process them during the incubation period of the next step.
   i. For total phage concentration (T1), dilute 10 µl from tube V10 for titration in triplicate on MNA with *E. coli* O157:H7 EC990298 as the indicator strain.
   ii. For free phage concentration (F1), dilute 50 µl of culture from the V10 tube in 450 µl of Lambda Diluent, filter the diluted sample through a 0.2 µm syringe filter and titrate

the filtrate in triplicate on MNA with *E. coli* O157:H7 EC990298 as the indicator strain.

iii. For viable bacterial concentrations (C1), take 10 μl from the V10 tube and 10 μl from the Negative tube for titration in triplicate on LB agar.

iv. Incubate the plates overnight at 37°C.

5. Pellet the bacterial cells in both tubes by centrifugation at 6,000 rpm for 10 min at room temperature. Remove and discard the supernatants from both tubes.
6. Resuspend each pellet in 0.5 ml of LB broth containing the antiserum. Incubate the tubes for 40 min at 37°C with shaking at ~200 rpm to allow expression of kanamycin resistance in the transductants.
7. After 40 min, determine the numbers of viable bacterial cells (C2), total phage (T2), and free phage (F2) in the tubes, as described in Step 4 above, leaving ~400 μl of each culture.
8. Spread 3 × 125 μl aliquots of the cultures from the V10 and Negative tubes on control plates of LB agar with antiserum, LB agar with kanamycin (100 μg/ml), and LB agar alone. Incubate the plates overnight at 37°C.
9. Examine the plates, record the results and ensure the controls gave expected results:
   i. LB agar with antiserum control: confluent bacterial growth indicating complete neutralization of phages, and no effect on bacterial growth when compared to growth on LB agar without antiserum.
   ii. LB agar with kanamycin control: no bacterial growth, indicating no unwanted contaminates affecting the results.
   iii. LB agar control: ensures viable growth of bacteria and serves as a control for the effect of the antiserum on bacterial growth.
10. Calculate the following:
    i. The titer of the recipient culture of *E. coli* O157:H7 EC990298 used in Step **3.3.1.** Use this titer and the known titer of the test phage to calculate the actual MOI in the assay.
    ii. The titers of total phages (T), free phages (F), and viable bacteria (C) per ml after 20 min adsorption (T1, F1, C1) and 40 min recovery (T2, F2, C2).
    iii. The titer of infected cells (IC) in cfu/per ml after 20 min adsorption (IC1) and 40 min recovery (IC2), by subtracting F1 from T1, and F2 from T2, respectively. This calculation assumes infection of each bacterial cell by only one phage.
    iv. The total number of infected cells (TIC) after 40 min recovery (TIC2), by multiplying the titer of IC2 by the volume of the sample in ml, e.g., IC2 cfu/ml × 5 ml = TIC2 cfu.

v. The total number of infected cells plated (TICP) onto LB agar with kanamycin and antiserum after 40 min recovery, with the formula: total number of infected cells after recovery (TIC2 cfu)× the volume (ml) plated to this agar (~4 ml) ÷ by the total volume (ml) during recovery (~ 5 ml).
vi. The number (cfu) of transductants (i.e., the mean of the counts of kanamycin-resistant colonies on triplicate LB agar plates with kanamycin and antiserum).

3.4 Generalized Transduction Assay with Lytic Phage

See **Fig. 25.2** for a flow diagram of this procedure.

1. Prepare an overnight culture of *E. coli* O157:H7 EC990298 in ~30 ml of LB broth. This culture is used as the recipient strain for TP. Assume that the culture contains ~10^9 cfu/ml, and take off a sample for determination of the titer by standard plate counts. Subculture the broth at 1/100–1/1000 in 30 ml

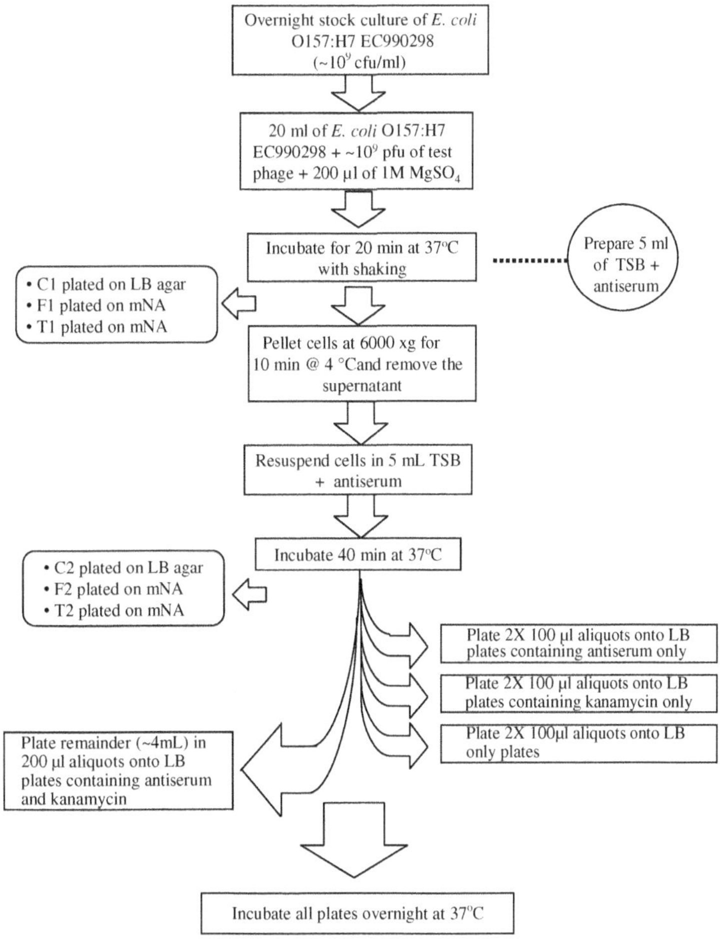

Fig. 25.2. Flow chart for testing lytic phage rV5 for generalized transduction of kanamycin resistance. The letters C, T and F refer to the concentrations of viable bacteria (C), total phages (T), and free phages (F) as explained in the text.

fresh LB broth and incubate with shaking at 37°C for 2–4 h to produce a fresh culture in exponential growth for titration of free and cell-bound phages (i.e., for calculation of the number of infected bacterial cells). It is preferable to conduct transduction experiments with lytic and temperate phage simultaneously.

2. Transfer 20 ml of the culture to a sterile centrifuge tube. Supplement the culture with 10 mM $MgSO_4$ by adding 200 μl of 1 M MgSO4.
3. Add ~10^{10} pfu of the test phage creating a MOI of 0.05–0.1. Mix the contents well and incubate the tube for 20 min at 37°C with shaking at ~200 rpm to allow adsorption of the phage to the cells.
4. During this incubation period, prepare 5 ml of LB broth containing the amount of antiserum required to neutralize the test phage in broth (see above).
5. After the 20-min incubation period, remove the following samples from the cultures for titration of total (cell-associated and cell-free) phages (T1), cell-free phages (F1), and viable bacterial cells (C1). Hold the samples on ice and process them during the incubation period of the next step.
6. For total phage concentration (T1), dilute 10μl of the culture for titration in triplicate on MNA with *E. coli* O157:H7 EC990298 as the indicator strain.
7. For free phage concentration (F1), dilute 50μl of the culture in 450 μl of Lambda Diluent, filter the diluted sample through a 0.2 μm syringe filter and titrate the filtrate in triplicate on MNA with *E. coli* O157:H7 EC990298 as the indicator strain.
8. For viable bacterial concentrations (C1), dilute 10 μl of the culture for titration in triplicate on LB agar. Incubate the plates overnight at 37°C.
9. Pellet the bacterial cells by centrifugation at 6000 rpm for 10 min at room temperature. Remove and discard the supernatant.
10. Resuspend the pellet in 5 ml of LB broth containing the antiserum. Incubate the tubes for 40 min at 37°C with shaking at ~200 rpm to allow expression of kanamycin resistance in the transductants.
11. After 40 min, remove samples of the culture for determination of the numbers of viable bacteria cells (C2), total phage (T2), and free phage (F2) in the tubes, as described in Steps 6, 7 and 8, above, leaving ~ 5 ml of the culture.
12. For additional controls, spread 100 μl aliquots of the culture onto duplicate plates of LB agar with antiserum, LB agar with kanamycin (100 μg/ml), and LB agar alone. Incubate the plates overnight at 37°C.
13. Spread plate the rest of the culture (~ 4 ml) in 200 μl aliquots onto LB agar with kanamycin (100 μg/ml) and antiserum

(i.e., ~20 plates). Incubate the plates overnight at 37°C. Record the volume plated.

14. Examine the plates, record the results and ensure the controls gave expected results:
    i. LB agar with antiserum control: confluent bacterial growth indicating complete neutralization of phages, and no effect on bacterial growth when compared to growth on LB agar without antiserum.
    ii. LB agar with kanamycin control: no bacterial growth, indicating no unwanted contaminates affecting the results.
    iii. LB agar only control. For ensuring viable growth of bacteria and as a control for the effect of the antiserum on bacterial growth.
15. Calculate the following:
    i. The titer of the recipient culture of *E. coli* O157:H7 EC990298 used in **Step 3.4.1.** Use this titer and the known titer of the test phage to calculate the actual MOI in the assay.
    ii. The titers of total phages (T), free phages (F) and viable bacteria (C) per ml after 20 min adsorption (T1, F1, C1) and 40 min recovery (T2, F2, C2).
    iii. The titer of IC in cfu/per ml after 20 min adsorption (IC1) and 40 min recovery (IC2), by subtracting F1 from T1, and F2 from T2, respectively. This calculation assumes infection of each bacterial cell by only one phage.
    iv. The TIC after 40 min recovery (TIC2), by multiplying the titer of IC2 by the volume of the sample in ml, e.g., IC2 cfu/ml × 5 ml = TIC2 cfu.
    v. The TICP onto LB agar with kanamycin and antiserum after 40 min recovery, with the formula: total number of infected cells after recovery (TIC2 cfu) × the volume (ml) plated to this agar (~4 ml) ÷ by the total volume (ml) during recovery (~ 5 ml)8.
    vi. The number (cfu) of transductants (i.e., the mean of the counts of kanamycin-resistant colonies on triplicate LB agar plates with kanamycin and antiserum).

3.5 Frequency of Transduction

Calculate the frequency of transduction with the formula:

Mean number of transductants on LB agar with kanamycin and antiserum (from previous calculation) ÷ the total number of infected cells plated to this medium (i.e., TICP).

Where there are no transductants, the result is expressed as a frequency greater than the total number of cells plated on LB agar with kanamycin and antiserum. Typical results obtained with tests of rV5 and V10 are:

Frequency of transduction with V10: 1×10^{-8}

Frequency of transduction with rV5: $< 1 \times 10^{-11}$

4 Notes

1. The availability of rV5 is restricted due to its commercial development as a phage-based therapeutic.
2. Before using the Top Agarose melt it in an isothermal autoclave by heating for ~30 min. Cool the medium to about 60°C in an incubator then add 1:100 (v/v) 1 M $MgSO_4$(be sure to mark on the bottle of medium that it contains $MgSO_4$). The medium may be stored at 60°C for up to a week. Cool the medium to 48°C the day the medium is used.
3. LB medium containing antiserum and/or kanamycin and $MgSO_4$ are made from pre-made bulk medium in bottles, antiserum, kanamycin and $MgSO_4$ stocks. Bulk medium is melted in an isothermal autoclave and then cooled to about 60°C. The required volume stocks are added, plates are poured and then stored at °C. Before use, the plates are warmed to room temperature and dried in a biohood to remove excess moisture that can interfere with titrations of bacteria and phage.
4. The number of host strain bacteria added to the tubes for the titration of temperate phage should be minimized because their plaques will be poorly visible if too many cells are present initially. The amount of culture to add is determined by trial and error.
5. The test phage initially may grow poorly in this strain, particularly if it is coming from a different background (i.e., a non-pathogenic strain belonging to a different serotype). If the phage grows poorly in *E. coli* O157:H7 298Kan-1, propagate it in *E. coli* O157:H7 EC990298 until a satisfactory titer is obtained. Then propagate it one time only in *E. coli* O157:H7 298Kan-1.

References

1. Petty, N.K., I.J. Foulds, E. Pradel, J.J. Ewbank, G.P. Salmond, N.K. Petty, I.J. Foulds, E. Pradel, et al. 2006. A generalized transducing phage (fIF3) for the genomically sequenced *Serratia marcescens* strain Db11: a tool for functional genomics of an opportunistic human pathogen. Microbiology *152*: 1701–1708.
2. Bendig, M.M. and H. Drexler. 1977. Transduction of bacteriophage Mu by bacteriophage T1. Journal of Virology *22*:640–645.
3. Drexler, H. 1977. Specialized transduction of the biotin region of *Escherichia coli* by phage T1. Molecular & General Genetics *152*:59–63.
4. Drexler, H. 1970. Transduction by bacteriophage T1. Proceedings of the National Academy of Sciences of the United States of America *66*:1083–1088.
5. Roberts, M.D. and H. Drexler. 1981. Isolation and genetic characterization of T1-transducing mutants with increased transduction frequency. Virology *112*: 662–669.
6. Wilson, G.G., K.Y. Young, G.J. Edlin, W. Konigsberg, G.G. Wilson, K.Y. Young, G.J. Edlin, and W. Konigsberg. 1979. High-frequency generalised transduction by bacteriophage T4. Nature *280*:80–82.
7. Young, K.K., G.J. Edlin, and G.G. Wilson. 1982. Genetic analysis of bacteriophage T4 transducing bacteriophages. Journal of Virology *41*:345–347.

8. Tianiashin, V.I., V.I. Zimin, A.M. Boronin, V.I. Tianiashin, V.I. Zimin, and A.M. Boronin. 2003. The cotransduction of pET system plasmids by mutants of T4 and RB43 bacteriophages. Mikrobiologiia *72*:785–791.
9. Taniashin, V.I., A.A. Zimin, M.G. Shliapnikov, A.M. Boronin, V.I. Taniashin, A.A. Zimin, M.G. Shliapnikov, and A.M. Boronin. 2003. Transduction of plasmid antibiotic resistance determinants with pseudo-T-even bacteriophages. Genetika *39*:914–926.
10. Cerquetti, M.C., A.M. Hooke, M.C. Cerquetti, and A.M. Hooke. 1993. Vi I typing phage for generalized transduction of *Salmonella typhi*. Journal of Bacteriology *175*:5294–5296.
11. Dzhusupova, A.B., T.G. Plotnikova, V.N. Krylov, A.B. Dzhusupova, T.G. Plotnikova, and V.N. Krylov. 1982. Detection of transduction by virulent bacteriophage f KZ of *Pseudomonas aeruginosa* chromosomal markers in the presence of plasmid RMS148. Genetika *18*:1799–1802.
12. Morgan, A.F. 1979. Transduction of *Pseudomonas aeruginosa* with a mutant of bacteriophage E79. Journal of Bacteriology *139*: 137–140.
13. Gorbunova, S.A., V.S. Akhverdian, L.V. Cheremukhina, V.N. Krylov, S.A. Gorbunova, V.S. Akhverdian, L.V. Cheremukhina, and V.N. Krylov. 1985. Effective method of transduction with virulent phage pf16 using specific mutants of *Pseudomonas putida* PpG1. Genetika *21*:872–874.
14. Daz, R., T.G. De, J.L. Canovas, R. Daz, G. De Torrontegui, and J.L. Canovas. 1976. Generalized transduction of *Pseudomonas putida* with a thermosensitive mutant of phage pf16h2. Microbiologia Espanola *29*:33–45.
15. Rheinwald, J.G., A.M. Chakrabarty, and I.C. Gunsalus. 1973. A transmissible plasmid controlling camphor oxidation in *Pseudomonas putida*. Proceedings of the National Academy of Sciences of the United States of America *70*:885–889.
16. Weiss, B.D., M.A. Capage, M. Kessel, and S.A. Benson. 1994. Isolation and characterization of a generalized transducing phage for *Xanthomonas campestris* pv. *campestris*. Journal of Bacteriology *176*:3354–3359.
17. Meile, L., P. Abendschein, T. Leisinger, L. Meile, P. Abendschein, and T. Leisinger. 1990. Transduction in the archaebacterium *Methanobacterium thermoautotrophicum* Marburg. Journal of Bacteriology *172*: 3507–3508.
18. Bender, R.A. 1981. Improved generalized transducing bacteriophage for *Caulobacter crescentus*. Journal of Bacteriology *148*: 734–735.
19. Petty, N.K., A.L. Toribio, D. Goulding, I. Foulds, N. Thomson, G. Dougan, and G.P. Salmond. 2007. A generalized transducing phage for the murine pathogen *Citrobacter rodentium*. Microbiology *153*:2984–2988.
20. Canosi, U., G. Luder, and T.A. Trautner. 1982. SPP1-mediated plasmid transduction. Journal of Virology *44*:431–436.
21. de Lencastre, H. and L.J. Archer. 1980. Characterization of bacteriophage SPP1 transducing particles. Journal of General Microbiology *117*:347–355.
22. Riska, P.F., Y. Su, S. Bardarov, L. Freundlich, G. Sarkis, G. Hatfull, C. Carriere, V. Kumar, et al. 1999. Rapid film-based determination of antibiotic susceptibilities of *Mycobacterium tuberculosis* strains by using a luciferase reporter phage and the Bronx Box. Journal of Clinical Microbiology *37*:1144–1149.
23. Waddell, T.E., A. Mazzocco, J. Pacan, R. Johnson, R. Ahmed, C. Poppe, and C. Khakhria. 2002. Use of bacteriophages to control *Escherichia coli* O157 infections in cattle, United States Patent No. 6,485,902.

Index

A

B

C

D

E

F

G

H

I

L

M

N

O

P

R

S

T

U

V

Y

GPSR Compliance

The European Union's (EU) General Product Safety Regulation (GPSR) is a set of rules that requires consumer products to be safe and our obligations to ensure this.

If you have any concerns about our products, you can contact us on ProductSafety@springernature.com

In case Publisher is established outside the EU, the EU authorized representative is:

Springer Nature Customer Service Center GmbH
Europaplatz 3
69115 Heidelberg, Germany

Batch number: 10762579

Printed by Printforce, the Netherlands